RADIO COMMUNICATIONS RECEIVERS

BY CORNELL DRENTEA

TAB TAB BOOKS Inc.
BLUE RIDGE SUMMIT, PA. 17214

FIRST EDITION

FIRST PRINTING

Printed in the United States of America

Library of Congress Cataloging in Publication Data

Drentea, Cornell.
Radio communications receivers.

Bibliography: p.
Includes index.
1. Radio—Receivers and reception. I. Title.
TK6563.D73 1982 621.3841'36 82-5836
ISBN 0-8306-2393-0 AACR2
ISBN 0-8306-1393-5 (pbk.)

Cover photo courtesy of Norlin Communications. Background in top half of cover courtesy of *EDN* magazine.

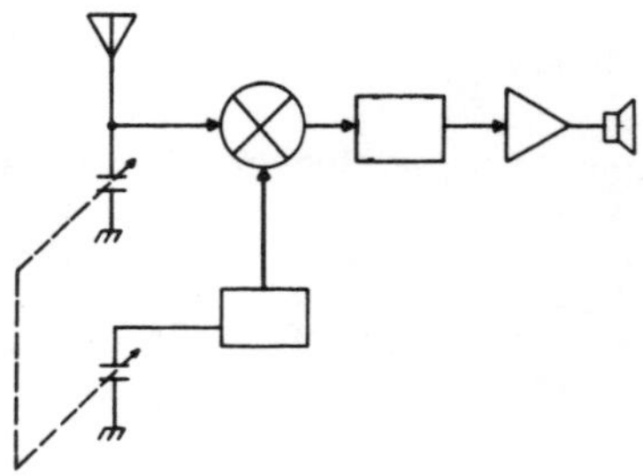

Contents

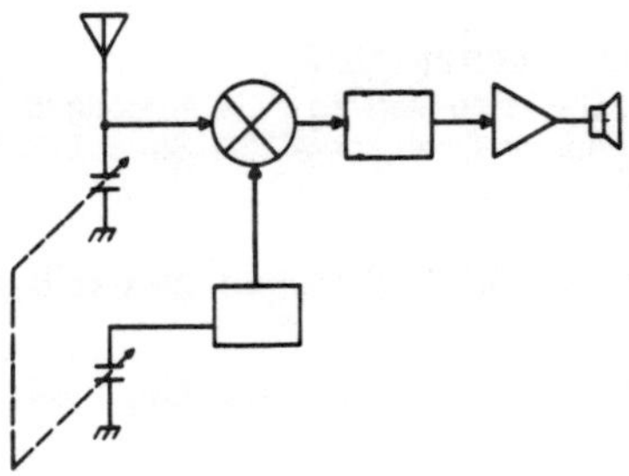

Introduction

As we approach the twenty-first century, radio technology is becoming increasingly complex, with the progress of the communication receiver being one of the most dynamic technical episodes of our century.

New stringent requirements such as good frequency stability, large signal handling capability, and improved selectivity make today's radio receiver a totally different apparatus than its predecessor of the 1920s. These new requirements have brought about the need for new technologies and make the "gap" of achieving technical excellence ever narrower.

The scope of this book is to give you a complete picture of what has been done and what can be done in receiver design, as well as to describe in a chronological way the historical technical facts that impacted radio technology over the past eighty years. This book will explain and expand on these facts showing present and future trends and finally present a receiver design from system through packaging. In this respect, the book is an accelerated design guide, filling a gap for the beginner while at the same time being a comprehensive work for the advanced professional.

Although portions of this work can be used for practical implementation, the material is not meant as a cookbook, but rather as a concentrated educational kit in radio receiver design.

Because the subject is so vast, in designing the book I chose to emphasize certain technical areas that cannot usually be found in one single publication and combine them with facts and experiences of my own. From a professional standpoint, this work should be good for anyone considering an rf signal processor (as well as radio) concept. It is hoped that the book will be found useful and entertaining by the beginner as well as the advanced reader.

—•••—

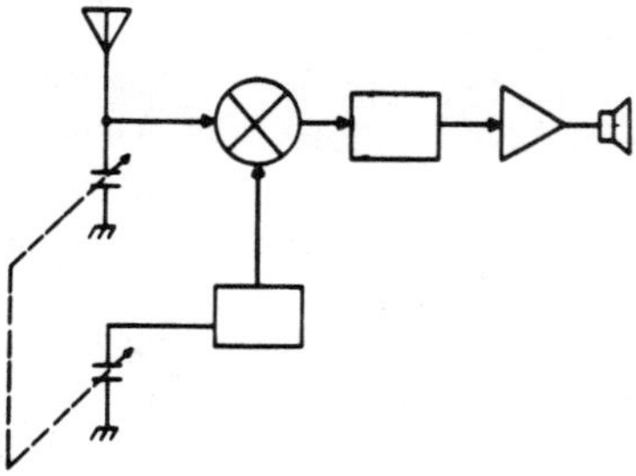

Acknowledgments

I wish to thank all of the reviewers for their input and constructive criticism during the many phases of completing this book. Special thanks go to the following individuals and/or companies who contributed to this work: James A. Andem of Norlin Communications; Frank A. Arneson of GTE-Sylvania; Roselyn Arsenault of National Astronomy and Ionosphere Center, Cornell University; Virginia L. Baltes of Honeywell Inc.; Walter Baronian of IntraAction Corp.; Jerry Behm of Disc Instruments; Glen G. Bruins of Honeywell Inc.; Jerry L. Carter of Rockwell International-Collins Telecommunications Products Division; Richard A. Coppock of GTE-Sylvania; Toni Deeb of Watkins-Johnson Company; Dominique Drentea; Oliver P. Ferrell of Gilfer Associates Inc.; Frank J. Ferrin of Honeywell Inc.; Pamela J. Goral of SGC Inc.; John Van Groos of True Time Division (Kinemetrics); Bruce Jaroski of Communication Coil Co.; Robert A. Johnson of Rockwell International-Electronic Devices Division; Tom Jorgensen of Honeywell Inc.; Don Junk of McCoy Electronics Co.; Carol Kiser of Vari-L Co. Inc.; Joseph L. Koproski of McCoy Electronics Co.; Nate Lewis of Teledyne Relays; Ron Maples of Cubic Communications; Rodney K. McDowell of Watkins-Johnson Company; R.L. Midkiff of R.L. Drake Co.; Maria J. Monusco of Honeywell Inc.; Roger Musolf of Honeywell Inc.; Mahlen Norton; Plessey Semiconductors; Susan Rabinovitz of EDN; Erna R. Rasmussen of Honeywell Inc.; Steve N. Raymer of Vintage Radio Service; Ulrich L. Rohde of Rohde & Schwarz; Frida Shubert of RCA Corp.; Marcus Staloff of Honeywell Inc.; Harry Stillwell of Honeywell Inc.; Baard Thue of Honeywell Inc.; Grant A. Wales of Honeywell Inc.; Wendel O. Walles of Honeywell Inc.; Lee R. Watkins of Martin Marieta Corp.; Samuel Weber of Electronics; William Werner of Honeywell Inc.; LeRoy Williams Jr. of Honeywell Inc.; and Jess C. Wright of Cubic Communications.

To the Memory of My Father

It is a beautiful Sunday morning in September 1946. I am going to be six years old in another week. With a pair of pliers in one pocket and a screwdriver in the other, I am patiently waiting for my parents to leave home. They are going to visit with relatives and will not return until late in the afternoon. I can finally take it apart to see what the little people inside look like, though something tells me that this cannot be true . . . but, how does it work?

"Good-bye Mama, . . . , bye Dad; I'll be good, I promise!"

The plan is to take it all apart, and put it back together before they get back at five o'clock. If everything works right, I should have enough time to find all there is to know about this magic box that my Dad got last Christmas. I know, I've been told not to touch it, but . . . I really have to find out what is inside. If my Dad ever found out about this, I would be in big trouble. It is a heavy radio to move, but it's on the table now. I better start working. What comes out first?

It is two o'clock and by this time I have lots of nuts and bolts laying around the table. I also have these funny-looking objects made out of glass, with things inside them . . . and no little people yet. There are also some tweaking things that I cannot take out, but something tells me not to touch them. So that is where the sound comes from, this big round looking thing with paper inside . . . and a magnet? But why? I hear noises . . . it can't be, it's only two-thirty. They're back. The door opens.

"What did you do?" I hear my Father's voice as I was unveiling another gray area of the radio, with the pliers in my right hand and the screwdriver in my mouth. Quick thinking tells me not to lie at this point.

"Well . . . I wanted to know how it works Dad, but I'll put it back together, I promise, I promise." I feel tears coming out of my eyes.

"Very well, if you put it back together, and it still works, I will forgive you. But it has to work."

I am scared. Will it still work? I know I was very careful to remember where everything came from, but I didn't plan on having so many little parts.

Where did this come from? Oh yes, here. And this funny-looking bottle, it doesn't want to go back . . . oh yes, it has a key . . . there.

It's seven o'clock and I have three screws left. I don't know what to do with them; I really don't remember where they came from. Will it work without them? I have to take a chance.

"I am ready Dad," I call as I am hiding the screws in my pocket. The big test is here, oh God help me. The set is plugged in and . . . suddenly, the little people are back again. What a relief!

Although I never found out where the three little screws went, I later went to school and learned a lot more about how things worked.

My father has long since passed away, but I will never forget my first contact with a radio. This book is dedicated to the memory of my Father, who taught me the laws of self discipline and perseverance.

Chapter 1

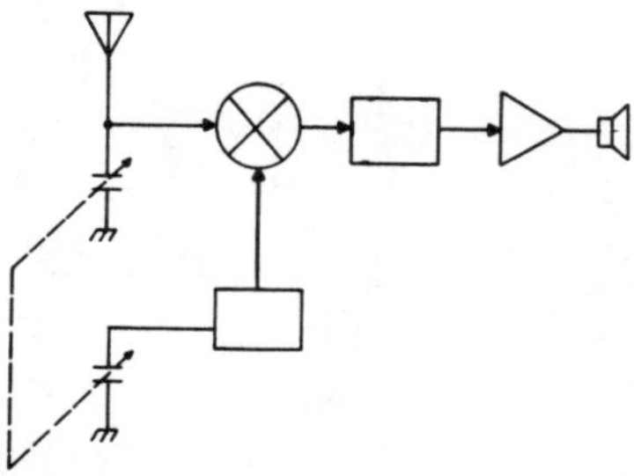

Introduction to Receivers

Whenever thinking of radio, we usually think of one man: Guglielmo Marconi. However, radio, as a technology, resulted from the work of many men. The following is a list of some of the many men involved in the development of radio.

Robert Adler
Ernest F. W. Alexanderson
Edwin H. Armstrong
Jones J. Berzelius
Edouard Branly
George Campbell
John Carson
Arthur A. Collins
Frank Conrad
William Crookes
Jacques and Pierre Curie
Amos E. Dolebear
R. L. Drake
William D. Duddel
H. H. Dunwoody
Thomas A. Edison
Albert Einstein
Robley Evans
Michael Faraday
Reginald A. Fessenden
John A. Fleming
Lee De Forest
Ben Franklin
L. Alan Hazeltine
Oliver Heaviside
Heinrich R. Hertz
Christian Huygens
Karl Jansky
Arthur E. Kennelly
Irving Langmuir
Oliver J. Lodge
James C. Maxwell
G. M. Minchin
Samuel F. B. Morse
Greenleaf W. Pickard
Alexander Popov
William H. Preece
Theodore Roosevelt
David Sarnoff
Nikola Tesla
Jules Verne
T. L. Wadley
Clemens Winkler

Fig. 1-1. Communications pioneers - a group of distinguished scientists visiting RCA's experimental Transoceanic Communications Station at New Brunswick, New Jersey in 1921. From left to right: (starting fourth from left) David Sarnoff, Thomas J. Hayden, Dr. E.J. Berg, S. Benedict, Prof. Albert Einstein, John Carson, Dr. Charles P. Steinmetz, Dr. Alfred N. Goldsmith, A. Malsin, Dr. Irving Langmuir, Dr. Anthony W. Hull, E.B. Pillsbury, Dr. Saul Dushman, R.H. Ranger, and Dr. G.A. Campbell (courtesy of RCA).

And also the many dedicated ham radio operators around the world.

Some of these men were science fiction dreamers. Their contribution to this invention was that of stimulating the other's imaginations. Still others were scientists and mathematicians. Their role was to pave the road for future developments. Some were politicians and businessmen, and some were inventors and technical practitioners, or what we usually refer to as engineers. See Fig. 1-1.

—•••—

Chapter 2

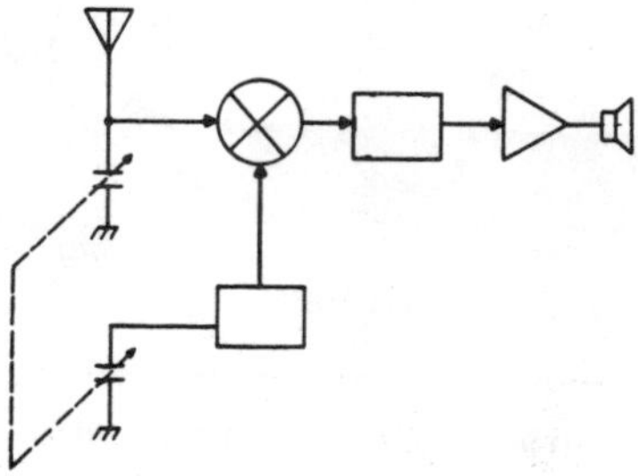

History of Radio

The following is a brief history of the development of radio.

THE COHERER

The radio receiver appeared as a consequence of an invention called the *coherer* (Fig. 2-1). Edouard Branly, of France, discovered that a glass tube with two silver electrodes, filled with loose iron particles, will conduct dc electricity better in the presense of so-called *Hertzian waves* generated with conventional spark generators.

THE FIRST RADIO RECEIVER

Branly would not see the real application for this device, but Marconi developed a receiver on this principle. Apparently, a Russian by the name of Alexander Popov claimed to create a storm receiver which rang a bell every time an electrical storm would approach within several miles. This receiver had no antenna and was of limited range.

Marconi discovered that by adding two wires with large metal plates at the ends to his receiver, and/or transmitter, the range could be increased considerably. He called these wires "catch wires" in 1895, and not much later he buried one of these wires in the ground, while elevating the other, thus discovering the antenna system as we know it today.

It appears that the two inventors were in correspondence for a while. While Popov did not see any further application for his receiver, the brilliant and enterprising Marconi perfected the machine and demonstrated its usefulness. He noticed that once conducting, the coherer would stay in that state, so he invented the decoherer.

THE DECOHERER (PRACTICAL COHERER/DECOHERER RECEIVERS)

The *decoherer* was nothing more than an electrical bell, slightly mod-

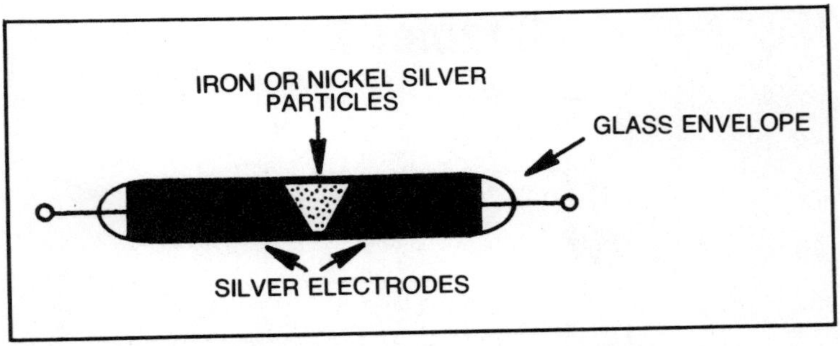

Fig. 2-1. The coherer was invented by Edouard Branly in 1891.

ified and connected in series with the coherer and/or the receiver relay as shown in Fig. 2-2. Every time a signal was received, the coherer would be set in an "on" state, triggering the relay and, therefore, the electrical bell, which in turn would knock on the coherer, resetting it for a new signal. Although practical receivers of this type required a decoherer, they were typically referred to as coherer receivers.

Marconi perfected this receiver and produced several versions of it. He left Italy and went to England where he improved the coherer, producing equipment that established maritime communications for the Royal Navy. By 1899, he established communications across the English Channel, and two years later he sent the letter "S" over the Atlantic Ocean*, by using a similar receiver (Fig. 2-3). He realized the limitations of the coherer receiver, but it wasn't until 1906 that any improvement came about.

*The transmitted signal came from Poldhu in Cornwall, England and the receiver was located in St. John's Newfoundland, Canada some 1800 miles away.

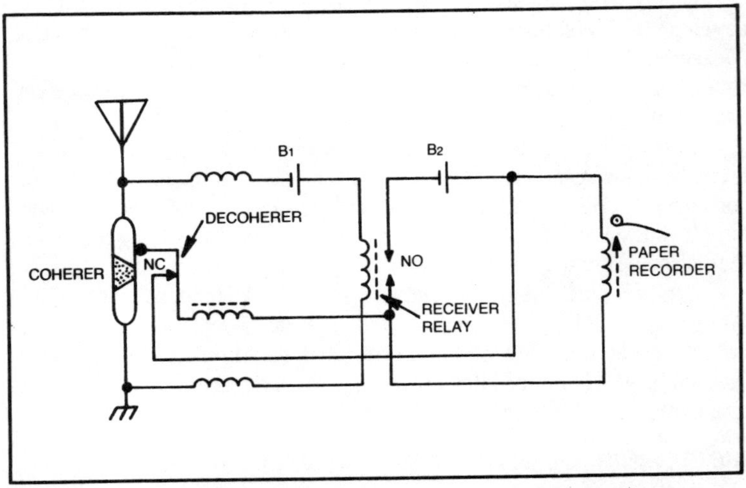

Fig. 2-2. Typical coherer receiver of 1901 had an automatic decoherer.

Fig. 2-3. Guglielmo Marconi and his famous receiver at St. John's, Newfoundland, where on December 21, 1901, the first transatlantic signal was heard (courtesy of RCA).

GALENA CRYSTAL DISCOVERY, THE FLEMING VALVE AND THE AUDION

General H. H. Dunwoody of the U.S. Navy discovered the *crystal detector* in 1906 (Fig. 2-4). This produced a new type of receiver which was more sensitive, the crystal, or galena, receiver (Fig. 2-5). The crystal receiver set didn't last long as a commercial receiver type because of *Fleming's valve* (1904) and the *De Forest audion* (1906).

As a result of these inventions, new and even more sensitive receivers evolved from the crystal receiver as shown in Fig. 2-6.

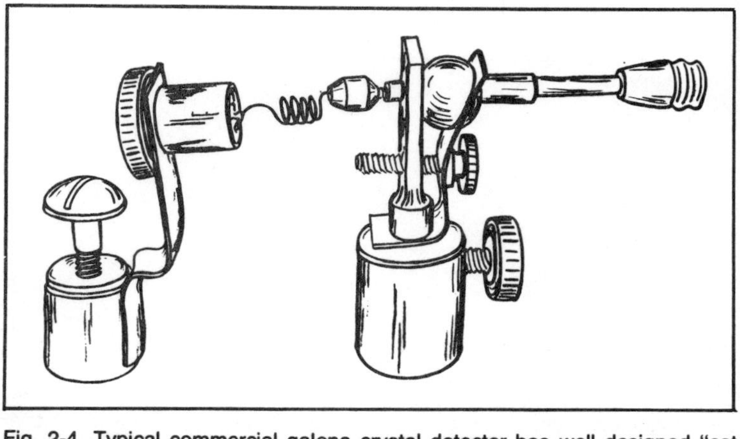

Fig. 2-4. Typical commercial galena crystal detector has well designed "cat whisker arm".

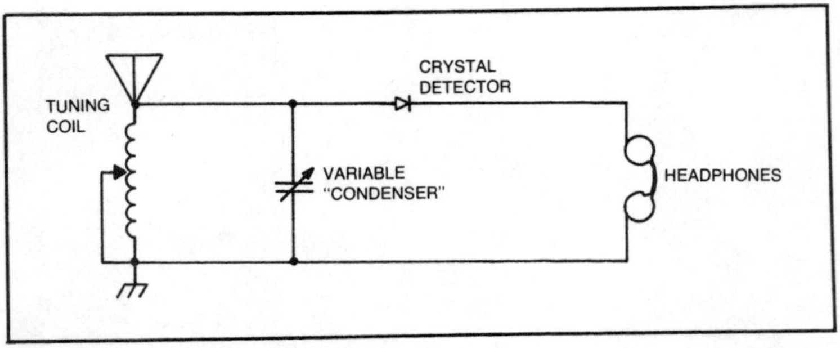

Fig. 2-5. A typical 1906 crystal-galena receiver.

THE AUDION AND THE REGENERATIVE RECEIVER

A new generation of radio receivers were born at this time, (i.e., 1906) regenerative sets. Edwin Howard Armstrong is responsible for the invention of the *regenerative receiver*. He understood that the Audion circuit used in Fig. 2-6B was not satisfactory. In this circuit, the Audion's grid was connected in a tuned circuit to the receiver's antenna, while the plate was connected in series with the headphones and the anode battery to ground. The circuit was closed through the electron-emitting filament, which in turn was activated by the filament battery.

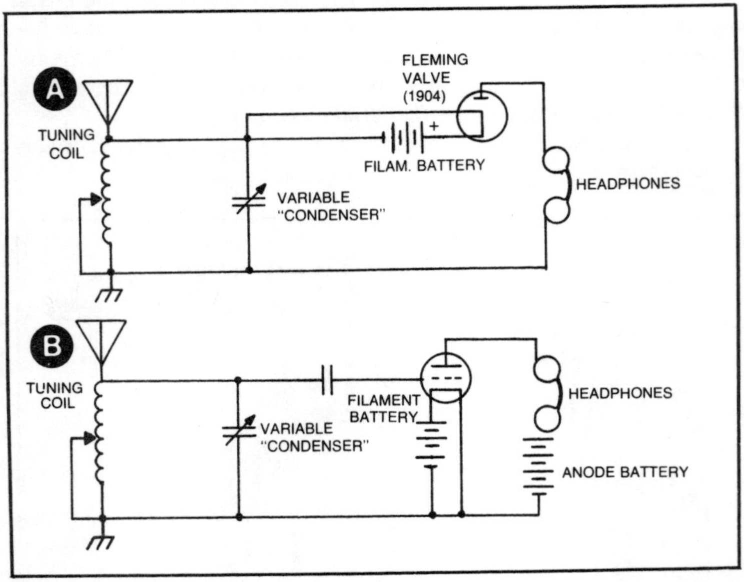

Fig. 2-6. The evolution of the crystal-galena receiver was a result of the Fleming valve and the Audion. The Fleming valve replaces the crystal detector in this 1907 receiver (A). This 1907 receiver uses De Forest's Audion in a detector/amplifier mode (B).

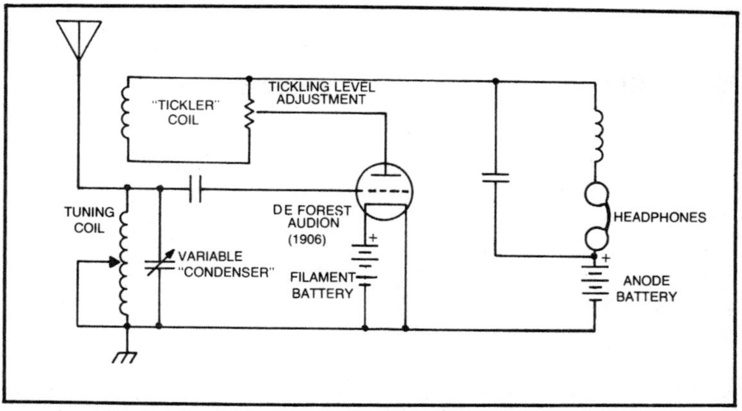

Fig. 2-7. Typical regenerative receiver using De Forest's audion.

Armstrong found that some alternating current was produced in the plate-headphone circuit (called the wing circuit) where it wasn't expected and attempted tuning it in much the same manner as the antenna-to-grid circuit, thus coupling some of the amplified output signal back to the grid circuit inductively, as shown in Fig. 2-7. This method was called *regeneration*, and the "tickling" level adjustment shown in Fig. 2-7 provided for a threshold allowing for large amplification, not previously possible with any other method, to be achieved. This newly discovered feedback amplifier circuit, made practical long distance reception a reality for the first time.

THE AUDION AND THE LOCAL OSCILLATOR

Both Armstrong and De Forest noted that when pressing the Audion to higher amplification (by increasing the "tickling" level, over the threshold point) in a regenerative set, an audible "hissing" would result. While De Forest dismissed this fact as an "irritating noise that hindered proper operation," Armstrong went on to prove that the Audion was not only a receiving device, but an oscillator of electromagnetic waves which would serve later as the basis for the local oscillator to be used in the superheterodyne receiver.

THE TUNED RADIO FREQUENCY (TRF) RECEIVER

Another form of radio receiver characteristic of this era was the *TRF receiver*, which stands for "tuned radio frequency receiver." The TRF receiver was nothing more than a chain of individually tuned amplifiers. This radio took advantage of the audion valve in an amplifying mode. Frequently, many dials were present on these early radios and it took patience to tune one of them. Better versions employed mechanical tracking for slaving several variable capacitors to the motion of a single knob in order to provide identical tuning as shown in Fig. 2-8. These radios were popular until 1922.

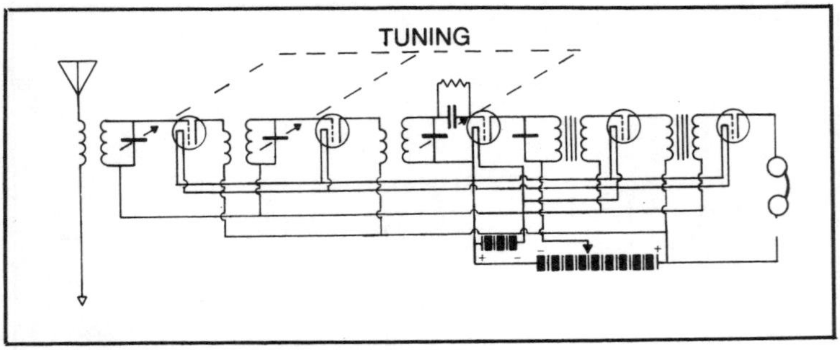

Fig. 2-8. Typical TRF receiver has triple-tuned circuit.

PROGRESS OF RADIO RECEIVERS

As a result of the Audion invention, ***amplitude modulation*** (AM) was born, and before long the spectrum was crowded with voices and music, and electromagnetic interference (EMI) finally began.

The type of radios described earlier were quite sensitive, but had very poor selectivity. Many stations were received at the same time. It is interesting to look at some statements made during this time, statements that probably define the radio receiver as we know it today.

"Imagination is better than knowledge" said Albert Einstein, and scientists certainly had imagination during this time in history. William

Fig. 2-9. David Sarnoff receiving traffic at the John Wanamaker store in New York City in 1912. He stayed at his post for 72 hours to report the Titanic disaster, demonstrating the importance of radio, and generating new interest in radio technology (courtesy of RCA).

Crooks, an English physicist who missed discovering x-rays, envisioned the progress of radio receivers. He said in 1892: "More delicate receivers which will respond to wavelengths between certain defined limits and be silent to all others remain to be discovered."

David Sarnoff (Fig. 2-9) a self-educated technical genius, entrepreneur, and a ham radio operator famous for receiving the Titanic distress message, said in 1916: "The receiver can be designed in the form of a simple radio music box and arranged for several different wavelengths, which should be changeable with the pulling of a single switch or pressing of a single button."

The realization of such selective receivers, did not come true until 1918, when Professor Armstrong invented the superheterodyne receiver.

—•••—

Chapter 3

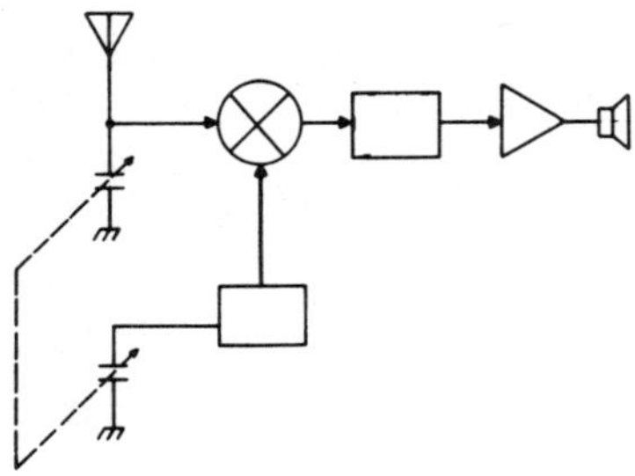

The Superheterodyne Receiver

The *superheterodyne* is a type of radio receiver that uses a process in which the incoming signal is mixed with a local oscillator signal in a manner generating new signal components which are equal to the sum and the difference of the original frequencies. One of these products is designated as the *intermediate frequency* (i-f) and is passed by a tuned circuit which rejects the undesired products as well as the original incoming frequency while still maintaining the information contained in it. This low level signal (caused by the natural losses along the propagation path between the transmitting point and the receiver's antenna) is further amplified before detection takes place.

In the amplitude modulation (AM) mode, the audio frequencies riding on the carrier are detected through the process of rectification, directly from the i-f, while code signals (cw) or single sideband signals (SSB) can be recovered by another mixing process which produces audible heterodynes with the help of a *beat frequency oscillator* (BFO).

In the case of *frequency modulation* (FM) where the carrier frequency moves around a center point at the rate of the audio frequency, a *frequency discriminator* or a *phase-locked loop* are used for detection. The final step after detection is audio amplification.

In the Armstrong superheterodyne, the *audion tube* performed several functions as shown in Fig. 3-1. It first acted as a local oscillator which produced high frequency currents of a different frequency than the signal to be received. It was used again as a mixer when this energy was heterodyned with the incoming frequency from the antenna, producing the intermediate frequency which was amplified again by the audion. Amplitude modulation (AM) detection was performed with the help of the Fleming valve, while final audio amplification used the audion again. Today there are

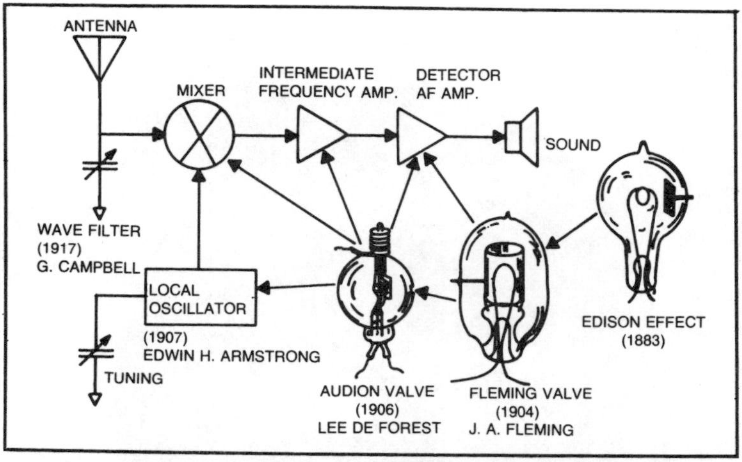

Fig. 3-1. The impact of earlier inventions on the superheterodyne which was invented in 1918 by Edwin H. Armstrong.

several types of superheterodynes, and their mixing schemes dictate their terminology, as shown in Table 3-1. Also, see Fig. 3-2.

SINGLE CONVERSIONS

A *single conversion receiver* can down-convert from the received frequency or up-convert, or both down-convert and up-convert can be used in a single conversion superheterodyne, when the range to be covered is so wide that the intermediate frequency falls somewhere in the middle of it, as in the case of many general coverage radio receivers with an i-f at 5.5 MHz, 9 MHz, or 10.7 MHz. There are advantages and disadvantages in these schemes, as we will see later.

MULTIPLE-CONVERSIONS

Another superheterodyne type is the *double conversion*, and as a growth from the double conversion, the *triple* and *quadruple conversion*.

Table 3-1. Superheterodyne Receiver Types.

1. Single-conversions	Down conversion Up Conversion Combination of the above
2. Multiple - conversions	Double-conversion Triple-conversion Quadruple-conversion
3. Direct-Conversion	

Fig. 3-2. The Philco model 81 is an example of an early superheterodyne receiver, intended for the reception of the broadcast and police bands in the early 1930s.

DIRECT-CONVERSION (ZERO I-F OR SYNC HROYNE)

The last and the simplest form of superheterodyne is the *direct conversion* type. We will analyze all these types of radio receivers, and examine their advantages and disadvantages.

—•••—

Chapter 4

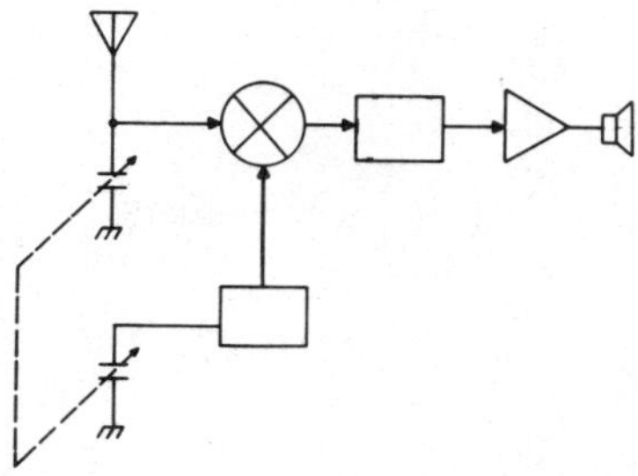

Implementing Single Conversion Superheterodynes

The most popular superheterodyne ever realized was a single conversion type with an i-f of 455 kHz. The block diagram in Fig. 4-1 shows a typical single conversion superheterodyne which tunes the frequency range from 500 kHz to 30 MHz. The i-f amplifier is tuned to 455 kHz.

The incoming signal anywhere between 500 kHz to 30 MHz passes through a tuned circuit called the *preselector*. The preselector has the role of filtering out all frequencies coming from the antenna, except for the one that is needed. This filtered signal goes to the *mixer*, which can take several forms, where it combines with the frequency coming from the *local oscillator* (in our case a variable frequency oscillator) and produces an i-f signal at 455 kHz.

As you have probably noticed, this is a down-conversion scheme because the i-f falls below all of the frequencies to be received within the range of 500 kHz to 30 MHz. The Vfo is operating 455 kHz above the incoming frequency at all times. For example, if we want to receive 500 kHz, the Vfo must generate 955 kHz. And, if we want to receive 30 MHz, the Vfo should be 30.455 MHz.

As a result of the interaction between the Vfo and the signal, there are two signals (referred to as *products*) coming out of our mixer, the sum and the difference (other products are generated as a result of the mixing and they will be reviewed later in this book). The higher product is rejected by the i-f filter, in this case, a transformer tuned to 455 kHz.

Note that no matter where we tune the receiver, the i-f is always the same, making processing of the signal identical for all frequencies to be received. This is why the superheterodyne has lasted until our day as the best design.

The 455 kHz signal, which also contains the information that came

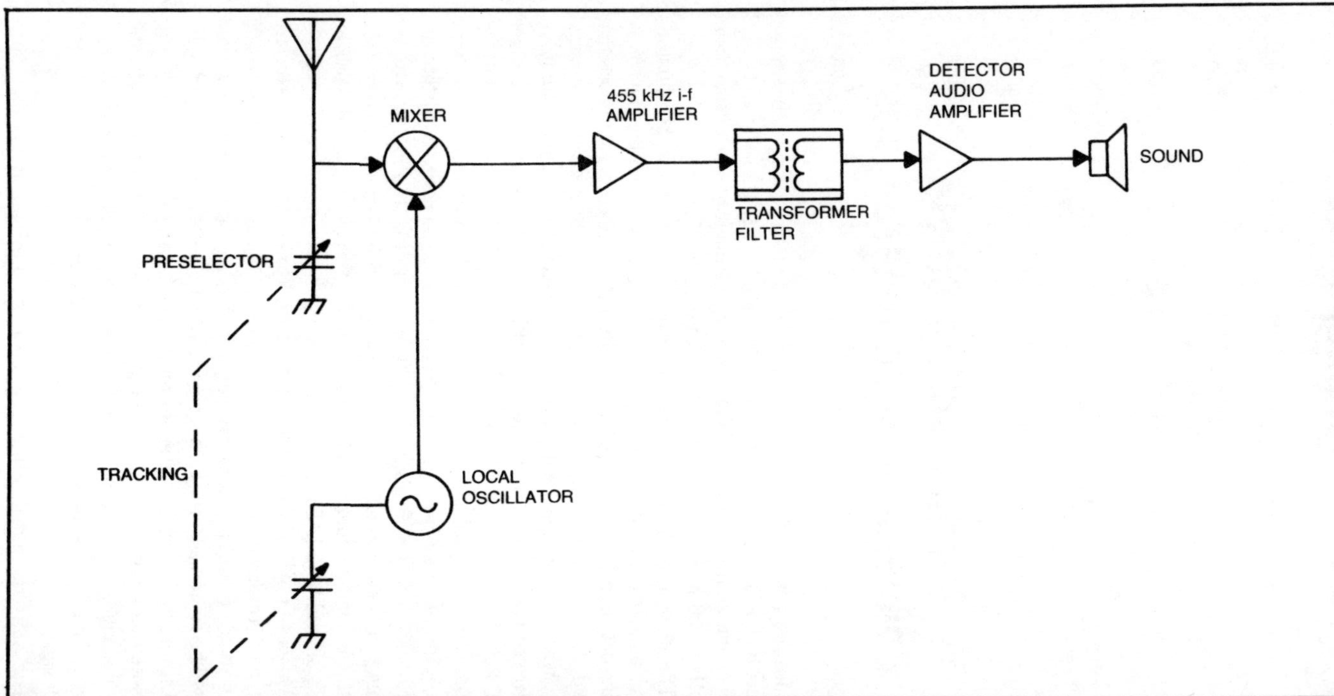

Fig. 4-1. General coverage single conversion superheterodyne receiver block diagram.

from the antenna, is filtered and amplified in the i-f amplifier, and is finally detected and amplified again at audio frequency before being delivered to the speaker. This scheme remains, to date, a basic approach of the single-conversion superheterodyne. Note the tracking mechanism of the scheme (Fig. 4-2). In this type of radio, it is imperative that the preselector and the Vfo track together.

In a basic single conversion hf receiver covering 500 kHz to 30 MHz, the incoming signal does not really go to just one filter preselector. Several filters are switched in and out for different ranges*, one at the time, allowing for several bands to be covered. The reason for this is that one tuned circuit cannot cover the entire range of 29.5 MHz properly. The Vfo is switched in similar bands, which exactly correspond in coverage to the preselector bands.

The tuning element of the preselector, the variable capacitor, is mechanically ganged with the Vfo capacitor, and they track together, so that when the preselector peaks at some frequency, the Vfo operates exactly 455 kHz above that frequency, as shown before. For example, if the preselector is tuned to exactly 1 MHz, it's capacitor is mechanically ganged with the Vfo capacitor, so that the Vfo produces exactly 1.455 MHz (Figs. 4-3 and 4-4). The rest of the circuitry remains the same, as described previously.

A practical receiver could be implemented as shown in the schematic diagram in Fig. 4-5. The signal from the antenna goes to switch S1-1 which couples it with any one of the five preselector tuned circuits, corresponding to the 5 bands in this case. Switch S1-2 which is mechanically ganged with S1-1 picks up the output of the preselector range chosen, and together with the variable capacitor which is always in the circuit, tracks the exact frequency to be received, which is exactly 455 kHz below the frequency of the Vfo.

The Vfo capacitor is mechanically coupled with the preselector capacitor. S1-3A and S1-4 are also mechanically connected with S1-1 and S1-2 in order to provide the right ranges for the Vfo to match the preselector. The Vfo signal couples through C18 into Q2 where it combines with the proper preselected signal and produces a difference frequency of exactly 455 kHz for any selected receiver frequency between 500 kHz and 30 MHz.

The signal is directed through a ceramic filter which eliminates all the unwanted mixing products and keeps only the 455 kHz signal which is routed through a chain of amplifiers, where it is detected and fed through a volume control, into an audio amplifier made of Q1, Q2, Q3, and Q4 (circuit board #2) and finally into the speaker.

By now, some of us probably wondered why was 455 kHz the choice for the i-f? Is there anything special about this frequency? There are two reasons for this choice.

- It was right below the broadcast band.

*Portion of the radio frequency spectrum designed to be received by a particular radio receiver, which for practical reasons can be briken down into "bands".

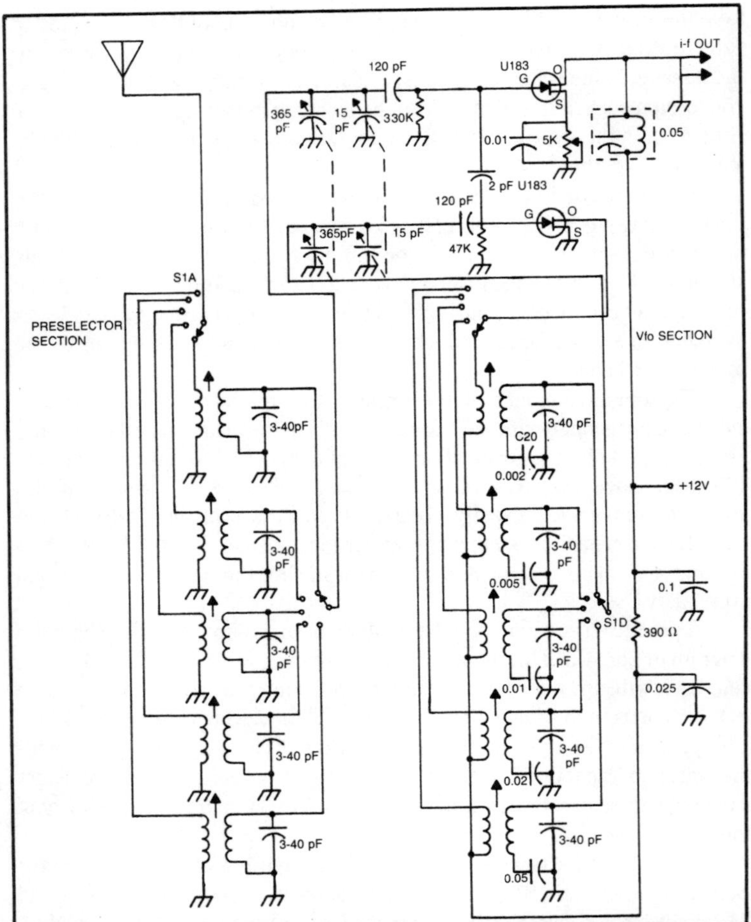

Fig. 4-2. Modern implementation of tracking mechanism for preselector and variable frequency oscillator, showing band selector.

■ It was low enough in frequency to allow adequate selectivity to be obtained with conventional inductors and capacitors (L/C).

Selectivity was not really a problem in the earlier days, but became one by the 1930s when the spectrum became crowded and because of it, the main disadvantage of the 455 kHz scheme became apparent. See Figs. 4-6 and 4-7.

THE IMAGE PROBLEM

We were all very happy with our radios and could hear a lot of stations with our general coverage receivers. Some of these stations were one and

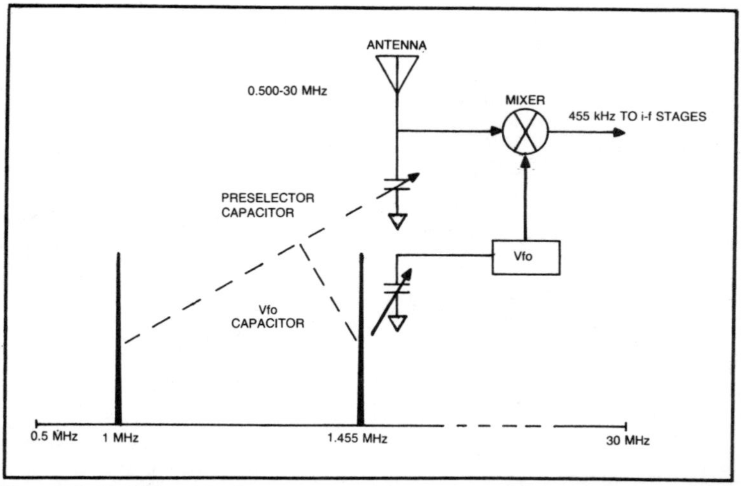

Fig. 4-3. Tracking mechanism of a general coverage single conversion superheterodyne.

the same (received in several places on the dials). Even though this phenomenon was discovered during the 1920s, it wasn't until the 1950s when manufacturers did something about it. Here is what the 1926 edition of the Radio News Superheterodyne Book says in regard to this problem, offering a solution: "There are many other instances where slight improvements may be made. Two straight-line frequency condensers, mounted on one shaft, for the oscillator and tuner circuits are recommended. This not only simplifies tuning, but eliminates the objectionable feature of every superheterodyne—the reception of all stations on more than one setting of the oscillator dial". This phenomenon is called *image*.

Fig. 4-4. Modern variable capacitors featuring as many as four tracking sections are being used in superheterodyne receivers.

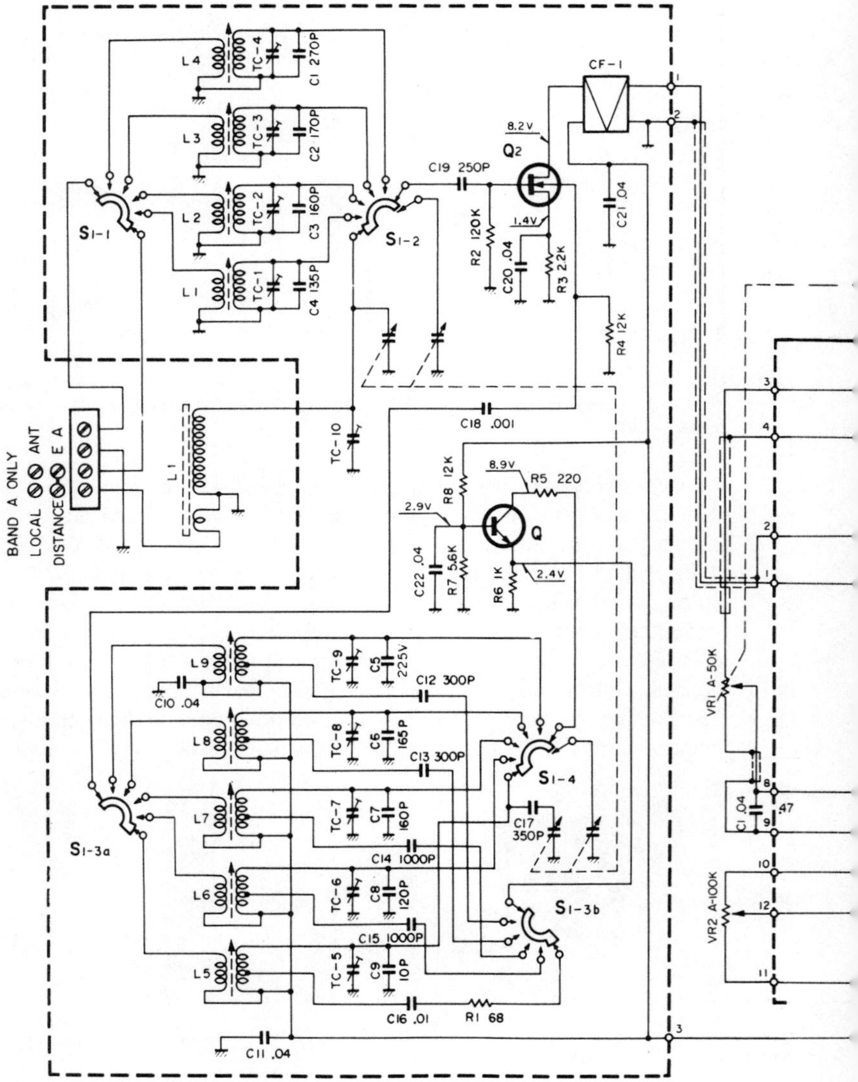

Fig. 4-5. Modern implementation of a general coverage 455 kHz single conversion superheterodyne receiver.

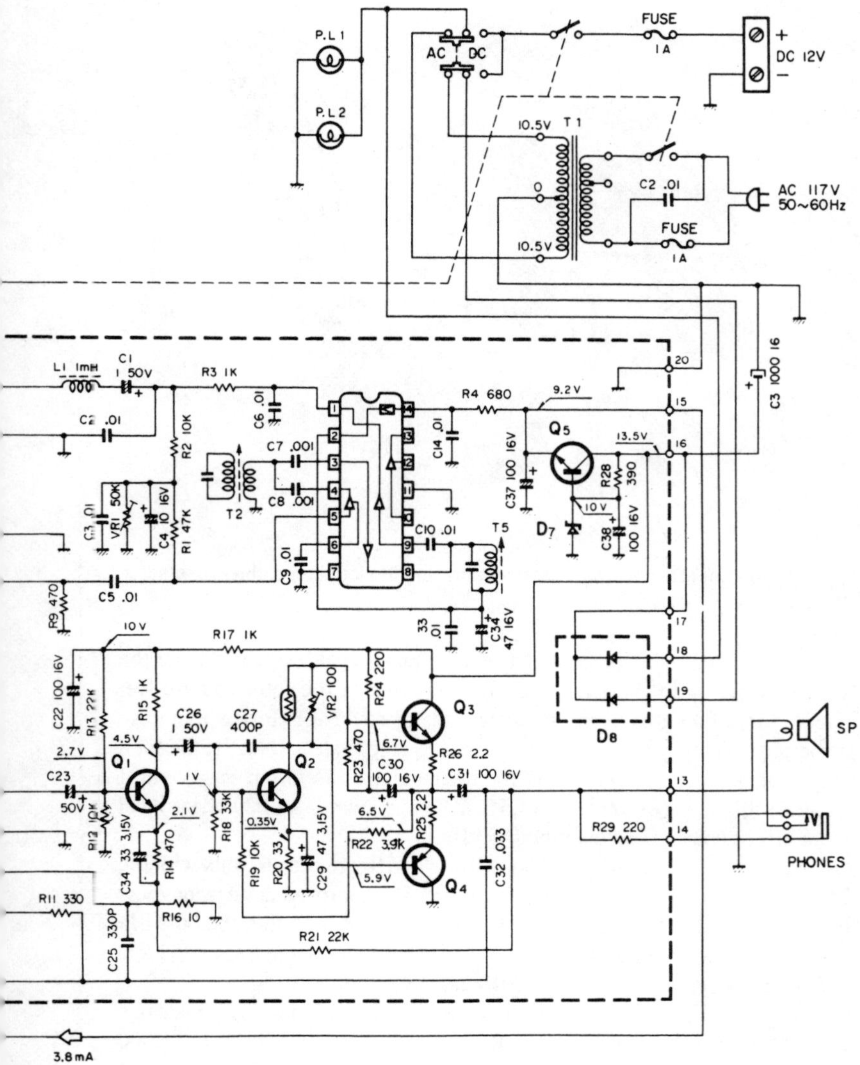

P.L 1
P.L 2
AC
DC
FUSE
1A
DC 12V
10.5V
T1
0
C2 .01
AC 117V
50~60Hz
L1 1mH
C1 1 50V
R3 1K
R4 680
9.2V
Q5
13.5V
C3 1000 16
R2 10K
C6 .01
C7 .001
C8 .001
T2
VR1 50K
R1 47K
C14 .01
C37 100 16V
R28 390
10V
D7
C38 100 16V
C10 .01
T5
C9 .01
C5 .01
R9 470
C34 47 16V
R17 1K
C22 100 16V
R13 22K
R15 1K
C26 1 50V
C27 400P
VR2 100
R24 220
Q3
D8
SP
Q1
Q2
R23 470
6.7V
R26 2.2
C30 100 16V
C31 100 16V
C23 50V
R12 10K
R18 33K
R19 10K
R20 33
C29 47 3.15V
R22 3.9K
R25 2.2
6.5V
5.9V
Q4
C32 .033
R29 220
PHONES
R14 470
R11 330
R16 10
C25 330P
R21 22K
3.8mA

Fig. 4-6. The general coverage receiver showing the location of the preselector section.

It has been established that the incoming signal frequency from the antenna beats with the local oscillator to produce the intermediate frequency for further amplification. This oscillator could be below or above the incoming frequency by the intermediate frequency value. Usually, the total oscillator is above the incoming frequency signal, as in our case, and it would oscillate, for example, at 14.545 MHz to convert a 14.090 MHz signal to an intermediate frequency of 455 kHz.

Another station, such as WWV at 15.000 MHz, would also mix with the 14.545 MHz local oscillator producing the 455 kHz intermediate frequency. If the front-end selectivity of the receiver is poor, and the 15.000 MHz station is strong, it will also be heard with the receiver tuned to 14.090 MHz and interfere with a real 14.090 MHz signal.

To dramatize this, consider the case in Fig. 4-8. Station A broadcasts at 14.090 MHz with 100 watts of power, and Station B broadcasts at 15.000 MHz with a power of 10 kilowatts. The relatively wide preselector is set at 14.090 MHz as shown. The Vfo oscillates at 14.545 MHz. Station A combines with the Vfo producing 455 kHz and this signal is processed further in the i-f chain and amplifier and is detected and heard in the speaker.

Station B at 15.000 MHz, falls within the pass band of the preselector, although somewhat attenuated by it. The preselector is of the type discussed previously. It involves one or two tuned circuits, and Station B is powerful enough to make up for the off-frequency rejection as shown in our exaggerated case here. Station B at 15.000 MHz also mixes with the 14.545 MHz oscillator producing the 455 kHz i-f signal which is processed along with the signal from Station A and is heard at the same time, thus interfering with the real signal.

The image is twice the intermediate frequency above the received signal if the oscillator frequency is above the desired signal frequency, or twice the intermediate frequency below the signal frequency if the oscillator is below the desired signal frequency. We can then say that in a typical 455 kHz receiver, images will be created by any stations 910 kHz above the frequency being received.

In such a system, a strong station from one end of the 28 MHz amateur band can interfere with the reception of a weak station at the other end. Most images are caused by other than amateur signals, and can be quite strong, presenting a great problem if such a receiver would be used aboard a

Fig. 4-7. Overall view of the 455 kHz i-f general coverage receiver showing the vfo capacitor mechanically coupled with the preselector capacitor.

military ship or aircraft, where as many as thirty or more transmitters could be operating in close physical proximity.

Improving preselector selectivity would mean increasing the number of tuned circuits with additional complex mechanical switching. The complexity of such an arrangement is exemplified by Fig. 4-9 which shows the schematic diagram of an actual implementation of two-pole preselector and local oscillator in the Realistic DX-160 communications receiver. This arrangement provides an image rejection from – 20 dB to – 65 dB below the station being received, depending on the band of interest (measurements were made in the center of the bands).

Some attempts have been made to minimize the switching of multipole preselectors by using combinations of variable capacitors with very wide range, with magnetically or electromagnetically tuned ferrite coils. Figure 4-10 shows a list of such products made by Vari-L Corp. It is known that a ferrite coil changes its inductance characteristics over a wide range in the presence of a magnetic field. This phenomena is exploited in these products by applying a dc current bias to an electromagnet facing the main coil which is wound on a ferrite core as shown in Fig. 4-11. Changing this bias changes the inductance of the particular coil, providing the designer with a means of extending the preselected range with minimum switching. See Figs. 4-12 and 4-13.

This property was further developed in a preselector that uses opposing polarity magnets with a toroidal ferrite coil sandwiched between them as shown in Fig. 4-14 and Fig. 4-15. A set of four magnets polarized as shown are the stator of the device, while another set of four magnets are mounted in the rotor. The ferrite toroidal coil is mounted between the stator and rotor and forms the first pole of the preselector.

The second pole is made of another toroidal ferrite coil which is identical with the first one and is mounted on the other side of the rotor,

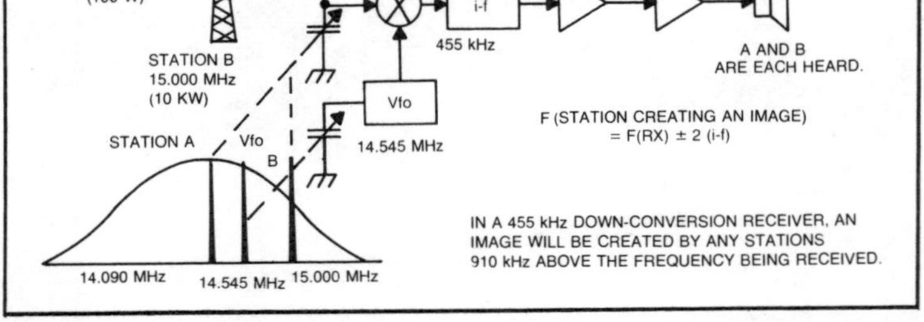

Fig. 4-8. The image in a 455 kHz i-f single conversion superheterodyne receiver.

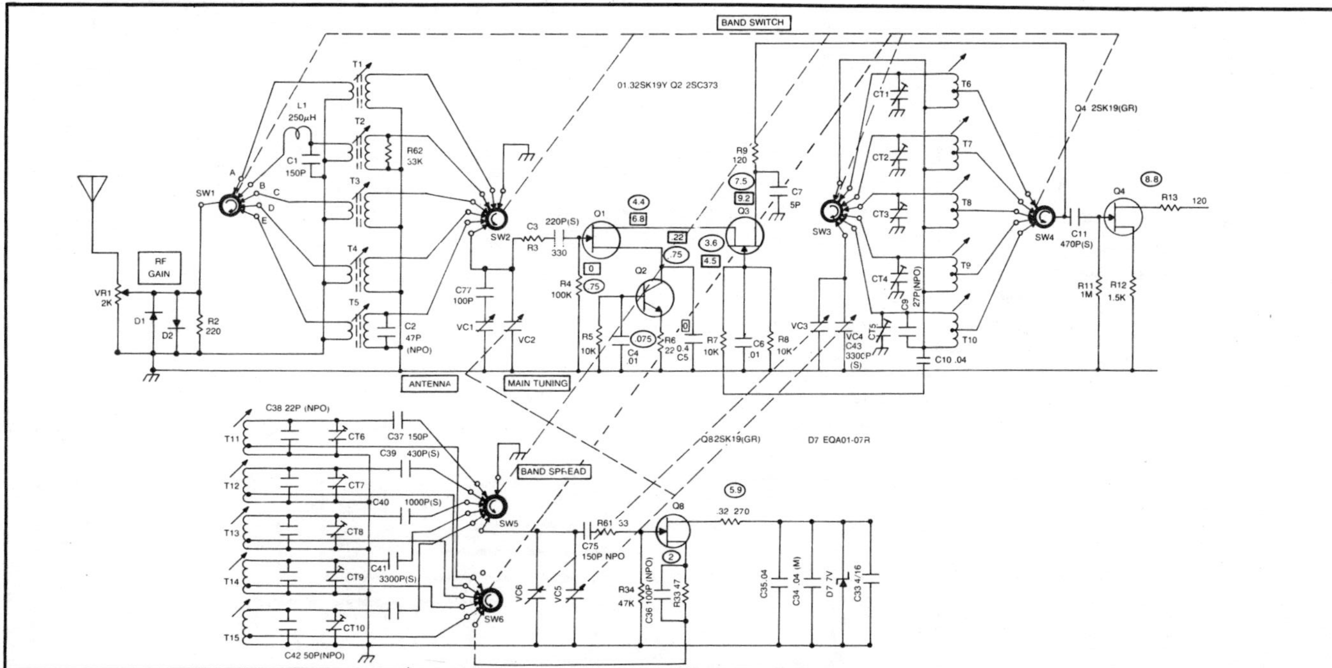

Fig. 4-9. The preselector-local oscillator mechanism of the Realistic DX-160 general coverage communications receiver, involves a six-wafer, five-position band switch, combined with a five-section mechanically-tracked variable capacitor.

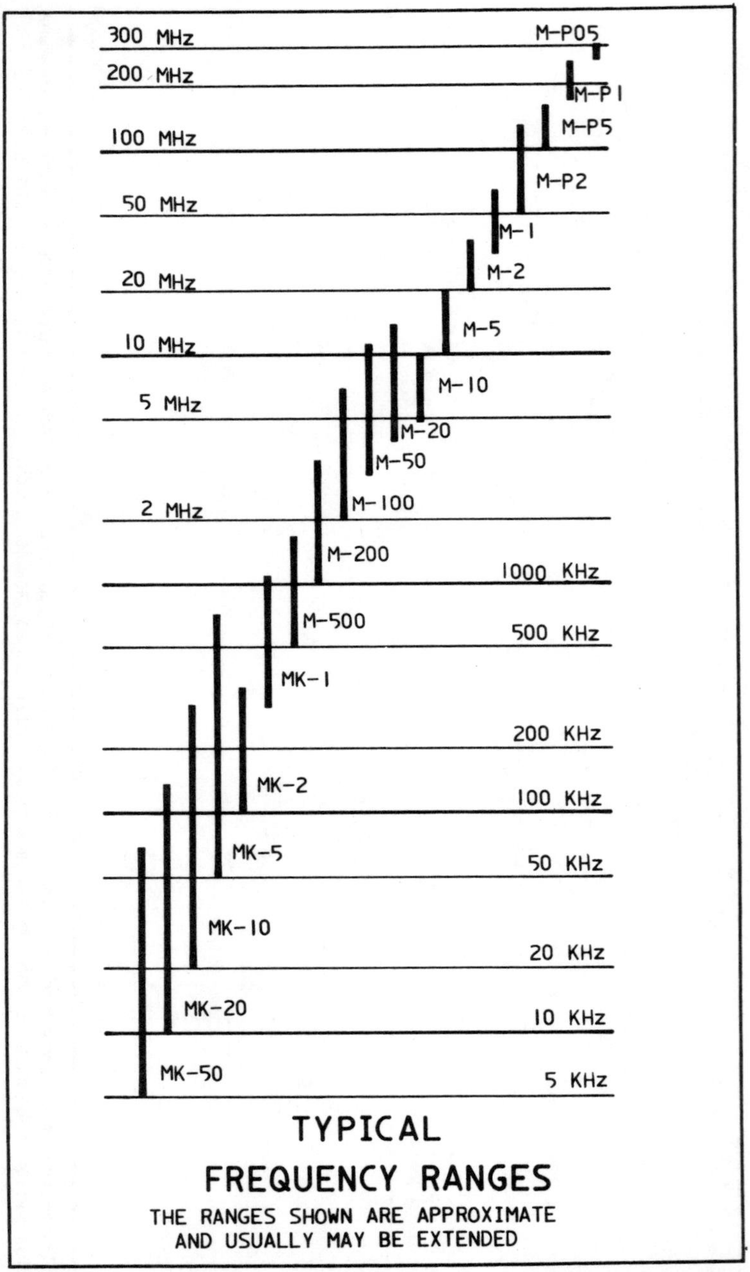

Fig. 4-10. Typical frequency ranges of electrically variable inductors. The ranges shown are approximate and usually may be extended (courtesy of Vari-L Corporation).

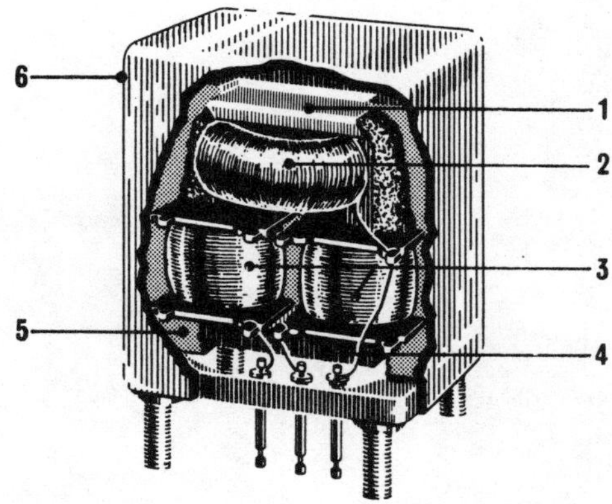

Fig. 4-11. Cut-away of a Vari-L inductor (courtesy of Vari-L Corporation).

which is now common to both poles. A third set of magnets polarized the same way as the first part of the stator completes the arrangement.

The specially-arranged magnetic field created by the rotor can then permeability-tune both toroidal ferrite coils proportionally, making for an extremely wide tuning non-switching preselector. The number of poles could be increased providing for sharper response.

Fig. 4-12. This aircraft radio receiver is typical of the ARC-5/BC type. Several such receivers were used in order to cover a greater range of frequencies.

Fig. 4-13. ARC-5 receivers used in the cockpit of an aircraft during WWII (Photo taken by Roger Musolf at the National Air and Space Museum, Washington, D.C.).

This preselector was designed to cover from 0.5 to 30 MHz by only switching the capacitors across once. Several problems occur with such an approach, however. Too much magnetic field will drive ferrites into saturation and the point of saturation is not always predictable since it depends on temperature. Secondly, unless the preselector is magnetically shielded, ac magnetic fields resulting from power supply transformers and other sources can easily modulate the ferrite coils. In reality, this preselector would become impractical. The real fix for supressing the image problem is

Fig. 4-14. A two-pole permeability tuned preselector.

to choose the i-f at a much higher frequency. Mathematically, this proves to be preferred.

If the frequency of a strong station creating an image equals F(RX) + 2•F(i-f), and the range to be covered by the receiver is for example 500 kHz to 30 MHz, a rule of thumb is that the intermediate frequency should be chosen above the highest frequency to be received (in our case 30 MHz) by about 35% (or higher), which happens to be 41 MHz as shown in Fig. 4-16. This is easy to establish mathematically but there was another element that impacted this design consideration, that being selectivity.

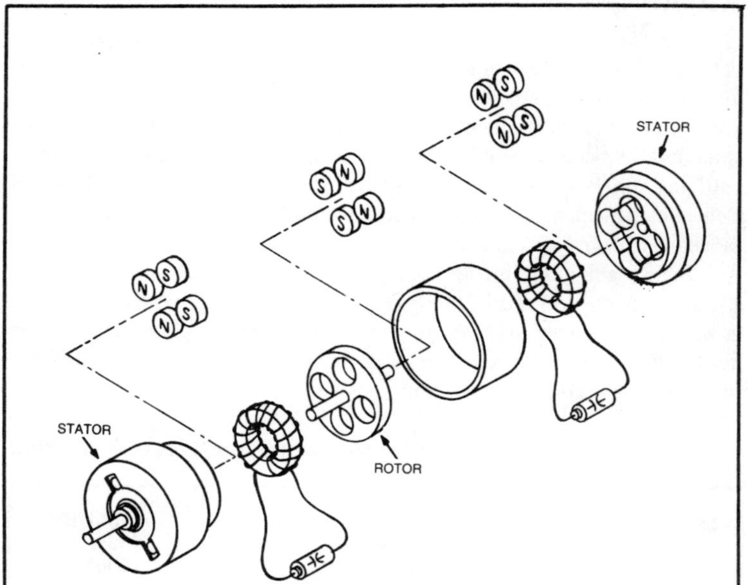

Fig. 4-15. A mechanical arrangement of parts in the permeability-tuned preselector.

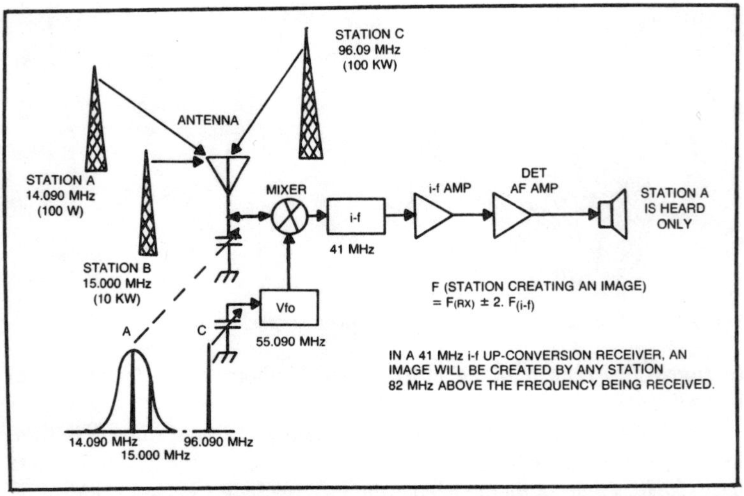

Fig. 4-16. Image with a 41 MHz i-f up-conversion receiver is 82 MHz above the received frequency (96.090 MHz) and can be easily filtered out at the preselector. The former image causing station at 15 MHz is not a problem with this arrangement.

SELECTIVITY

The bandwidth of the information that we are trying to capture varies between about 8 kHz for amplitude modulation signals to as little as 300 Hz for continuous wave signals (Table 4-1). This selectivity is much easier to accomplish at lower frequencies such as the 455 kHz (the adjacent channel selectivity is accomplished in the i-f filter). We were simply not capable until the 1950s of providing i-f filters with this kind of bandwidth at frequencies much higher than 455 kHz. Humanity stumbled along for about 30 years with the 455 kHz scheme.

Before going any further, it should be mentioned that there is one advantage to the 455 kHz i-f, it is simple to implement. The *image response*, however, remains a problem to the present. Another approach to the front-end preselector which eliminates image problems while using a low i-f scheme will be presented later.

Table 4-1. Selectivity Requirements for Different Modes of Radio Communication.

MODE OF OPERATION	BANDWIDTH
1) AMPLITUDE MODULATION (AM)	8 kHz
2) FREQUENCY MODULATION (FM)	15 kHz
3) SINGLE SIDEBAND (SSB)	2.1 kHz
4) INDEPENDENT SIDEBAND (ISB)	6.1 kHz
5) FREQUENCY SHIFT KEYING (FSK)	1 kHz
6) CONTINUOUS WAVE (KEYED) (CW/KCW)	≤ .4 kHz

Fig. 4-17. Modern high-Q rf transformers used in commercial equipment.

INTERMEDIATE FREQUENCY FILTERS

The trend of i-f filter design started with high-Q rf transformers such as those shown in Fig. 4-17, and gradually progressed to mechanical, ceramic, and crystal filters, and also to *surface-acoustic-wave* (SAW) *filters*.

MECHANICAL FILTERS

The advent of i-f filters began in 1947 when Robert Adler of Zenith designed and built a six-element electro-mechanical filter for 455 kHz (Figs. 4-18 and 4-19). He writes: *"Intermediate frequency currents, upon entering the filter, are converted into mechanical vibrations of the same frequency. These vibrations then pass through a structure resembling a ladder, consisting of several mechanically resonant metal plates coupled to each other by wires which act as springs.*

*This structure forms a bandpass filter for mechanical vibrations. Width of the pass band, as will later be shown, is determined by the design of each individual section and, because the several sections are all alike, bandwidth does not depend upon the number of sections. Attenuation outside the band limits, however, increases with the number of sections.**

To understand the operation of the filter most easily, let us first consider a familiar electrical filter, composed of inductors and capacitors; the type shown

Author's note: *This is a theoretical case and assumes perfect square response.

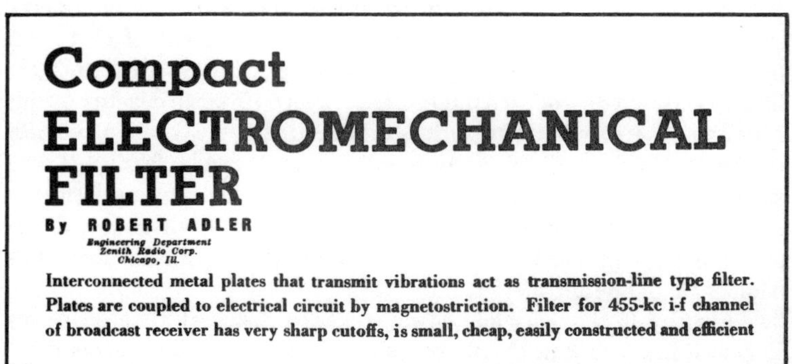

Compact ELECTROMECHANICAL FILTER

By ROBERT ADLER

Engineering Department
Zenith Radio Corp.
Chicago, Ill.

Interconnected metal plates that transmit vibrations act as transmission-line type filter. Plates are coupled to electrical circuit by magnetostriction. Filter for 455-kc i-f channel of broadcast receiver has very sharp cutoffs, is small, cheap, easily constructed and efficient

Fig. 4-18. Title reprinted with permission from Electronics, vol. 20, pp. 100-105, copyright April, 1947.

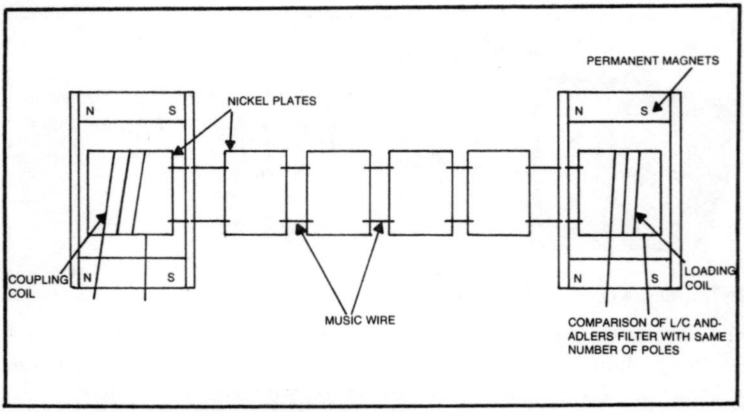

Fig. 4-19. The first 455 kHz mechanism filter-1946. Reprinted with permission from Electronics, vol. 20, pp. 100-105, copyright April, 1947.

in Fig. 1A (figure not shown) *is rather fundamental and can be found in many books on electrical wave filters.*

*The analogy between electrical and mechanical network elements, with masses substituted for inductors and springs for capacitors, is well known; and it appears quite feasible to build a mechanical filter structure, which is fully equivalent to the electrical filter shown in Fig. 1A, by combining masses and springs in analogous fashion."*** Adler applied his filter to a 455 kHz i-f receiver. He notes:

"It was first suspected that such a set would be hard to tune, but tests with a number of lay listeners did not bear this out. Change in tone quality caused by incorrect tuning sets in at two clearly defined points much more abruptly than in conventional receivers; listeners seem to find it quite easy to tune between these points". **

Adler made several tests showing clear improvement against adjacent channel interference with his newly invented filter.

Although L/C filters are still being used, mechanical filters offered superior performance as shown in Fig. 4-20. Mass production of this type of filter was pioneered by the Collins Radio Company in 1947. By 1952, the first 455 kHz single sideband filters were produced in small quantity by this company. The popularity of such filters continues today when complex mechanical filters with twelve or more resonators can be fabricated with shape factors of as low as 1.25 (Method of measuring filter performance represented by the ratio between filter bandwidth at – 60 dB and – 3 dB, as shown in Fig. 4-21) and passband ripple as low as ±0.1 dB. The lower this number is, the better the filter is.

Figure 4-22 shows a modern twelve-disc mechanical filter produced by Rockwell International. The theory of operation for this filter is not much different than that of Adler's which was presented earlier. The technologi-

**Excerpts reprinted with permission from Electronics, Vol. 20, pp. 100-105, Copyright April 1947.

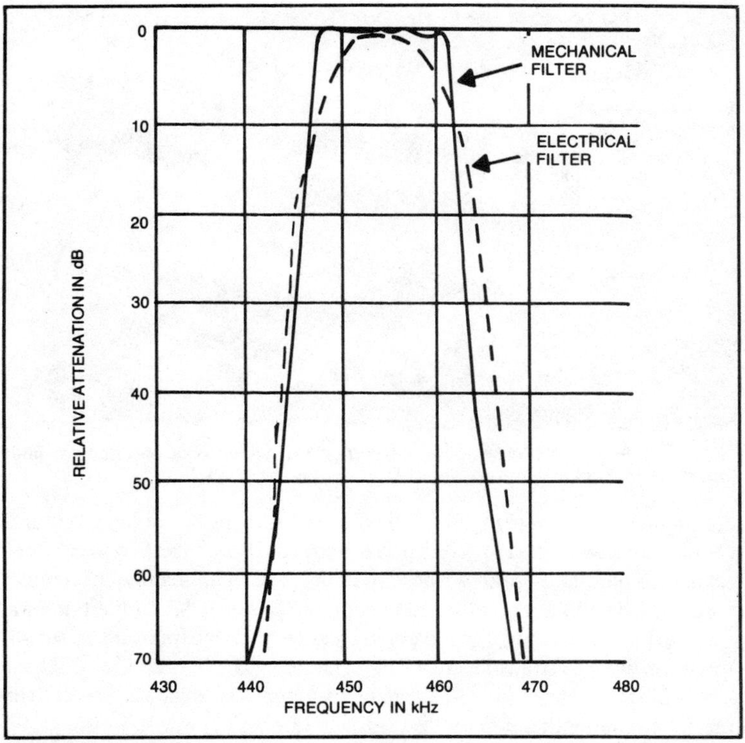

Fig. 4-20. Comparative frequency response curves between electrical and mechanical filters used in 455 kHz i-f's.

cal implementation, however, departs considerably. While Adler's filter was using inefficient and lossy magnetostrictive-wire transducers*, this modern disk-wire mechanical filter uses piezotechnology (PZT) transduc-

---

*A device which transforms one kind of energy into another. In our case electrical signals are transformed into mechanical signals, or the other way around.

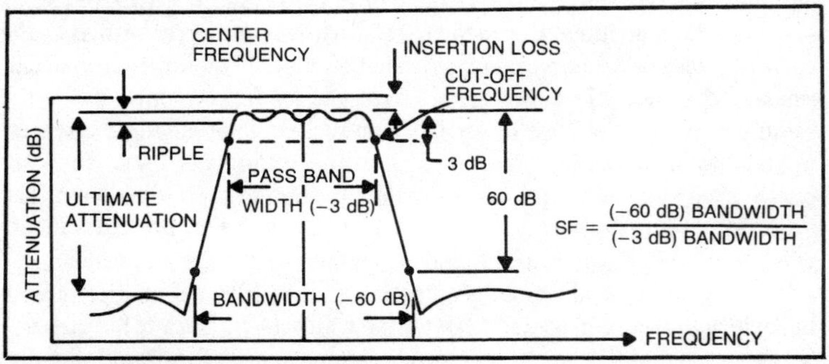

Fig. 4-21. Determining shape factor.

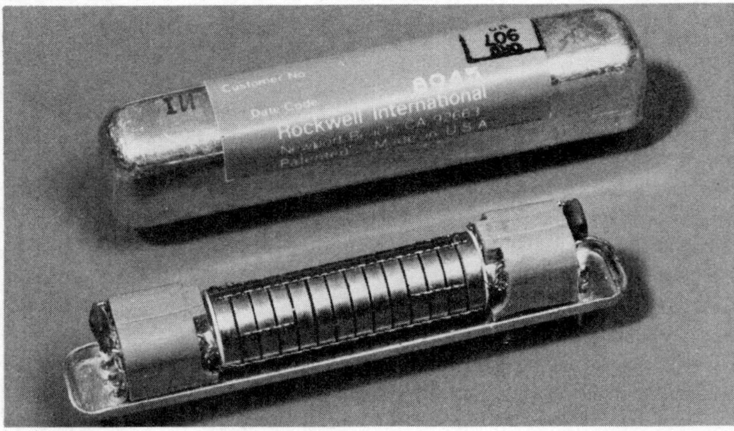

Fig. 4-22. Actual implementation of a modern twelve-disk mechanical filter (courtesy of Rockwell International Corporation).

ers in combination with ferrite cup cores and negative temperature-coefficient capacitors (Fig. 4-23) in order to reduce the insertion loss to less than 3 dB and the passband ripple to under 2 dB, and maintain frequency stability over the temperature range of −55°C to +85°C. (If the temperature performance is not a requirement, ferrite transducers could be substituted for the PZT transducers.) The bridging wires in both Fig. 4-22 and 4-23, allow for phase inversion needed to obtain attenuation poles, symmetrically, on both sides of the center frequency.

While Adler's filter was produced by laboratory methods, this modern implementation is the result of a highly technological production process which starts with six-foot long cylindrical bars made of a mechanically-stable iron-nickel alloy, which was ground to a precise diameter, cut into disks of a certain thickness, then heat treated, and ground within .0002 inches accuracy. The individual disks are then frequency tuned, which is accomplished by drilling out a small amount of metal at their geometrical center. This operation is performed automatically in only seven seconds with the help of a computer-controlled drillpress equipped with frequency sensors which continuously measure and anticipate the resonance of the particular disk and then automatically shut off the drill motor when the right resonant frequency is achieved. The transducers follow somewhat of the same pattern. In the case of ferrite, iron-nickel oxide powders are wet-mixed and extruded into rods similar to the iron-nickel bars. They are finally fired and cut to size before being attached to the metal disks. The final assembly incorporates the right number of disks, the transducers which are now attached to the ends, together with coils and magnets.

The bridging and coupling wires are spot welded automatically with the help of a computer-controlled machine which keeps track of the distance between the disks as well as the precise consistency of the weld. Finally, shock-absorbing rubber grommets are added before the final inspection is performed with the help of network analyzers, and the cover is put on by heat staking or automatic soldering.

Mechanical filters are easy to apply in i-f circuits as shown in Fig. 4-24 and Fig. 4-25.

CRYSTAL FILTERS

Today there are several types of i-f filters (Fig. 4-26), with the crystal lattice being by far the most widely used. Crystal technology improved considerably in the past 20 years; but one thing remains, it was harder to make filters into the MHz range than it was to make mechanical filters at 455 kHz.

A crystal filter usually consists of one to as many as twenty-eight individual quartz crystals, arranged together electrically by means of inductors, capacitors, and resistors, so that a certain bandpass characteristic is achieved. To best understand how this is implemented, we will consider a single individual crystal at resonance, as being a *high Q* series resonant circuit, as shown in Fig. 4-27. *Q* is a figure of merit for selectivity, indicating how sharp the frequency response is. Q's greater than 50,000 have been observed.

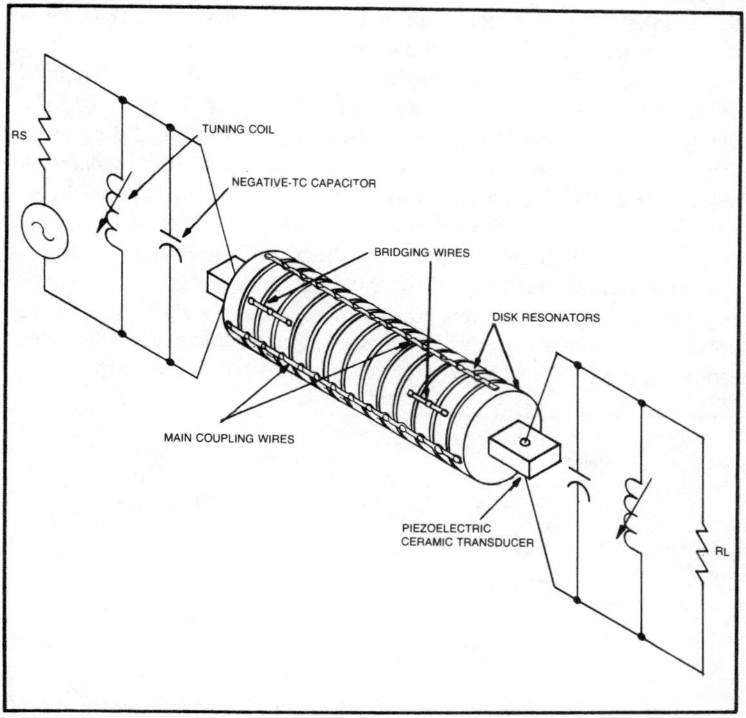

Fig. 4-23. Internal configuration of the disk-wire mechanical filter.

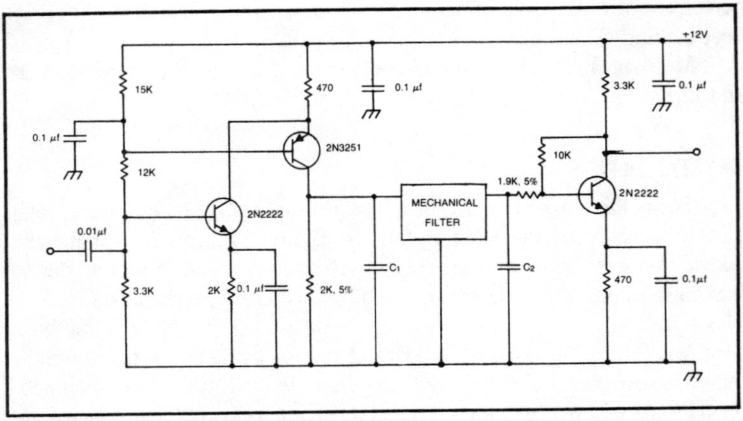

Fig. 4-24. Electrical implementation of a 455 kHz mechanical filter.

If the signal generator was swept slowly over a range of interest, a sharp output would be observed on the spectrum analyzer as we approach the resonant frequency of the crystal. The response characteristics of the one-pole crystal filter is shown in Fig. 4-28. (Quartz crystals exhibit other modes of resonance resulting in spurious outputs which are not considered in this example for reasons of simplicity.)

It can be seen from this example that this pole has a very narrow band width, which can vary from a few hundred Hz to as much as 80 kHz, depending on the crystal frequency. The addition of carefully balanced series and paralled elements such as other crystals with slightly different frequencies and inductances and capacitances can create additional poles, and eventually a passband response with steep skirts (small shape factor) can be achieved. Figure 4-29 shows the schematic diagram of a two section crystal-lattice filter. In this example, the shunt inductances for the crystals are part of the input and output transformers. The crystals Y1 and Y2 are slightly different in frequency from crystals Y3 and Y4. In reality the series resonant crystals (Y1 and Y2) are tuned to the parallel resonance frequency

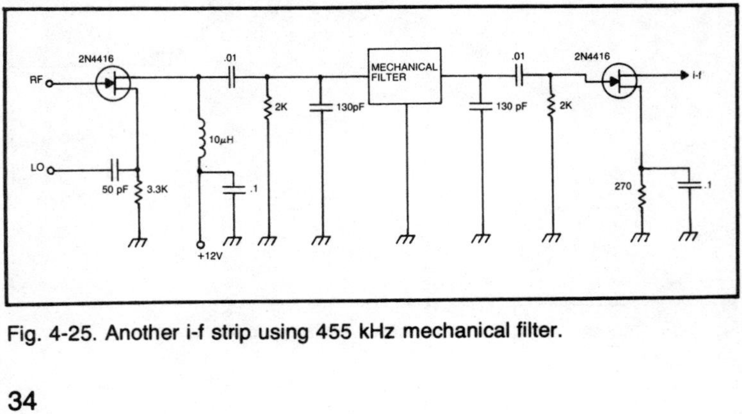

Fig. 4-25. Another i-f strip using 455 kHz mechanical filter.

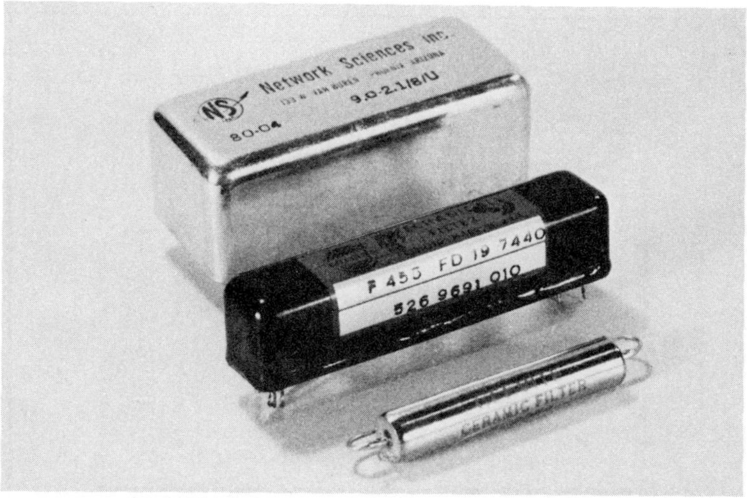

Fig. 4-26. Three types of modern i-f filters: Crystal lattice (top), mechanical (center), and ceramic (bottom).

of the other two crystals (Y3 and Y4) providing for a bandpass characteristic as shown in Fig. 4-30.

Modern commercial and military crystal filters involve several individual crystals as shown in Fig. 4-31. The theory and design of such filters is several orders of magnitude and more difficult than that of an individual crystal. In such an electromechanical arrangement, quartz crystals can exhibit several modes of resonance, as we discussed earlier, resulting in complex interaction of such parameters as center frequency, bandwidth, shape factor, group delay, transient response, phase linearity, insertion loss, temperature, and aging, making for exhaustive mathematical analysis which is usually performed by computers. Networks with as many as thirty-two quartz crystals have been theoretically modeled by filter engineers, however, implementing such filters obeys natural laws not understood yet by these computers. Complex quartz crystal filters are being manufactured by several companies. Figure 4-32 shows examples of such filters.

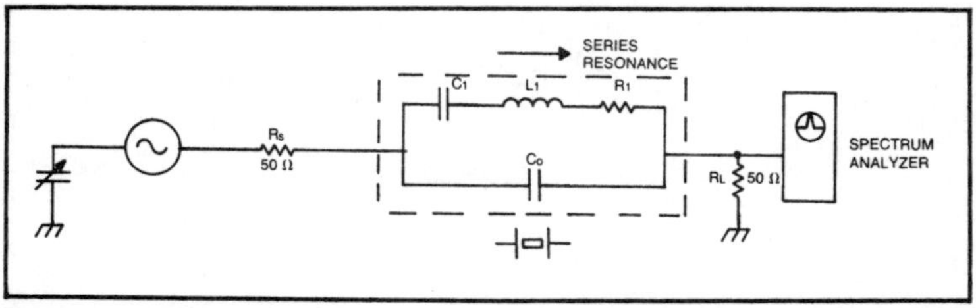

Fig. 4-27. Test set-up for a quartz crystal, showing its equivalent high Q series resonant circuit.

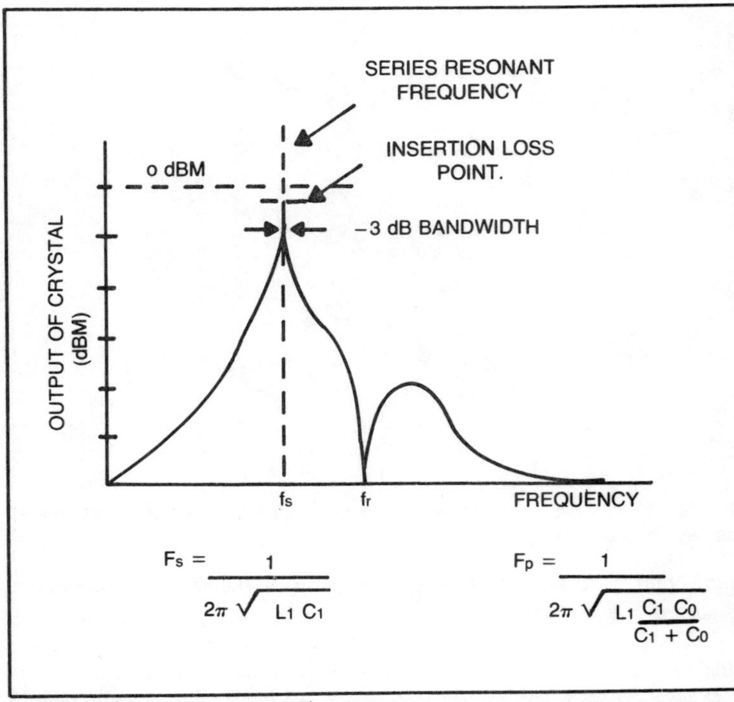

Fig. 4-28. Response characteristics of a quartz crystal. The highest response is achieved at the series-resonant frequency (fs).

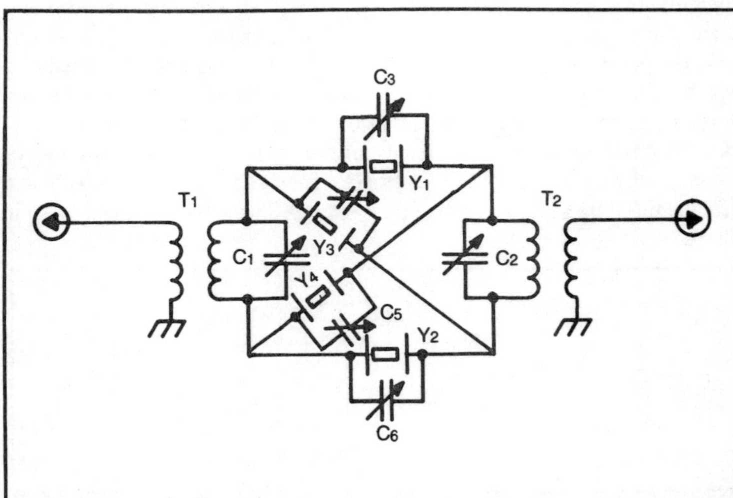

Fig. 4-29. A two section crystal-lattice filter. The transformers are used as shunt inductances.

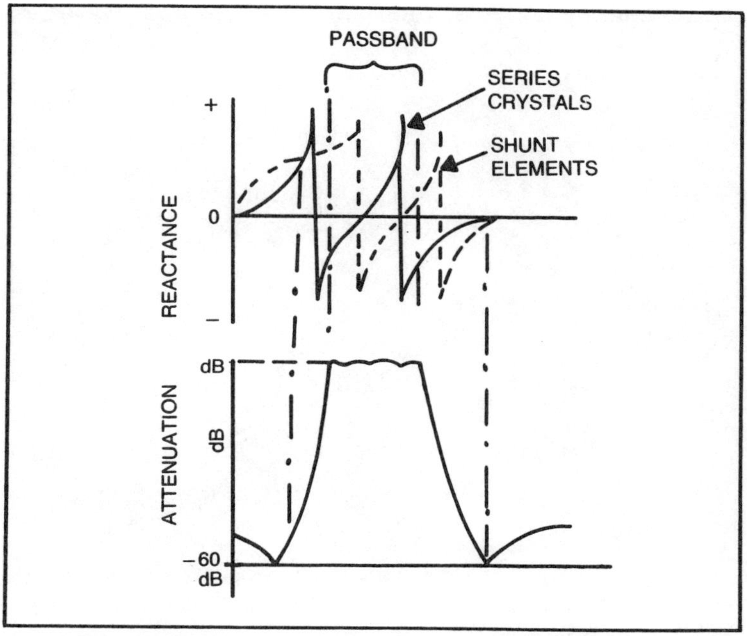

Fig. 4-30. Response characteristics of the two-section lattice filter.

MONOLITHIC CRYSTAL FILTERS (MCF)

A simpler way of achieving a quartz filter is the monolithic approach. In a *monolithic filter*, there are no discreet adjustments such as inductances and capacitors. Two electrode pairs are deposited on a quartz wafer (Electrochemical methods of depositing aluminum, silver, or gold are used in conjunction with AT cuts of quartz crystals) as shown in Fig. 4-33.

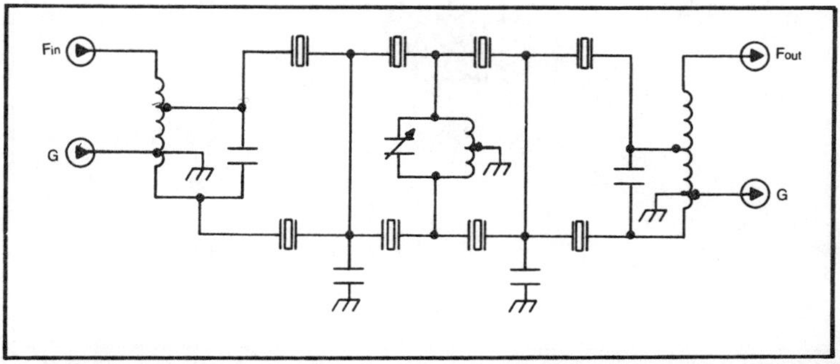

Fig. 4-31. Internal configuration of a commercial eight-pole crystal-lattice filter (XF-9B) available from KVG* (Kristallverarbeitun Neckarbischofsheim, West Germany; available from Spectrum International). Each stage is made of tuned half-lattice sections with one crystal in each branch.

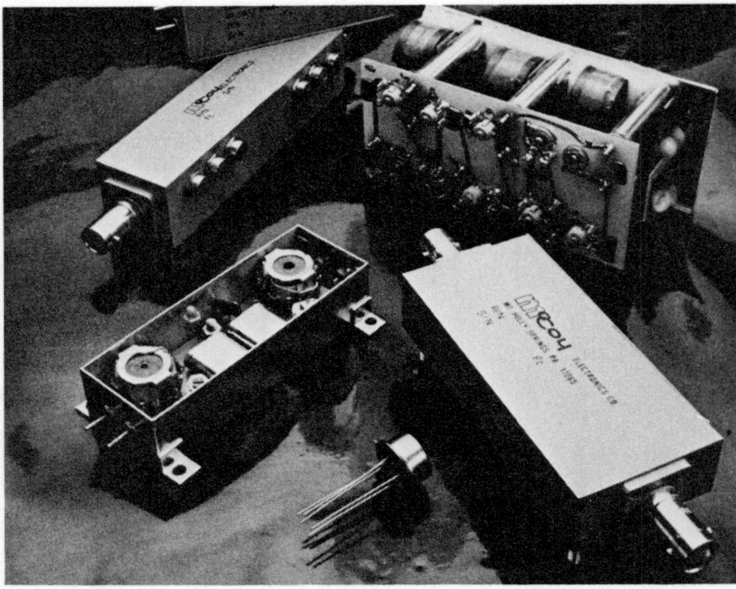

Fig. 4-32. Quartz crystal filters manufactured by McCoy Electronics (courtesy of McCoy Electronics).

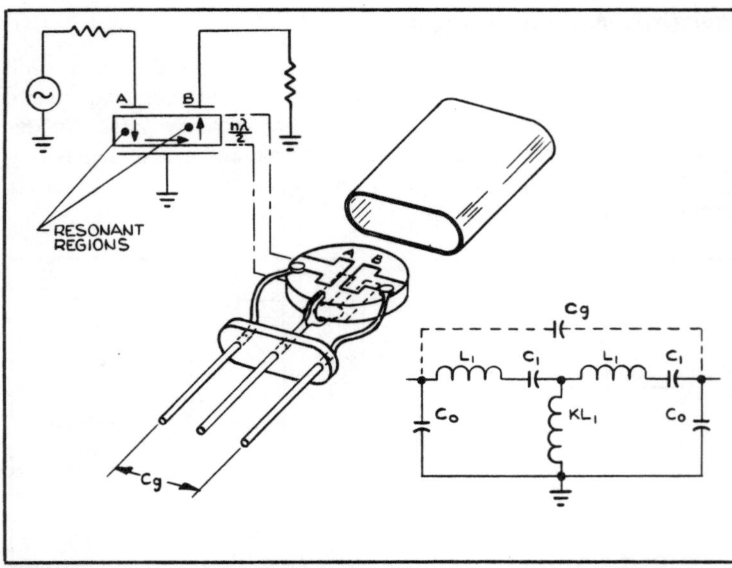

Fig. 4-33. A two resonator monolithic filter and its electrical equivalent. The thickness of the crystal wafer equals $\frac{n\lambda}{2}$, where λ is the operating wavelength and n is an odd integer. C_g is the gap capacitance between leads, a limiting factor for the stop band rejection of the monolithic filter.

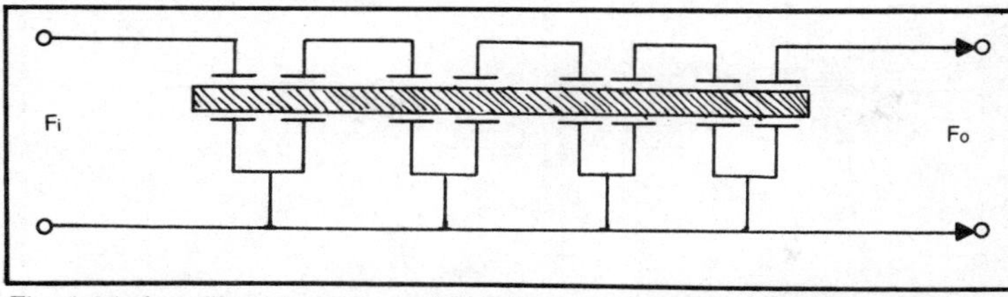

Fig. 4-34. A multiresonator monolithic filter with eight electrode pairs on a single substrate.

A sound wave generated at electrode A travels through the crystal wafer to electrode B, at the two resonant regions as shown in our example. The thickness of the wafer at the two points determines the band-pass characteristic of the monolithic filter.

This "folded" approach to acoustically coupled resonators was not fully understood until 1965 when R. A. Sykes at Bell Telephone Laboratories, and M. Onoe in Japan discovered (simultaneously) the mathematics governing this mechanism.

This event led to the development of multiresonator monolithic filters such as the one shown in Fig. 4-34. However, the inability to tune out unwanted elements, such as in a conventional quartz filter, limits the practical number of resonators in a monolithic to four pairs. While in a conventional quartz filter, the interaction of unwanted parameters could be minimized by adjusting variable elements such as capacitors and inductances, in a monolithic filter, this interaction is a fixed function of the mechanical arrangement, and is further aggravated by indirect sound paths of unpredicted nature between the input and output, producing spurious responses and degrading the stop-band attenuation of the filter.

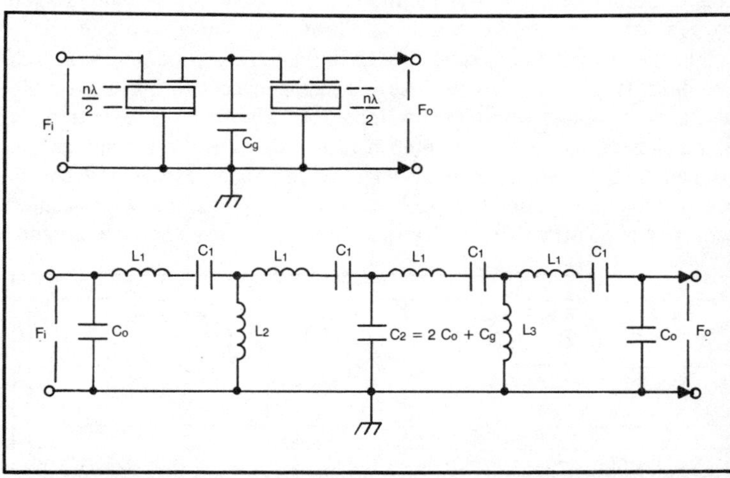

Fig. 4-35. Two two-pole monolithics are connected in tandem for increased stop-band attenuation and ease of manufacturing.

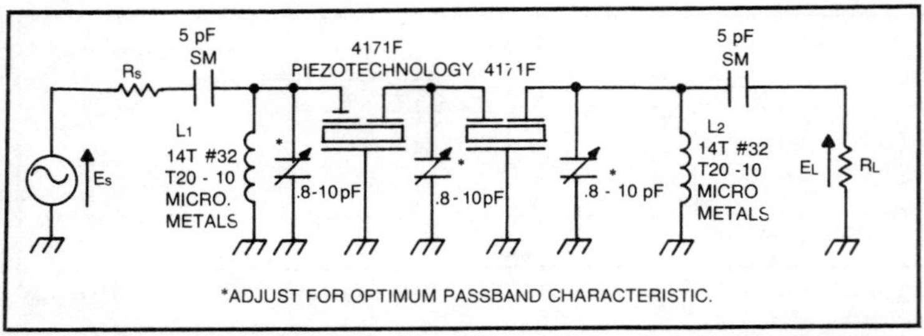

Fig. 4-36. A practical implementation of eight-pole tandem monolithic filter, centered at 75 MHz (−6 dB bandwidth is ±13 kHz) such a filter could be used in the first i-f of a double conversion superheterodyne. For reasons of simplicity the 4171 F filter is shown at a two-pole monolithic. An amplifier can also be used between the filters in order to provide better isolation.

The control of factors such as the quality of etching for the electrodes, or the polishing of the quartz, play an important role in cancelling these effects, making it very hard for such filters to be produced economically. As manufacturing techniques improve, this type of filter will become increasingly popular.

New technological breakthroughs, such as ion etching combined with sophisticated polishing techniques, as well as the use of other than quartz materials such as lithium niobate, lithium tantalate, and bismuth germanium oxide, will allow for increasingly complex monolithic filters to be built on the same substrate.

THE TANDEM MONOLITHIC

One way of achieving greater selectivity without the calibration problems of the multiresonator monolithic is to connect two or more simple two-resonator monolithics in a tandem fashion as shown in Fig. 4-35.

In this arrangement, control over the electrical parameters of each monolithic is guaranteed by the less demanding manufacturing process and the tandem design provides for increased stop-band attenuation by minimizing the effect of unwanted modes. Filters with as many as ten resonators have been manufactured by this technique. Figure 4-36 shows an actual implementation of an eight-pole tandem monolithic with a center frequency of 75 MHz. Two 4171F monolithic filters from Piezo Technolo-

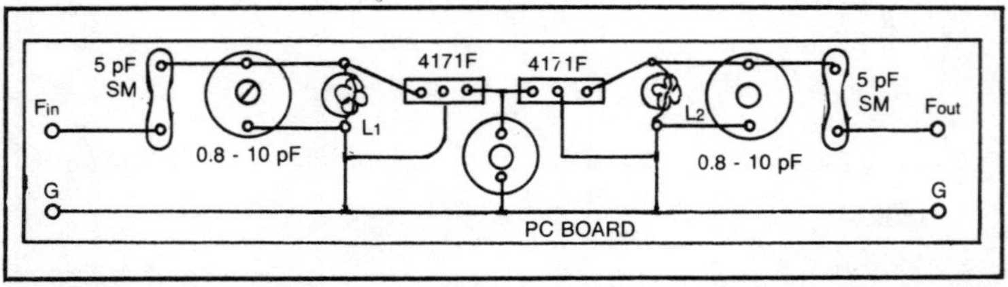

Fig. 4-37. Preferred layout for the 75 MHz tandem monolithic filter.

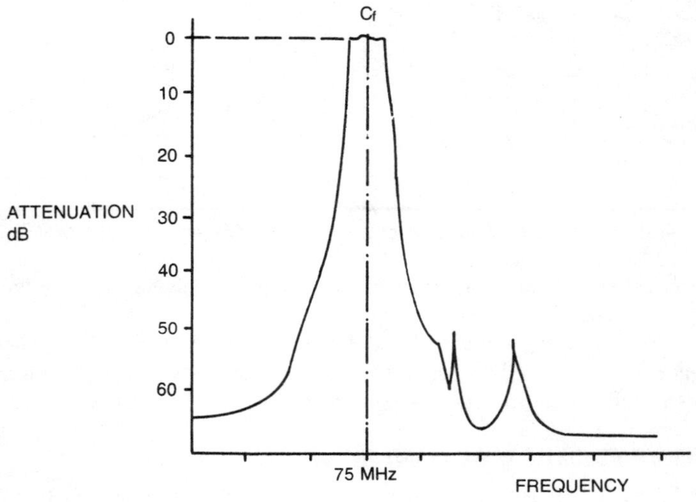

Fig. 4-38. Frequency response of the 75 MHz tandem monolithic filter.

gy, Inc. are used in this example. A preferred layout is also given (Fig. 4-37) for minimum interaction of the elements. The frequency response of the filter is shown in Fig. 4-38.

CERAMIC FILTERS

Where size and cost are important in a receiver design, the ceramic-piezoelectric filter is used in the i-f. Such filters usually operate at 455 kHz, although higher frequencies have been achieved with this technology. The electrical performance of the ceramic filter resembles that of the quartz-crystal filter, however, piezoelectric ceramics do not exist in natural form.

The complicated manufacturing process starts with ceramic compositions such as lead zirconate-titanate and lead metaniobate. These powders are mixed with water in order to produce a paste which is further formed into discs of a certain size. The discs are then dried out, before being baked at high temperatures, much like bricks in a ceramic factory. It is in the baking process that a high voltage "poling" field is applied in order to give them the piezoelectric property.

This shock distorts the physical shape and electrical properties of the ceramic structure creating a permanent piezoelectric element, like a powerful magnet would magnetize a ferric material. Silver electrodes are then deposited on each side of the discs, and a ladder arrangement can be accomplished by stacking as many as fifteen or more elements in a tubular package. Figure 4-39 shows the electrical equivalent of a ceramic ladder

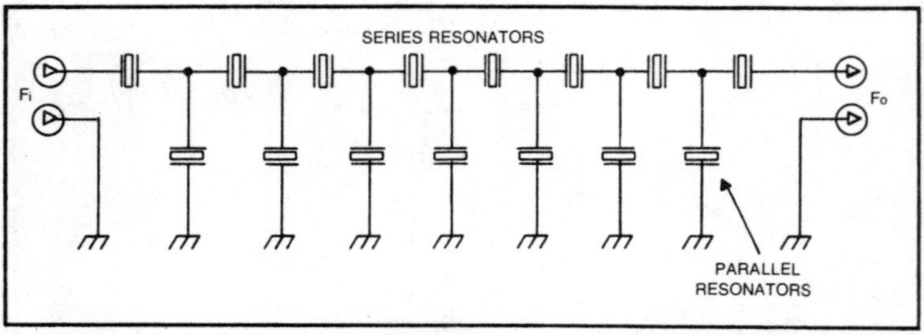

Fig. 4-39. Electrical configuration of a fifteen disc ceramic ladder filter.

filter. Other configurations such as monolithics are also achievable with this technology.

Ceramic filters are easy to implement as shown in Fig. 4-40. Their biggest drawback is their relative frequency instability with temperature which could be an important requirement in certain applications.

SURFACE ACOUSTIC WAVE (SAW) FILTERS

The *SAW filter* takes advantage of the acoustic wave propagation in piezoelectric materials, but unlike the quartz monolithic filter, which uses transversal propagation properties, the SAW uses mechanical waves that propagate at the surface of the solid substrate, via actual motion of the particles, much like a water wave in an ocean. This is shown in Fig. 4-41. A surface-acoustic-wave (SAW) filter usually consists of three sets of transducers which are electrochemically deposited on the surface of a piezoelectric crystal as shown in Fig. 4-42. These thin aluminum electrodes (chosen for minimum mechanical mass) are spaced ¼ or ½ wavelength from each other.

If transducer A is excited at the frequency of interest, a surface wave will be generated simultaneously toward the intercepting transducers B and C providing the same output at both locations. If only one receiving transducer was used, half the energy would be lost, thus a higher insertion loss. The two outputs are electrically paralleled in order to minimize the

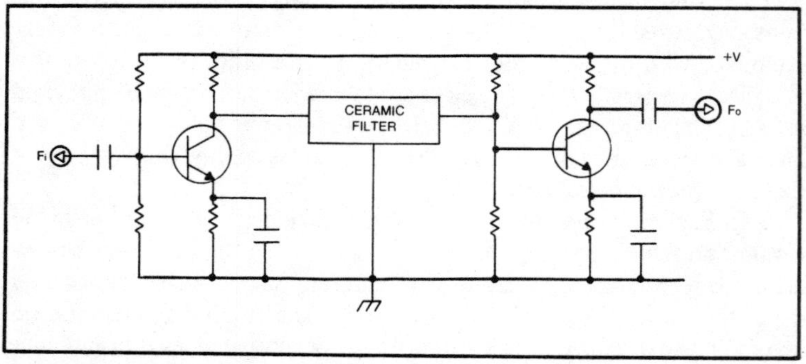

Fig. 4-40. Typical implementation of a ceramic filter in an i-f.

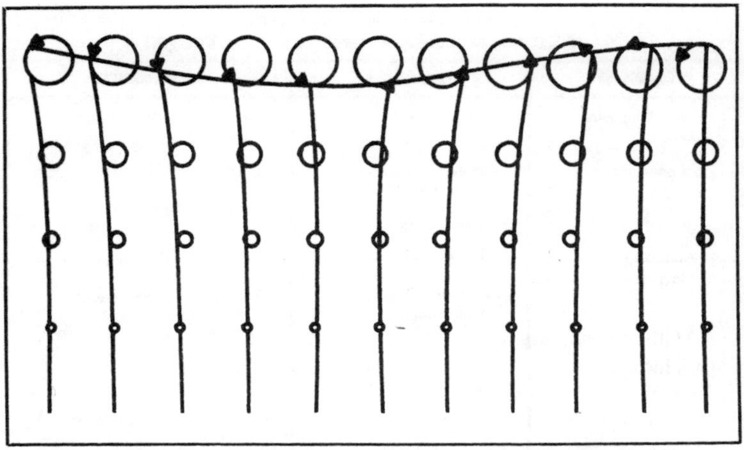

Fig. 4-41. Propagation of a surface acoustic wave across a quartz crystal substrate showing actual motion of particles. Secondary vertical propagation is also shown by the smaller circles spaced at about a tenth wavelength from each other in the vertical plane.

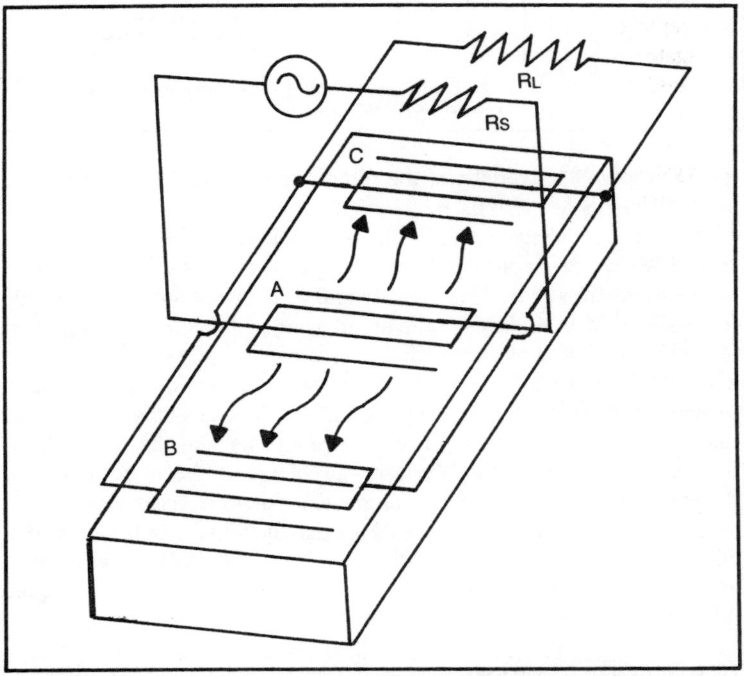

Fig. 4-42. A typical implementation of a SAW filter. Transducer A emits surface waves in both directions to be intercepted by transducers B and C, which are electrically coupled together for minimum loss.

Table 4-2. Advantages and Disadvantages of Several Technologies Used in I-f Filter Design.

TECHNOLOGY	ADVANTAGES	DISADVANTAGES
- Inductor/Capacitor L/C	- Inexpensive	- Does not provide a square response.
- Electromechanical	- Approaches a square response. - Very stable with temperature.	- Limited in frequency to below 1 MHz. - Relatively expensive
- Crystal/Lattice Monolithic	- Very high Q - Relatively stable with temperature.	- Limited in frequency to about 150 MHz - Narrow bandwidths cannot be achieved at higher frequencies.
- Ceramic Ladder	- Smaller in size than its electromechanical equivalent. - Inexpensive	- Generally limited in frequency to below 1 MHz - Unstable with temperature and time.
- Surface Acoustic Wave (SAW)	- Higher in frequency than crystal/lattice - Relatively inexpensive - Small in size	- Narrow bandwidths cannot be achieved. High insertion loss. - Poor ultimate rejection
- Optical	- Higher in frequency than surface acoustic wave.	- Very limited bandwidth - Very unstable with temperature and time. - Experimental.
- Digital	- Easily adaptive, versatile square response	- Experimental, very expensive.

insertion loss characteristic of the filter. Other methods of focusing the acoustic energy are also used.

The main advantage of the surface-acoustic-wave (SAW) filters is the relatively slow propagation rate of these types of waves, compared with electromagnetic waves. This velocity is 3×10^3 m/sec or about 100,000 times slower than the speed of light, allowing technological and physical feasibility for filters at very high frequencies, typically from 10 MHz to 2,000 MHz. Practical surface-acoustic-wave (SAW) filters have been realized at frequencies from 20 to 600 MHz. They are inherently small, rugged, and reliable, and require no adjustments like the monolithics. However, they suffer from the same problems as the monolithics do. (Inferior ultimate rejection due to interaction of unpredicted modes of wave propagation with the surface wave.) Although the initial cost is higher than other technologies, surface-acoustic-wave (SAW) filters are ideally suited for mass production, because of the lithographic manufacturing techniques that can be used with this type of filter.

TECHNOLOGICAL TRADEOFFS IN INTERMEDIATE FREQUENCY (I-F) FILTERS

We discussed several methods of achieving selectivity in the intermediate-frequency (i-f) portion of a superheterodyne receiver. New ways of intermediate frequency (i-f) filtering are being developed today and among them are the optical and digital filters. Although a thing of the future, these filters will provide certain advantages over the previously used technologies. A look at Table 4-2 will show a list of technological trade-offs between the various methods discussed, and provide us with a better understanding of their merits and their limitations, as we will apply them to communication receivers in the following chapters.

—•••—

Chapter 5

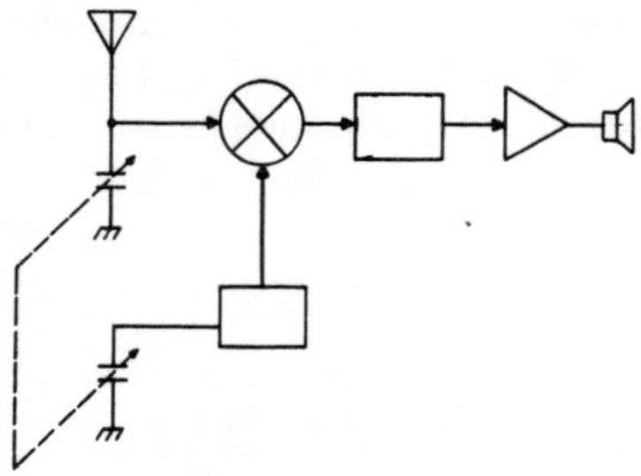

Implementing Double Conversions

The double-conversion scheme converts twice, just as its name implies (Fig. 5-1). First, it uses a fixed oscillator to convert to a relatively higher frequency than our previous 455 kHz, typically 5.5, 9, 10.7, or 21.4 MHz in order to place the image high enough to be readily attenuated in the preselector. This first i-f is filtered with the help of an electrical or crystal filter. To achieve proper selectivity, a second conversion is used at 455 kHz where already developed filters could be used. The second oscillator is a variable one, allowing for a typical 1 MHz (or 500 kHz) span.

It is important to know that the bandwidth of the first i-f has to be equal to that of the coverage of the second oscillator, in our case 1 MHz. The first i-f sometimes involves a manual preselector. This is usually called a variable bandpass i-f. In some more sophisticated receivers, this variable bandpass can be tuned electrically with the help of back-biased diodes which act as variable capacitors. The right amount of bias voltage is supplied to the varactors (variable capacitance diodes) to track the frequency being received. This voltage can be commanded from the Vfo through some rather fancy digital-to-analog techniques. A third mixer is used as the product detector, which is necessary for receiving cw and SSB signals. The third oscillator is called the bfo (Beat Frequency Oscillator).

—•••—

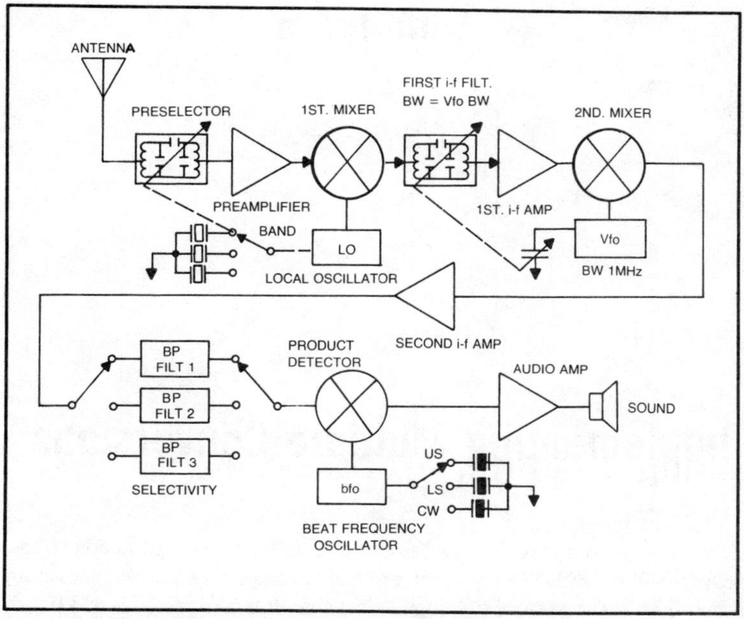

Fig. 5-1. Typical double-conversion block diagram.

Chapter 6

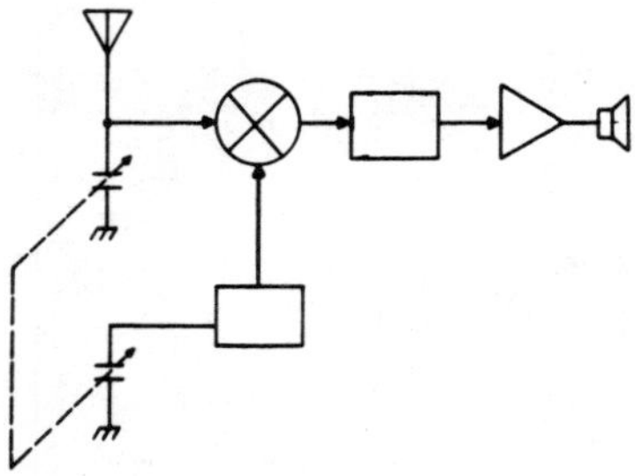

Implementing Multiple Conversions

Triple conversion receivers and even quadruple conversion receivers appeared in the 1960s. These receivers had schemes like the one shown in Fig. 6-1 and went so far as to bring the last conversion to as low as 85 kHz or even 50 kHz, in order to achieve good selectivity economically with L/C filters.

The Drake 2C was a famous example of such a receiver, and it used a triple-conversion scheme with the first i-f between 3.5 and 4 MHz, the second i-f of 455 kHz, and a third i-f of 50 kHz. For economical reasons, the selectivity was not achieved at 455 kHz but at 50 kHz with the help of L/C circuits. It consisted of four tuned circuits of very high Q that were capacitively coupled in cascade as shown in Fig. 6-1. A characteristic of such coupled circuits is that the bandwidth or selectivity is largely dependent on the degree of coupling, being greatest with light coupling. Thus, by switching in different coupling capacitors, the overall selectivity could be varied accordingly. As the coupling was increased, however, a dip in the pass-band tended to appear. R1 was then inserted (This tended to lower the element Q on the first resonator, thereby changing the pass-band.) or varied as needed to provide a flatter response in the pass-band at the broader selectivity portions. Also, as the coupling was increased, the response curve spread out from both sides of the initial resonant frequency. The set-up was, therefore, also arranged so that resonating capacitances were altered for the various selectivity positions, in order that the low frequency side of the response curve in each case remained at a fixed point.

Another receiver that used L/C coupling to achieve selectivity, was the Collins R-392 (Fig. 6-2). The 392 had variable bandwidth in the second i-f at 455 kHz by using a rather interesting electromechanical approach (Fig. 6-3). By mechanically varying the position of the coils in the i-f transformers, different degrees of coupling and, therefore, selectivities

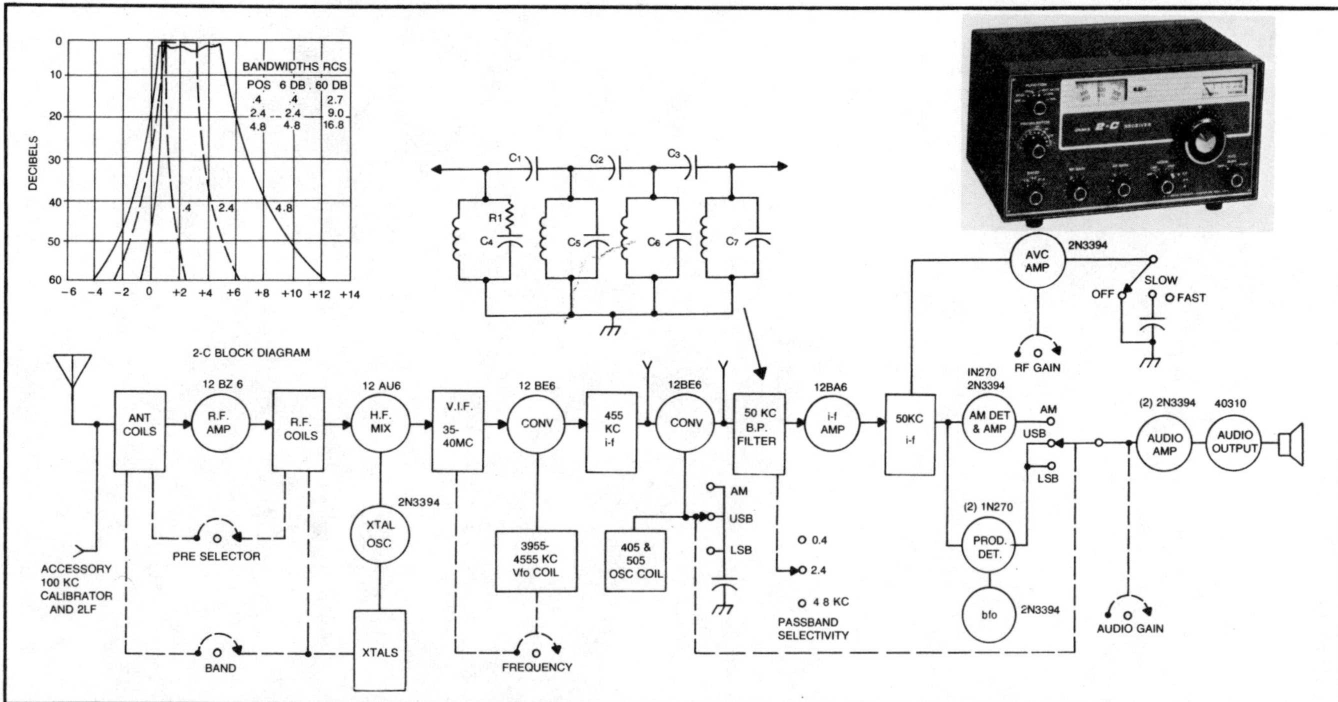

Fig. 6-1. Drake 2-C receiver block diagram and selectivity.

Fig. 6-2. Internal view of an R-392 receiver manufactured by Collins Radio Company.

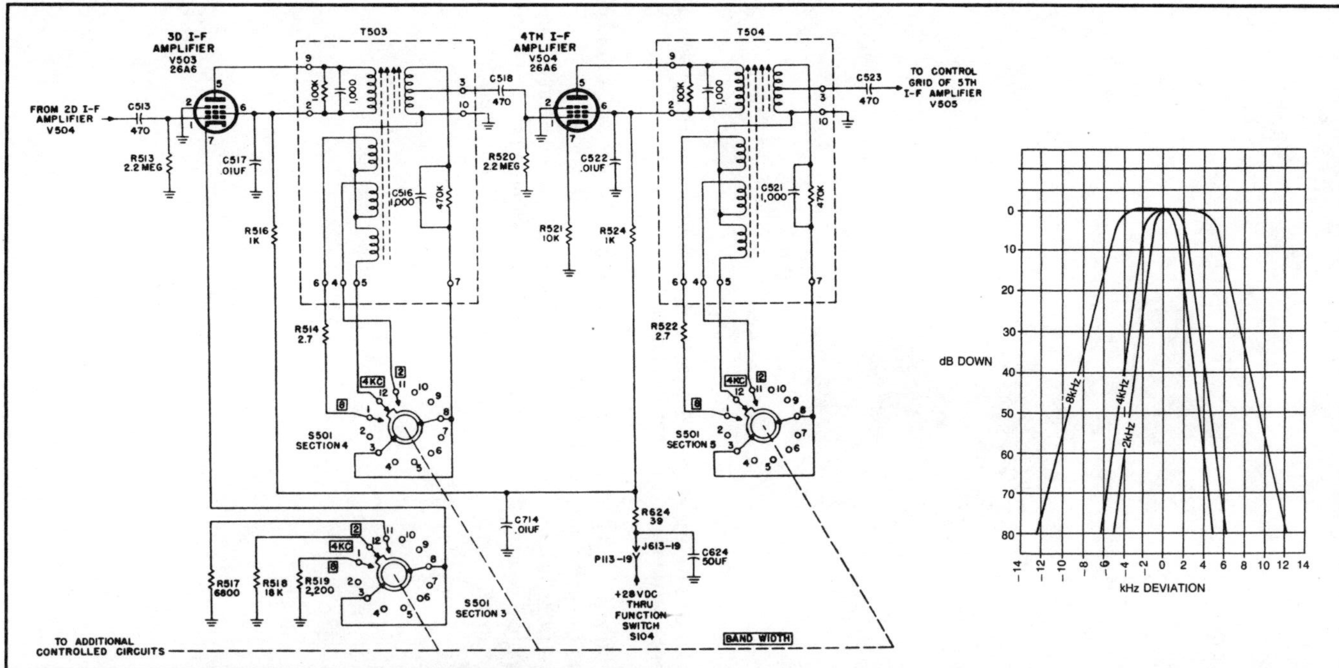

Fig. 6-3. In the R-392 receiver, several degrees of selectivity were achieved by switching in differently positioned coils in the i-f transformers T503 and T504.

Fig. 6-4. An ARC-5 i-f transformer. Selectivity was achieved at 85 kHz by varying the distance between two windings as shown.

were achieved. To illustrate this, we took an i-f transformer out of a radio that uses the same approach. Figure 6-4 shows how it works.

The R-392 used some of the most complex electro-mechanical implementation ever used to this date. While still considered to be a good receiver, even by today's standards, it remains a symbol of the past.

—•••—

Chapter 7

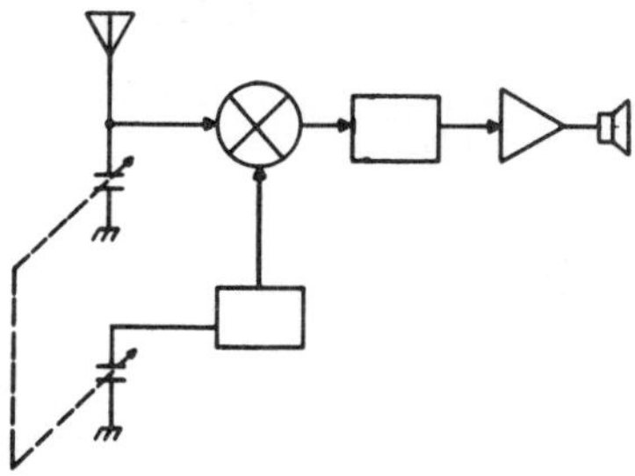

Direct Conversion Implementation

Another kind of superheterodyne is the well-known direct conversion type. Figure 7-1 shows a block diagram for such a receiver. Its i-f is at audio frequencies since the Vfo runs practically at the received frequency, offset only by the center of audio frequency range of interest, in our case 2 kHz.

Disadvantages of the direct-conversion receiver are primarily the very close image which cannot be eliminated and the hum that comes with the high level of amplification in the af stage. Hum is also caused by the so-called common-mode syndrome. That is, local oscillator energy can be radiated to the ac power supply where it mixes with 60 and 120 Hz in the rectifier and is then radiated back to the receiver front end.

The direct conversion receiver (also known as the synchrodyne, or the zero i-f receiver) is not usually used in practical communications except as a monitoring device in some military or commercial applications. Figure 7-2 shows a radar altimeter which uses such a receiver to process signals reflected by the ground, in order to determine the altitude of an aircraft.

As a communication receiver, the direct conversion approach is not suitable for single sideband reception since a signal present on one side of the zero beat center will interfere with the desired side, unless a phasing circuit is used.

Direct conversion receivers are not suitable for agc. One means of overcoming this would be to implement an audio compressor and that would approach an effect similar to that of the agc. Another deficiency of this type of receiver is microphonics, (Technical term used for an audio effect, noticed by knocking on a radio. It is caused by mechanical modulation of the rf frequency, generated by the Vfo.) especially at higher rf frequencies. Figure 7-3 shows a practical design for a dc receiver. Before going into the anatomy of a modern receiver, we'll look at another type of superheterodyne which is a very interesting case.

—•••—

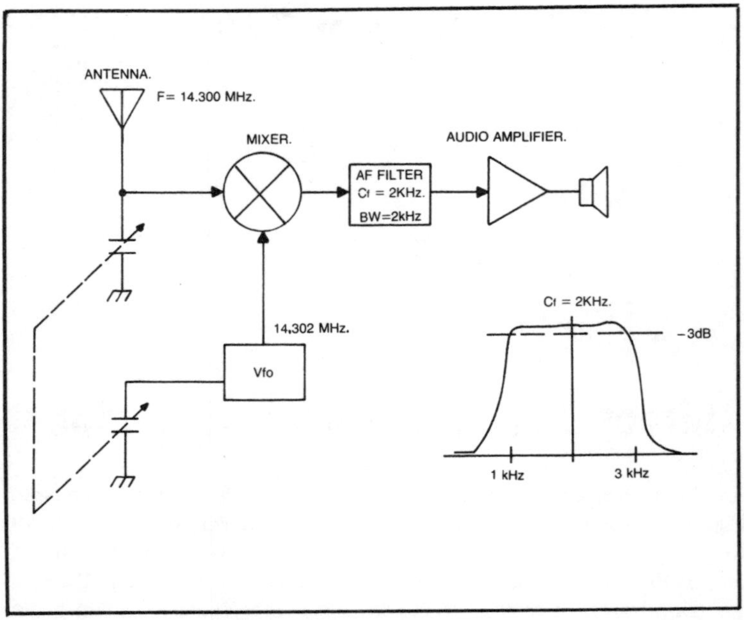

Fig. 7-1. Direct conversion receiver block diagram.

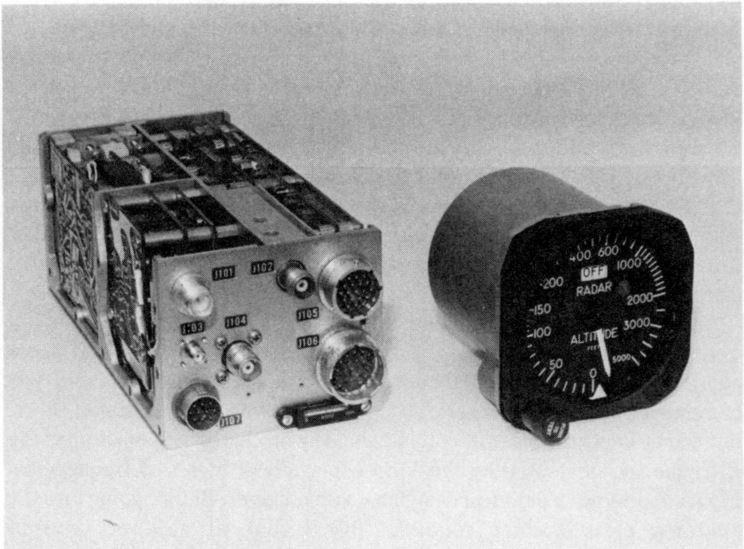

Fig. 7-2. The APN-194 radar altimeter uses a direct conversion (zero i-f) receiver in its signal path (courtesy of Honeywell, Inc.).

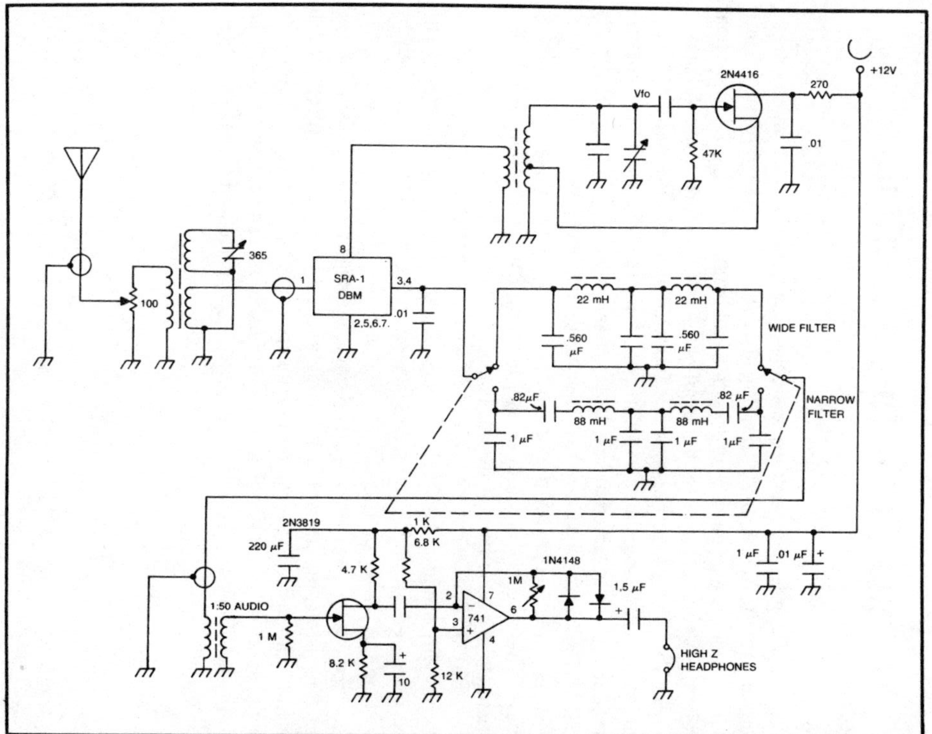

Fig. 7-3. Schematic diagram of a typical direct conversion receiver.

Chapter 8

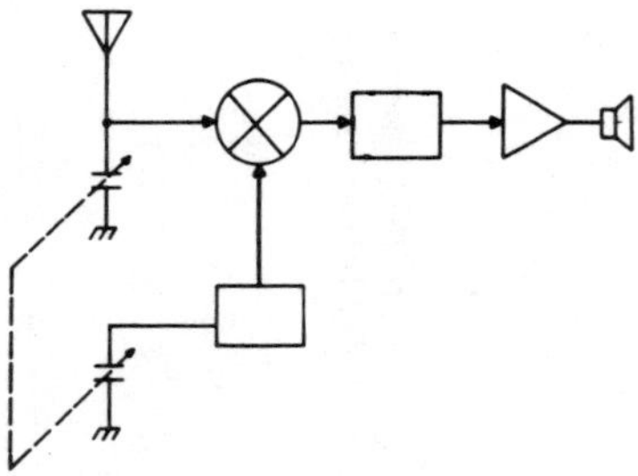

Special Conversions and Their Implementation

Some companies manufacture a single-conversion down-conversion receiver with the i-f around 16 kHz, (Fig. 8-1). The antenna is directly fed into a multi-pole crystal-lattice filter with a bandwidth of 2 kHz, allowing final selectivity to be achieved at the input of the radio, before any mixing takes place. This filter is very lossy and is followed by a very low noise amplifier and mixer which translates it to a 16 kHz i-f where additional selectivity and amplification is obtained before detection takes place. The image is greatly reduced by the crystal filter and a minimum amount of products are created. This is really a superbly performing radio from a signal handling point of view. The receiver is intended for maritime communication where fixed channels are usually allocated to users.

Physically, the receiver measures approximately 1×2×4 inches with the filter taking most space. About 12 separate receivers could be packaged in a box, and frequency change could be accomplished by turning the power on to the desired unit.

If we were to create a general-coverage receiver (0.5-30 MHz) based on this principle, we would be using about 14,750 crystal filters which would tend to cost several million dollars. Very impractical, but it has its application. If someone could make a tunable filter of this nature for the front end, this would probably be an ideal approach for a superheterodyne receiver.

—•••—

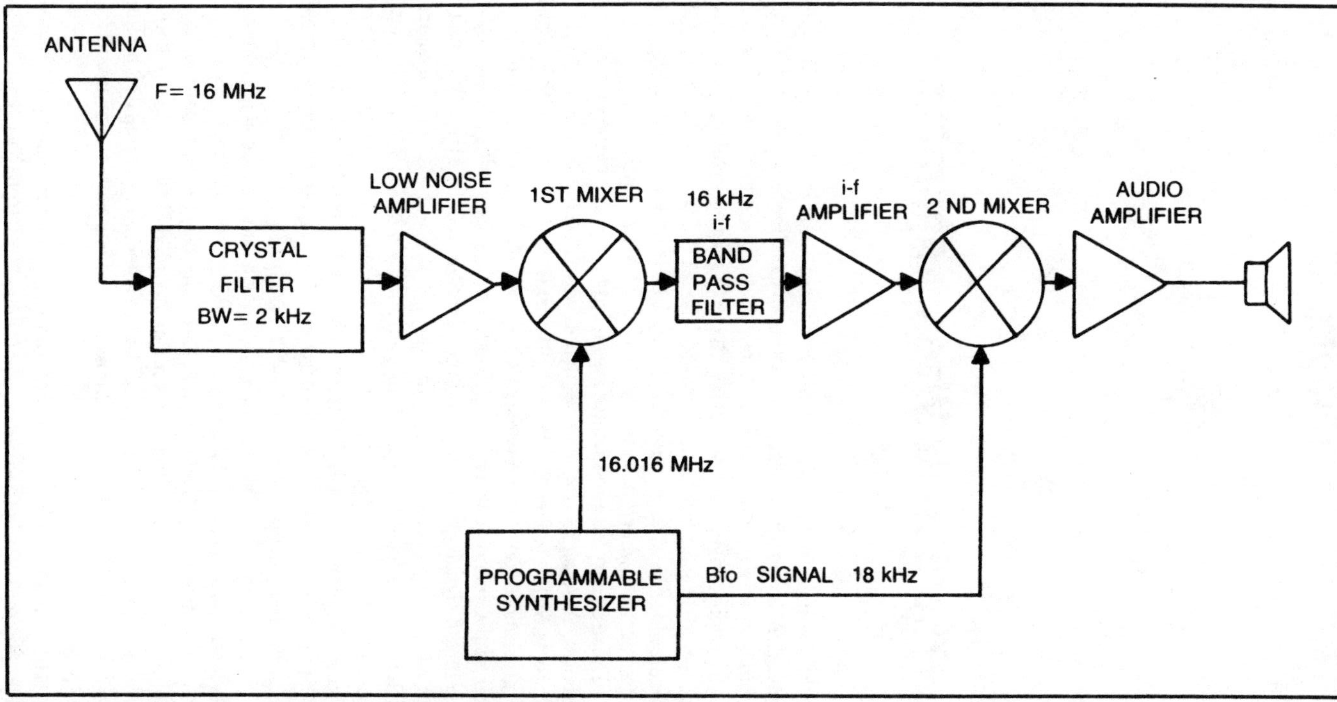

Fig. 8-1. Block diagram of the special conversion receiver.

Chapter 9

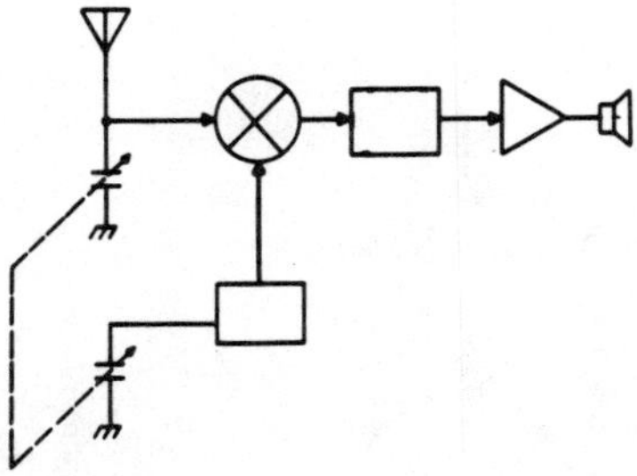

The Barlow Wadley Approach

One of the most ingenius receivers ever created is the Barlow Wadley receiver, invented by Dr. Wadley of South Africa. Figure 9-1 shows several examples of this type of receiver. A look at the block diagram in Fig. 9-2 shows a triple conversion approach with the first i-f at 55 MHz, the second i-f at 2.5 MHz, and the third i-f at 455 kHz. A 1 MHz crystal oscillator generates local oscillator signals at exactly 1 MHz intervals (this is a type of synthesizer) anywhere between 3 MHz to 32 MHz (Fig. 9-3).

These twenty-nine harmonics are all fed to one side of the loop mixer, as shown. A separate variable oscillator which oscillates between 55.5 MHz to 84.5 MHz (called the MHz oscillator) is manually controlled by the operator, and is fed simultaneously to the loop mixer, as well as to the first mixer of the receiver where it combines continuously with the incoming preselected signal, generating the first i-f centered at 55 MHz. This information is further filtered and amplified through the first i-f filter. The signal which now has a bandwidth of one megahertz (54.5 MHz to 55.5 MHz) and contains all of the rf signals present at the antenna within one megahertz of the tuned frequency is finally fed to one side of the second mixer. The other side of the second mixer is powered from the loop mixer, through a narrow bandpass filter amplifier centered at 52.5 MHz.

It can be seen that the 52.5 MHz signal will only be true at precisely 1 MHz intervals as a result of the selective mixing process which takes place in the loop mixer, as shown in Table 9-1. The biggest advantage of this scheme is the drift cancelling mechanism provided for the MHz oscillator by the double mixing approach. The drift is completely eliminated by the subtraction process in the second mixer, providing stable conversion for the second i-f.

The second i-f fulfills the bandwidth range requirements for a multi-conversion receiver (as discussed earlier) and tunes from 3 MHz to 2 MHz

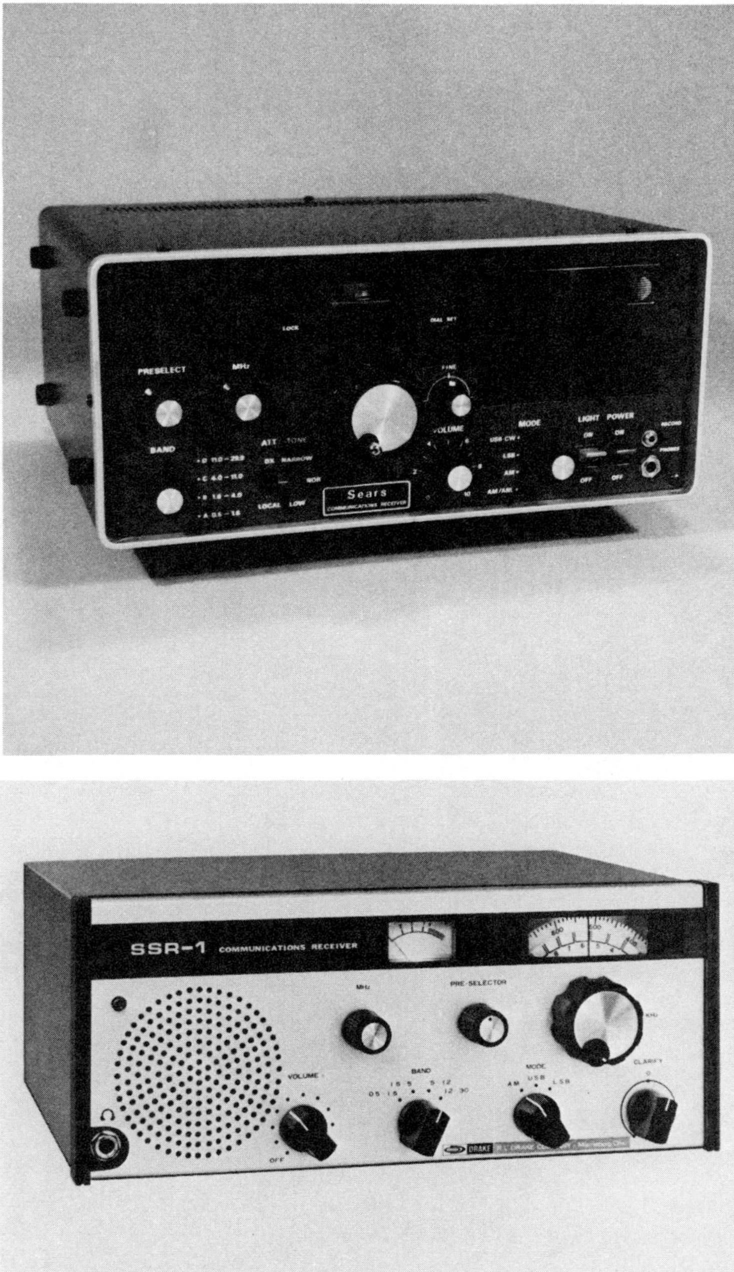

Fig. 9-1. Barlow Wadley type receivers (courtesy of Gilfer Associates and R.L. Drake Co.).

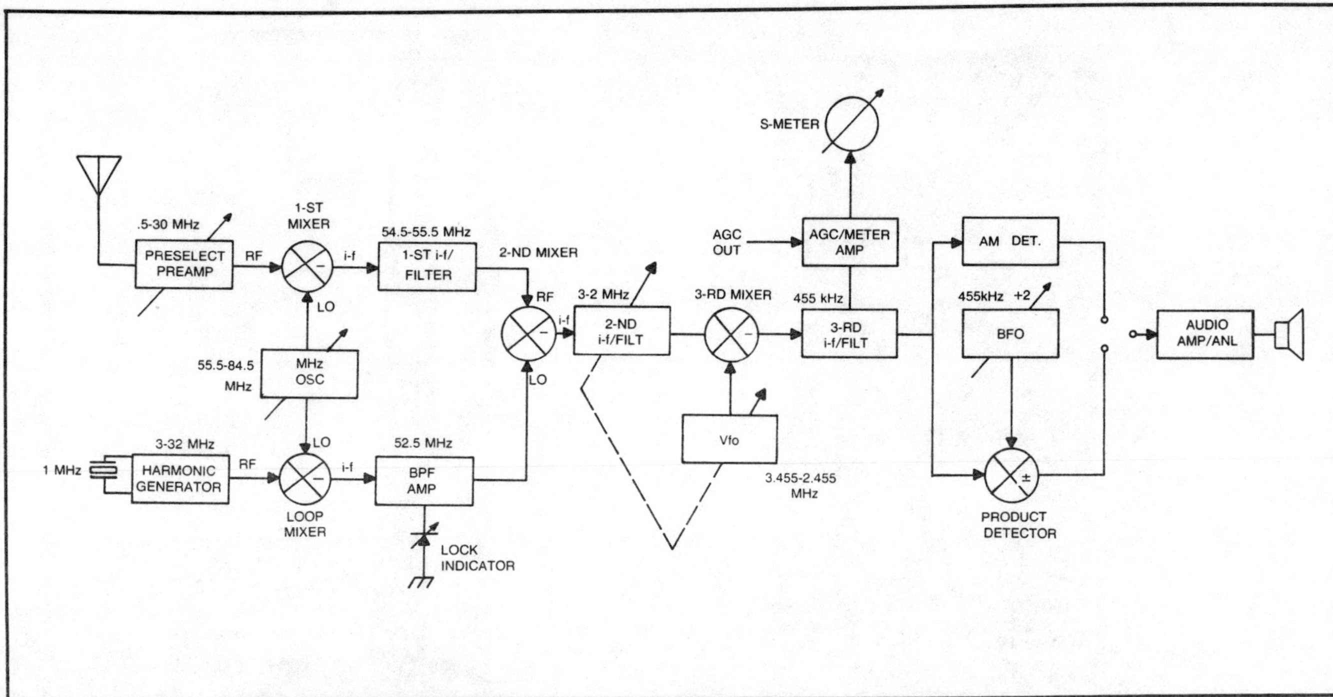

Fig. 9-2. Block diagram of a typical Barlow Wadley receiver (YAESU, FRG-7).

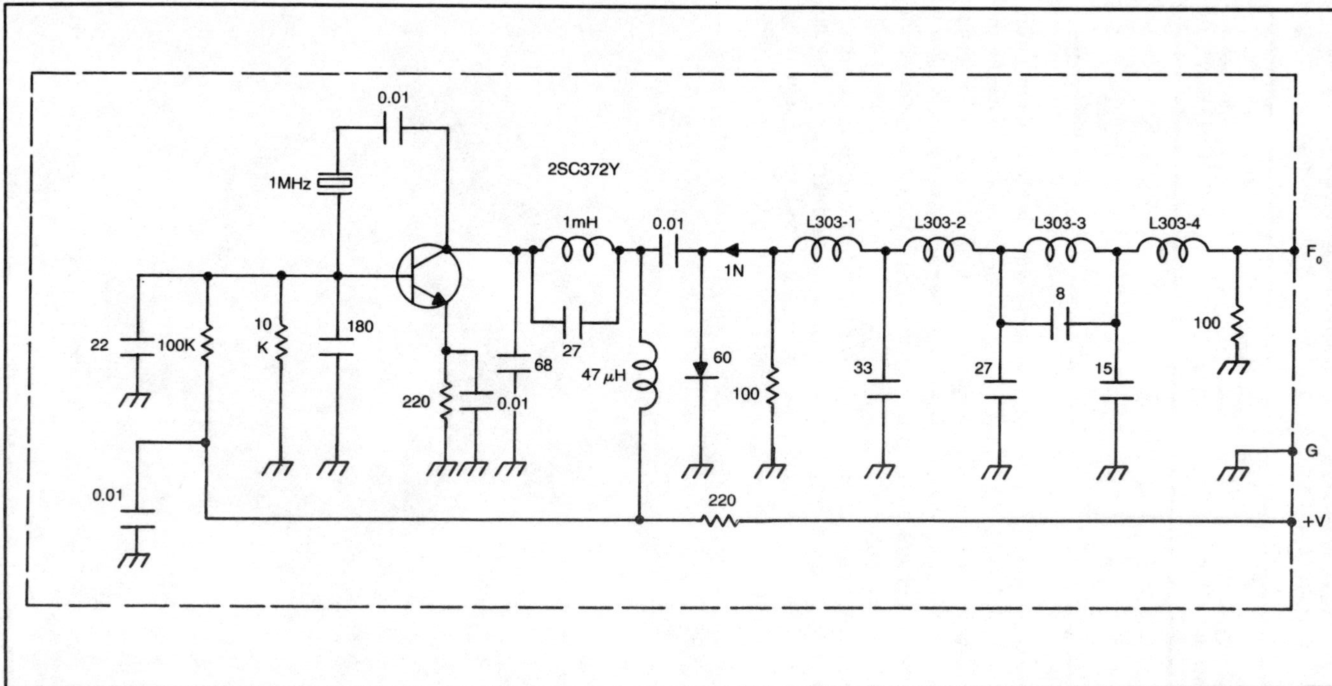

Fig. 9-3. Harmonic generator of the YAESU, FRG-7 receiver.

Table 9-1. Example of Selective Mixing Process in the Loop Mixer of a Typical Barlow Wadley Receiver.

MHz Oscillator Frequency (MHz)	Harmonic Number of Frequency Selected (MHz)	Local Oscillator Input to Second Mixer (MHz)
55.5	3	52.5
56.5	4	52.5
57.5	5	52.5

via a variable capacitor which tracks together with the Vfo, operating from 3.455 MHz to 2.455 MHz. This provides the fine-tuning conversion for the third i-f which is centered at 455 kHz within any of the 1 MHz bands previously selected. The 455 kHz i-f is further processed through narrow filters and amplifiers, and is finally detected and fed to the speaker via the audio amplifier (Fig. 9-4).

A disadvantage of the Barlow Wadley scheme is that the signals produced by the harmonic generator are usually heard at the beginning and the ends of each 1 MHz band as a result of system leakage, in spite of the fact that this generator is usually shielded and placed far away from critical

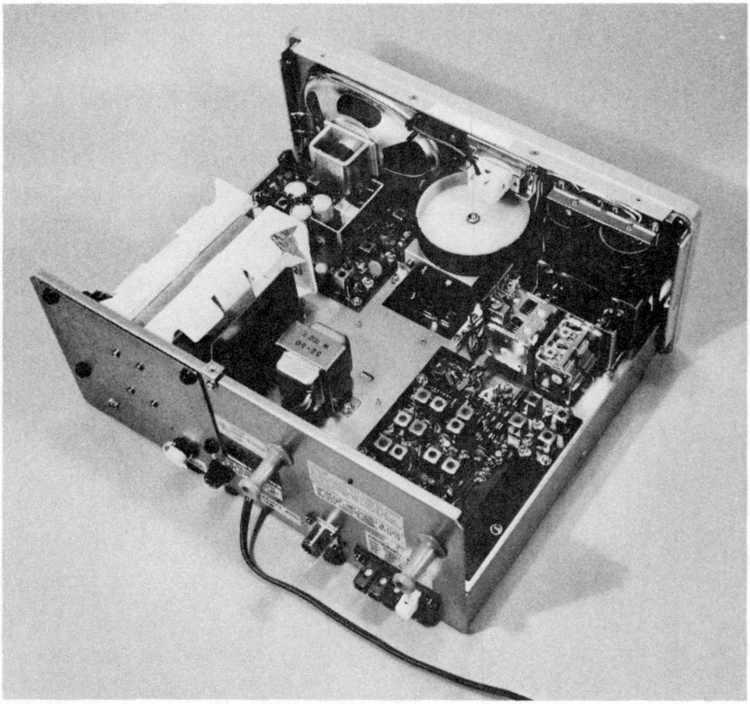

Fig. 9-4. Internal view of the YAESU, FRG-7 receiver.

Fig. 9-5. The harmonic generator (upper left) in a Barlow Wadley receiver is placed in a shielded box (shown with cover removed) far away from critical rf areas.

areas as shown in Fig. 9-5. This problem, however, can be considered as an advantage as it provides a means of dial calibration for the Vfo. The Barlow Wadley scheme remains as a widely used scheme and a symbol of technical ingenuity.

—•••—

Chapter 10

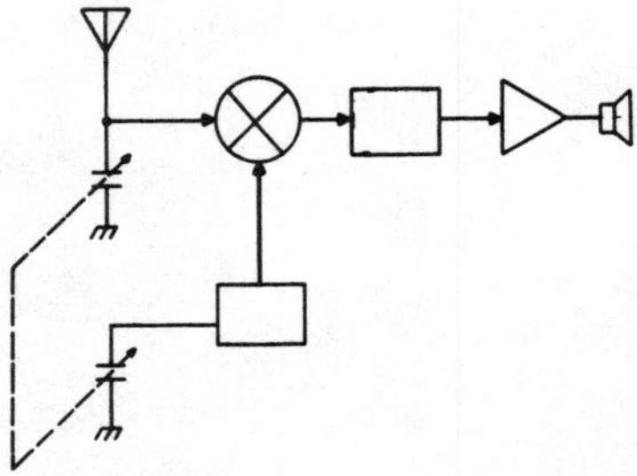

Modern Receivers

New types of receivers have appeared during the past decade. Figures 10-1, 10-2, and 10-3 show some examples. The advent of crystal filter design made up-conversion radios a reality. Today crystal filters can be fabricated at frequencies up to 150 MHz. However, bandwidths of 2 kHz or so cannot yet be obtained at frequencies above 30 MHz, because of crystal technology problems, making the dream of producing a single up-conversion receiver with an i-f above 30 MHz an impossibility (Fig. 10-4). As technology improves, this type of radio should be the preferred choice, over the double- and triple-conversions, because of the minimum number of mixers involved, as well as the minimized image problems.

—•••—

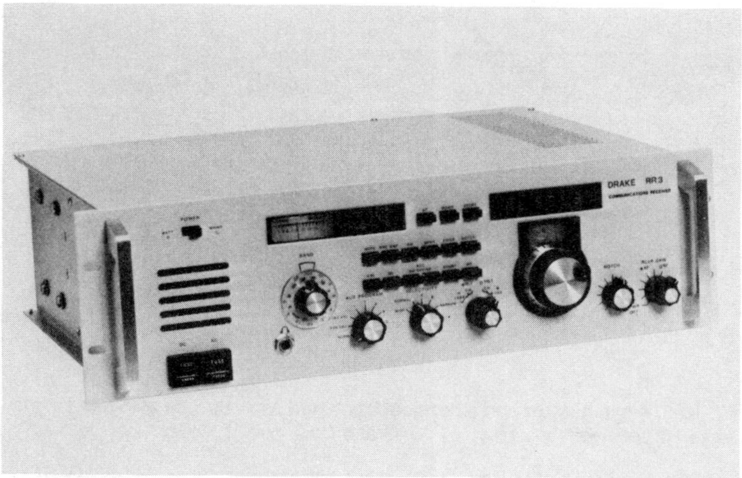

Fig. 10-1. The RR 3 communications receiver is a modern implementation of a triple conversion approach with the first i-f at 48.05 MHz, the second i-f at 5.645 MHz, and the third i-f at 50 kHz. This receiver was designed primarily for maritime communications (courtesy of R.L. Drake Co.).

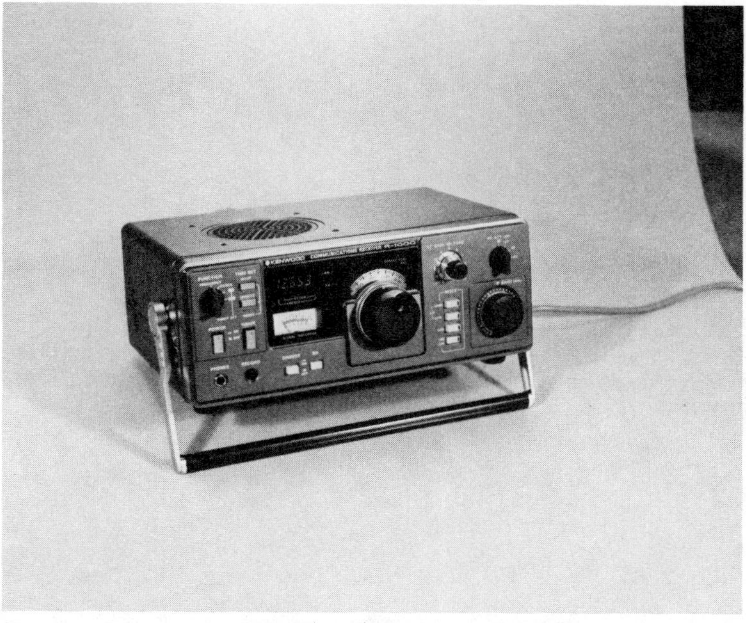

Fig. 10-2. The Kenwood R-1000 receiver is an example of modern receiver technology. It is synthesized in one MHz steps with the use of a digital phase-locked loop which also provides automatic switching for one of the six preselector filters. Fine tuning is obtained with a permeability tuned oscillator which is part of the loop.

Fig. 10-3. A synthesizer and a permeability tuned oscillator are also used in the Drake R7 receiver (courtesy of R.L. Drake Co.).

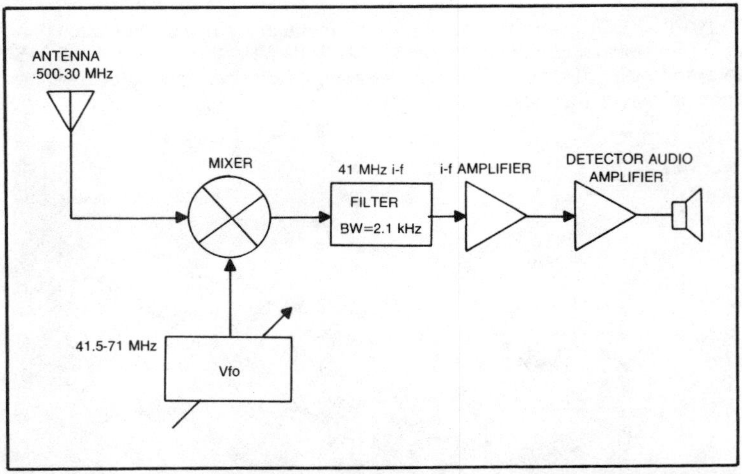

Fig. 10-4. Envisioned single conversion (up-convert) superheterodyne with i-f at 41 MHz. This scheme cannot be implemented presently because of the i-f filter limitations.

Chapter 11

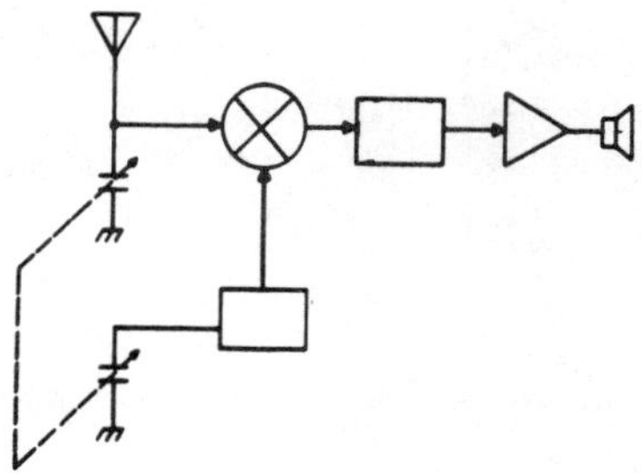

System Design Considerations for Modern Receivers

The realization of general coverage, multi-conversion receivers has brought about the need for a better understanding of their performance. In a radio receiver, mixers used to perform conversions are not only frequency adding and subtracting mechanisms, but they also exhibit a variety of undesired output products, in addition to their sum and difference frequencies. These are called *intermodulation products*. This phenomenon is complicated by the increased front-end bandwidth requirement (general coverage) as well as by the intermediate frequency i-f bandwidth requirement. If a multi-mixer situation exists, such as in a multiconversion receiver, the problem is further aggravated, as initial unwanted products from one mixer can combine with those of another, creating a multitude of "birdies"* at the final i-f output.

---

*Technical jargon used for unwanted interference which sounds like the whistling of a bird.

Fig. 11-1. This dedicated receiver is intended for WWV reception on one of the four selected frequencies. Automatic scanning of the receiving frequencies allows the unit to determine which frequency has the most usable signal and locks onto it. Careful system design shows that a double-conversion approach was used having the first i-f at 45 MHz followed by a second i-f at 4.6 MHz (courtesy of Kinemetrics, Inc. True Time Division).

Table 11-1. Some Internally Generated Spurious Frequencies in a Typical General Coverage Receiver with a 9 MHz I-f. For the Received Frequencies on the Left Column Interference from the Intermodulation Product is Created in the Mixers.

OPERATING FREQUENCY MHz	LOCAL OSCILLATOR FREQUENCY MHz	Bfo FREQUENCY MHz	LOWEST ORDER SPURIOUS PRODUCT EQUAL TO i-f	TYPICAL CARRIER LEVEL (μV) FOR 10 dB $\frac{S+N}{N}$		
				ADJACENT FREQUENCIES MAXIMUM	ADJACENT FREQUENCIES TYPICAL	SPURIOUS FREQUENCY TYPICAL
3.0025	12.0025	9.0025	3LO-3BFO	0.3	0.2	2.0
3.6030	12.6030	9.0025	5LO-6BFO	0.3	0.2	2.0
5.4035	14.4035	9.0025	5LO-7BFO	0.3	0.2	6.0
6.0033	15.0033	9.0025	3LO-4BFO	0.3	0.2	30.0
9.0025	18.0025	9.0025	LO-BFO	0.4	0.2	Receiver Blocked
11.2550	20.2550	9.0025	4LO-8BFO	0.4	0.2	8.0
12.0050	21.0050	9.0025	3LO-6BFO	0.4	0.2	300 for 3dB $\frac{S+N}{N}$
13.5000	4.5000	8.9975	2LO	0.5	0.3	Receiver Blocked
14.99916	5.99916	8.9975	3LO-BFO	0.5	0.3	8.0
18.0000	9.0000	8.9975	LO	0.5	0.3	Receiver Blocked
20.9975	11.9975	8.9975	3LO-3BFO	0.6	0.4	0.5
21.5970	12.5970	8.9975	5LO-6BFO	0.6	0.4	2.0
23.3965	14.3965	8.9975	5LO-7BFO	0.6	0.4	8.0
23.99667	14.99667	8.9975	3LO-4BFO	0.6	0.4	2.0
25.1960	16.1960	8.9975	5LO-8BFO	0.6	0.4	4.0
26.9975	17.9975	8.9975	LO-BFO	0.6	0.4	30.0
29.2450	20.2450	8.9975	4LO-8BFO	0.6	0.4	6.0

UNDERSTANDING INTERMODULATION PRODUCTS

The chart in Table 11-1 shows products generated at the output of a single-conversion general coverage communications receiver using a 9 MHz i-f. These products will interfere with received frequencies to the degree shown.

Whether a receiver is a dedicated type as the one shown in Fig. 11-1, or a general coverage type such as the one shown in Fig. 11-2, the problem of intermodulation products has to be carefully understood and weighed against system parameters, so a minimum number of "birdies" will be internally generated and heard within the pass-band of the receiver.

DESIGN TOOLS FOR PREDICTING INTERMODULATION DISTORTION (PRODUCT CHARTS AND THEIR USE)

Let's look at some analytical tools the system designer uses to determine these products. Let's assume we are going to design a fixed-frequency receiver for 70 MHz (Fig. 11-3). With a local oscillator of 90 MHz, the receiver will have a first i-f of 160 MHz if an up-conversion type is considered. The second conversion of this receiver is not discussed here in order to simplify this case. To find the two normalized* frequencies:

$$\frac{F_1}{F_2} = \frac{70}{90} = 0.778$$

$$\frac{F(\text{i-f})}{F_2} = \frac{160}{90} = 1\ 1.778$$

*Method of generalizing all cases so a single chart can be applied for finding a function.

Fig. 11-2. The Cubic model HF-1030 general coverage communications receiver is used in a dual frequency diversity mode in this rack mounted installation (courtesy of Cubic Communications).

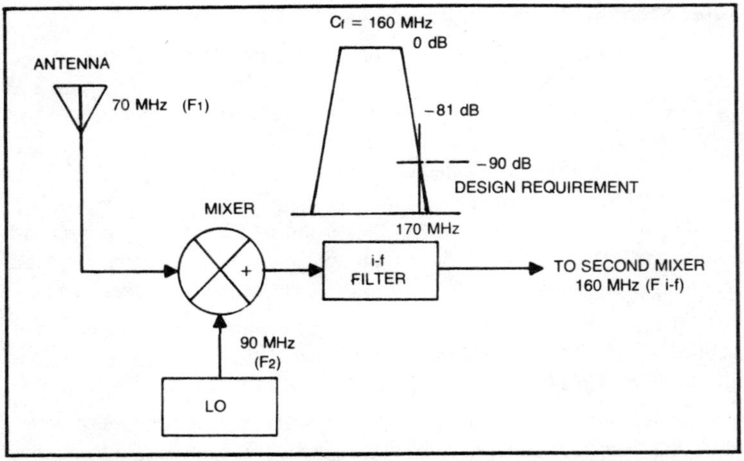

Fig. 11-3. Finding in-band intermodulation products dictates choice of i-f center frequency as well as the shape factor of the i-f filter, in order to meet system requirement.

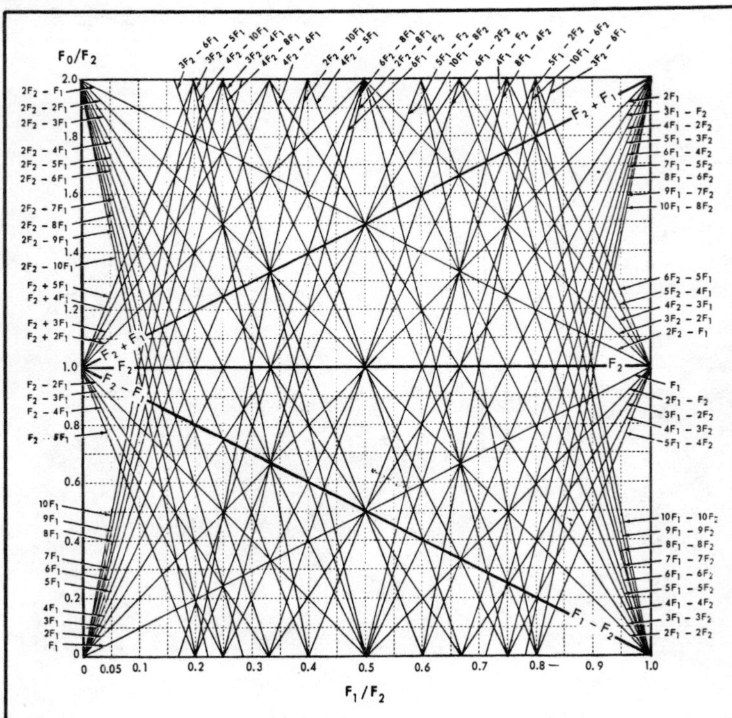

Fig. 11-4. Intermodulation products chart. Reprinted with permission from EDN Magazine, Copyright August 1967.

With this information and the mixer product chart in Fig. 11-4 find the locus point** for the two ratios, as shown in Fig. 11-5. The chart shows all products produced not only by the fundamentals, but also by multiples of the signal and oscillator frequencies which are present in the mixer stage, and correspond to the second, third, fourth, fifth, and sixth harmonics of the two mixing signals.

The order of the product is determined by the sum of the harmonic orders involved. For example, $5F_1 + 2F_2$ is a seventh order product (regardless of the mathematical operation involved) because it involves the fifth harmonic of F_1 combined with the second harmonic of F_2. Higher-order products are also present but they are usually of a sufficiently low level so as not to cause problems. Any line that crosses the locus point corresponds to a product which is identified on the edge of the chart.

**Intersection of a system of lines which satisfies one or more given conditions.

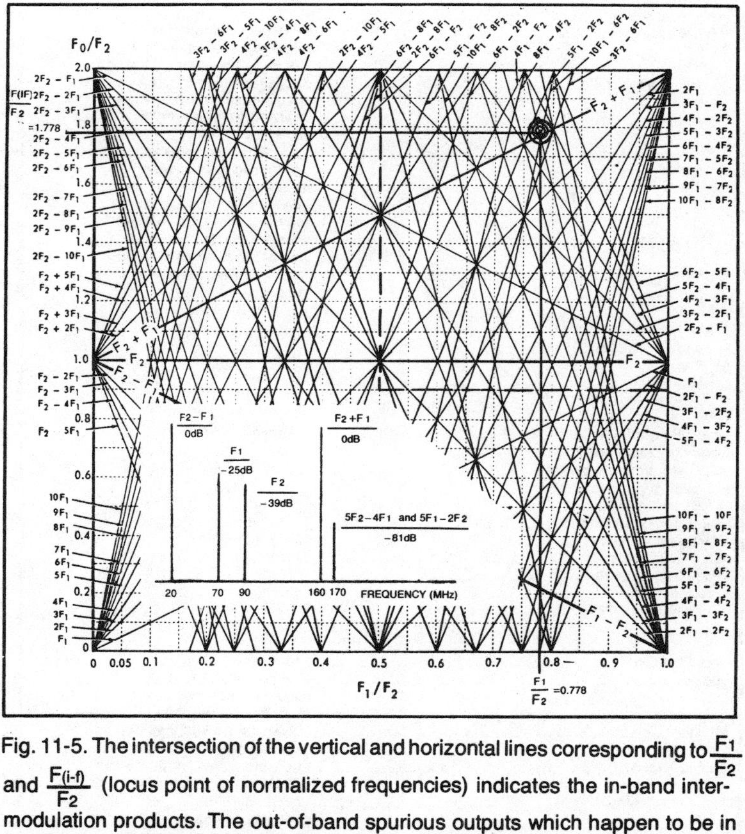

Fig. 11-5. The intersection of the vertical and horizontal lines corresponding to $\frac{F_1}{F_2}$ and $\frac{F_{(i\text{-}f)}}{F_2}$ (locus point of normalized frequencies) indicates the in-band intermodulation products. The out-of-band spurious outputs which happen to be in the vicinity of the i-f frequency can also be verified by looking at the products adjacent to the locus point.

Table 11-2. Specifications of the ESM 500 Vhf/Uhf Receiver.

Frequency Range	20 to 1000 MHz
Frequency Setting	a) quasi-continuous with rotary knob; the tuning speed increases with the speed of rotation b) from keyboard on front panel c) entered from internal memory d) entered from external computer
Resolution	1 kHz/10 Hz (SSB)
Readout, digital (can be shifted by 3 digits in SSB operation)	6-digit display for receive frequency, 6-digit display for frequency entered from keyboard or stored frequency value, 2-digit display for storage location
Error of frequency setting	$\pm 1 \times 10^{-8}$ (or external standard frequency, 10 MHz)
Antenna Input	$50-\Omega$, N socket
Oscillator reradiation with $50-\Omega$ termination	$<1\mu V$ corresponding to -107dBm
Input filters	tracking filters
Frequency setting store	
Store capacity	99 frequencies and their respective type of demodulation and i-f bandwidth
Loading of store	frequency entered from keyboard or current receive frequency including type of demodulation and i-f bandwidth
Scanning operation	up to 99 stored frequencies can be cyclically scanned; halts automatically if frequency is occupied; scanning operation continued after preselected period of time at the push of a button.
Scanning time	typically 50 ms per stored frequency
S/N ratio ($V_{in} = 1\mu V$, $f_{mod} = 1$ kHz, i-f bandwidth 30 kHz, AF filter ON)	
AM(m = 0.5)	$\geqq$ 10 dB
FM(deviation 10 kHz)	$\geqq$ 20 dB
Total noise figure (incl. AF section)	9 dB typ.
Oscillator phase noise (at 20 kHz from the carrier)	120 dB/Hz typ.
FM noise suppression (3 kHz deviation, $f_{mod} = 1$ kHz, $V_{in} = 1$ mV)	50 dB typ.
Intercept point 2nd order	50 dBm typ.
3rd order	12 dBm typ.
Image frequency rejection	>90 dB
i-f rejection	>90 dB
i-f bandwidth (3 dB)	2.3 kHz, 8 kHz, 15 kHz, 30 kHz, 100 kHz, 300 kHz, 2 MHz
Demodulation	AM, FM, SSB
Squelch	S/N ratio and adjustable carrier squelch circuits (both can be switched off)
AF filter	300 Hz to 3.3 kHz; can be switched out
Gain control AGC	i-f control for $V_{in} \leqq$ 80 dB (μV) rf /i-f control for $V_{in} \leqq$ 120 dB (μV)
MGC	i-f control 80 dB rf: 40 dB, can be switch-selected
AFC	digital tracking of signals of unstable frequency (can be switched off)

Table 11-2. Specifications of the ESM 500 Vhf/Uhf Receiver. (Continued from page 72.)

Indication	
Level	on moving-coil meter in dB (μV)
Frequency offset	on moving-coil meter; sensitivity of offset meter matched to bandwidth
Panoramic Display	
i-f Panoramic Display	
Sweep width	200 kHz
Resolution	4.5 kHz
Amplitude display	logarithmic approx. 80 dB
Screen area	4 cm × 3 cm
rf Panoramic Display and broadband i-f display	
rf sweep width	entire reception range (500 MHz, max.) and/or a particular section of it; superposition of frequency marker for receiver tuning
i-f sweep width	2 MHz, max.
Amplitude display	linear or logarithmic 80 dB (10 dB/cm)
Internal testing facilities	
Continual test	monitoring of subassemblies; error signalled with code number
Loop test	triggered by pressing a button; automatic testing of complete receive section including the AF section and all LED displays
Outputs	level, offset, AF (600ω), AM video, FM video, IF (10.7 MHz) 2 MHz broadband, 50Ω, 10 dB above input level, without AFC) i-f (10.7 MHz, narrowband, with AFC, 50Ω, 10 mV), inputs /outputs for Panoramic Adapter EZP, COR (Carrier Operated Relay): coupled with squelch; drop-out time internally adjustable
Inputs	external control voltage, squelch response threshold
Remote control (via IEC bus or RS 232 C interface)	all important functions; input and output

Values of F_1 and F_2 can be substituted and the interference can be anticipated and avoided. If the locus point is examined closely, one can be assured that there are no in-band products in this case, but analyzing the areas adjacent to the locus point indicates some out-of-band spurs (unwanted products) which will have to be suppressed by the i-f filter to the level specified in the requirement. By knowing their order given by the chart, their predicted amplitude can be found (in our case 170 MHz). The 7th and 9th product ($5F_1$-$2F_2$ and $5F_2$-$4F_1$) is predicted to be 81 dB below the i-f level (typical manufacturer prediction). The i-f filter would have to attenuate 9 dB at 170 MHz to accomplish a system requirement of 90 dB as shown in Fig. 11-3. Table 11-2 gives the specifications for the E5M 500 vhf/uhf receiver. Figure 11-6 shows this fine receiver.

A simpler method of finding these products can be achieved by using the charts in Tables 11-3 and 11-4. The chart shown in Table 11-3 is for mixers used in an adding mode (B + A) where A and B are the mixing

Table 11-3. Intermodulation Product Chart Used for Mixers in an Adding Mode (B+A) where A and B are the Mixing Frequencies and B > A. Ratios A Are Shown for Each Intermodulation Product Up to the 15-th Order.

If output frequency = B + A (condition B > A)

$R=\frac{A}{B}$	PRODUCT ORDER														
	1	2	3	4	5	6	7	8	9	10	11	12	13	14	15
0.000	B	B−A	B±2A	B±3A	B±4A	B±5A	B±6A	B±7A	B±8A	B±9A	B±10A	B±11A	B±12A	B±13A	B±14A
0.072															2B-13A 15A
0.077														2B-12A 14A	
0.083													2B-11A 13A		
0.091												2B-10A 12A			
0.100											2B-9A 11A				
0.111										2B-8A 10A					
0.125									2B-7A 9A						
0.143								2B-6A 8A							
0.154															3B-12A 14A-B
0.167							2B-5A 7A							3B-11A 13A-B	
0.182													3B-10A 12A-B		
0.200						2B-4A 6A						3B-9A 11A-B			
0.222											3B-8A 10A-B				
0.250					2B-3A 5A					3B-7A 9A-B					4B-11A 13A-2B
0.273														4B-10A 12A-2B	
0.286									3B-6A 8A-B						
0.300													4B-9A 11A-2B		
0.333				2B-2A 4A				3B-5A 7A-B				4B-8A 10A-2B			
0.364															5B-10A 12A-3B
0.375											4B-7A 9A-2B				
0.400							3B-4A 6A-B							5B-9A 11A-3B	
0.429										4B-6A 8A-2B					
0.445													5B-8A 10A-3B		
0.500			2B-A 3A			3B-3A 5A-B			4B-5A 7A-2B			5B-7A 9A-3B			6B-9A 11A-4B
0.555														6B-8A 10A-4B	
0.571											5B-6A 8A-3B				
0.600								4B-4A 6A-2B							
0.625													6B-7A 9A-4B		
0.667					3B-2A 4A-B					5B-5A 7A-3B					7B-8A 10A-5B
0.715												6B-6A 8A-4B			
0.750							4B-3A 5A-2B							7B-7A 9A-5B	
0.800									5B-4A 6A-3B						
0.833											6B-5A 7A-4B				
0.858													7B-6A 8A-5B		
0.875															8B-7A 9A-6B
1.000		2B 2A		3B-A 3A-B		4B-2A 4A-2B		5B-3A 5A-3B		6B-4A 6A-4B		7B-5A 7A-5B		8B-6A 8A-6B	

frequencies and B>A. The chart shown in Table 11-4 is for mixers used in a subtracting mode (B-A) with the same conditions applying. If using the same example, and substituting F_1 for A and F_2 for B the same ratio can be obtained.

$$\frac{F_1}{F_2} = \frac{A}{B} = \frac{70}{90} = 0.778$$

If output frequency = B − A (condition B > A)

$R = \frac{A}{B}$	ORDER														
	1	2	3	4	5	6	7	8	9	10	11	12	13	14	15
0.000	B	B+A	B±2A	B±3A	B±4A	B±5A	B±6A	B±7A	B±8A	B±9A	B±10A	B±11A	B±12A	B±13A	B±14A
0.063															15A
0.067														14A	
0.072													13A		
0.077												12A			
0.083											11A				2B-13A
0.091										10A				2B-12A	
0.100									9A				2B-11A		
0.111								8A				2B-10A			
0.125							7A				2B-9A				
0.133															14A-B
0.143						6A				2B-8A				13A-B	
0.154													12A-B		
0.167					5A				2B-7A			11A-B			
0.182											10A-B				3B-12A
0.200				4A				2B-6A		9A-B				3B-11A	
0.214															13A-2B
0.222									8A-B				3B-10A		
0.231														12A-2B	
0.250			3A				2B-5A	7A-B				3B-9A	11A-2B		
0.273												10A-2B			
0.286							6A-B				3B-8A				
0.300											9A-2B				4B-11A
0.308															12A-3B
0.333		2A				2B-4A 5A-B				3B-7A 8A-2B				4B-10A 11A-3B	
0.364													10A-3B		
0.375									7A-2B				4B-9A		
0.400					4A-B				3B-6A			9A-3B			
0.416															11A-4B
0.429								6A-2B				4B-8A			
0.445											8A-3B				5B-10A
0.455														10A-4B	
0.500	A			3A-B	2B-3A		5A-2B	3B-5A		7A-3B	4B-7A		9A-4B	5B-9A	
0.545															10A-5B
0.555												8A-4B			
0.571									6A-3B				5B-8A		
0.600						4A-2B				4B-6A				9A-5B	
0.625											7A-4B				6B-9A
0.667			2A-B				3B-4A	5A-3B				5B-7A	8A-5B		
0.700															9A-6B
0.715										6A-4B				6B-8A	
0.750					3A-2B				4B-5A			7A-5B			
0.778														8A-6B	
0.800							4A-3B				5B-6A				
0.833									5A-4B				6B-7A		
0.858											8A-5B				7B-8A
0.875													7A-6B		
1.000				2B-2A		3B-3A		4B-4A		3B-5A		6B-6A		7B-7A	

Table 11-4.
Intermodulation Product Chart Used for Mixers in a Subtracting Mode. Same Procedure is Used as in Table 11-3 for Finding Intermodulation Products to the 15-th Order.

We then use Table 11-3 since the mixer in our example operates in the adding mode, and find the corresponding products as indicated in Fig. 11-7 (5A - 2B and 5B - 4A). If the numerical values of A and B are inserted in these formulas, the same resultant values can be obtained with this method, which is usually preferred. If this receiver were not designed for a fixed frequency you can imagine what a job it could be to evaluate all the high-order products generated within such a scheme.

USING COMPUTERS TO SUPPORT THE PRODUCT CHARTS (A TI-59 PROGRAM)

Today, computer programs are used successfully to predict these problems. The following example will describe the use of a TI-59 programmable calculator (Fig. 11-8) to perform this tedious task for our example. A listing of the program is also included for those wishing to try their own problems.

This program finds all combinations of (m × LO) ± (n × rf) and prints those frequencies that fall in the center of the i-f by actually indicating "=i-f" in the printout. Those frequencies that fall within the predetermined i-f bandwidth but are not exactly in the center are also reported in the printout by the indication "*BW" (Fig. 11-9). The sample program in Fig. 11-10 shows how our 70 MHz fixed-frequency receiver may be analyzed using this method.

If the 90 MHz LO is entered into the user-defined key A, the 70 MHz rf into B, the 160 MHz i-f into C and the i-f bandwidth we wish to analyze into D (50 MHz) a report is obtained by depressing key E′, indicating that we are ready to run the program with the entered data as shown in Fig. 11-9. If a mistake occurred in the process of entering the information, new data can be entered by repeating the above process with no alteration to the actual

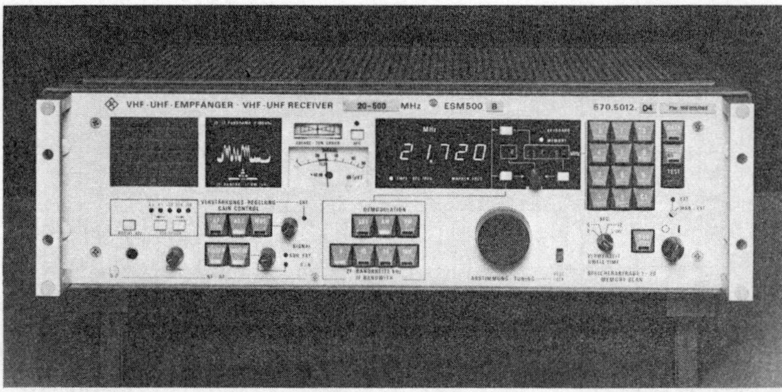

Fig. 11-6. The ESM 500 is a general purpose vhf/uhf receiver designed for radio monitoring. This highly digital receiver features a fast synthesizer with a resolution of 1 kHz, a memory that can store up to 24 frequencies, and a built in i-f panoramic display. Careful system and circuit design is visible throughout the receiver (courtesy of Rohde & Schwarz).

EXAMPLE:

$R = \frac{A}{B}$	i-f OUTPUT FREQUENCY = B + A (condition B >A) ORDER														
	1	2	3	4	5	6	7	8	9	10	11	12	13	14	15
0.000	B	B−A	B±2A	B±3A	B±4A	B±5A	B±6A	B±7A	B±8A	B±9A	B±10A	B±11A	B±12A	B±13A	B±14A
0.072															2B−13A 13A
0.077														2B−12A 14A	
0.083													2B−11A 13A		
0.091												2B−10A 12A			
0.100											2B−9A 11A				
0.111										2B−8A 10A					
0.125									2B−7A 9A−B						
0.143								2B−6A 8A−B							
0.154															3B−12A 14A−B
0.167							2B−3A 7A							3B−11A 13A−B	
0.182													3B−10A 12A−B		
0.200						2B−4A 6A						3B−9A 11A−B			
0.222											3B−8A 10A−B				
0.250					2B−3A 5A					3B−7A 9A−B					4B−11A 13A−2B
0.273														4B−10A 12A−2B	
0.286									3B−6A 8A−B						
0.300													4B−9A 11A−2B		
0.333				2B−2A 4A				3B−5A 7A−B				4B−8A 10A−2B			
0.364															3B−10A 12A−3B
0.375											4B−7A 6A−2B				
0.400							3B−4A 6A−B							3B−9A 11A−3B	
0.429										4B−6A 3A−2B					
0.445													3B−8A 10A−3B		
0.500			2B−A 3A			3B−3A 5A−B			4B−5A 7A−2B			3B−7A 9A−3B			6B−10A 11A−4B
0.555														6B−8A 10A−4B	
0.571											3B−6A 8A−3B				
0.600								4B−4A 8A−2B							
0.625													6B−7A 9A−4B		
0.667					3B−2A 4A−B					3B−3A 7A−3B					7B−8A 10A−3B
0.715												6B−6A 6A−4B			
0.750							3A−2B							7B−7A 9A−3B	
0.778															
0.800									5B−4A						
0.833											6B−5A 7A−4B				
0.858													7B−6A 8A−5B		
0.875															8B−7A 9A−6B
1.000		2B 2A		3B−A 3A−B		4B−2A 4A−2B		3B − 3A 5A − 3B		5B − 4A 6A − 4B		7B−3A 7A−3B		8A−6A 8A−6B	

Fig. 11-7. The same results can be obtained as with our previous example by using the intermodulation chart at A shown here. Ratio $\frac{A}{B}$ = 0.778 which points to a 7-th and 9-th order product (5A-2B and 5B-4A).

program. We can now run our analysis, by depressing Key E. A complete list of products will be automatically printed as shown in Fig. 11-10. The process will take approximately four minutes to analyze all cases of m and n within the bandwidth of 50 MHz. This may seem to be a long time, but not if we compare the effort with that of the product chart in Fig. 11-4.

Fig. 11-8. The TI-59 programmable calculator and a PC-100 printer-plotter are used to analyze the intermodulation products that may exist in a radio receiver.

This program can be recorded on two magnetic cards and could be used over and over again as necessary. For those equipped with a TI-59, Table 11-5 shows the actual listing of the intermodulation products program. Partitioning (OP 17) is 479.59. Table 11-6 lists all necessary procedures for running the program.

The amplitude of the undesired products identified depends on their particular order number (m+n). Most products of the seventh order or higher will be at least 60 dB down from the i-f level, and are usually not considered to cause problems. Unless instructed differently, the program will automatically calculate all products to the twelfth order (6 × LO) ± (6 × rf) (no user inputs to A′ and B′ are required). If a different resolution is

```
   CONDITIONS
   6.      × LO
   6.      × RF
  90.        LO
  70.        RF
 160.        IF
  50.        BW
```

Fig. 11-9. Depressing Key E′ will print out all conditions entered by the operator.

```
   3.      × LO
   6.      × RF
 150.      ÷ BW

   2.      × LO            0.     × LO
   5.      × RF            2.     × RF
 170.      ÷ BW          140.     ÷ BW

   5.      × LO            1.     × LO
   4.      × RF            1.     × RF
 170.      ÷ BW          160.     = IF

   4.      × LO            2.     × LO
   3.      × RF            0.     × RF
 150.      ÷ BW          180.     ÷ BW
```

Fig. 11-10. Depressing Key E on the TI-59 will run the intermodulation product program within the specified conditions indicating the same problem areas 5RF-2LO and 5LO-4 RF as previously determined with the intermodulation charts. Other problems are also indicated within the i-f bandpass but are not considered in our example for reasons of simplicity.

desired the two keys should be addressed accordingly. The execution time of this program is a direct function of the product order and the i-f bandwidth required by the user.

SYSTEM ANALYSIS FOR A GENERAL COVERAGE COMMUNICATION RECEIVER (A DESIGN CASE)

In the following pages we will consider a system design for a general coverage receiver with a wide input bandwidth (28 MHz). Unlike the dedicated single frequency receiver analyzed in the previous pages, this wideband receiver presents an immense product analysis problem, which is due to the many different cases that could be created within the input bandwidth. The problem is further increased if a double-conversion approach is used, as products generated in the first i-f can multiply in the second i-f, as we discussed at the beginning of this chapter. The designer should use good judgment in the initial choice of frequencies, since no computer or chart can take the place of good engineering procedures, as we will see in our example. Assume that a communication receiver is to be designed that will cover the range of 2 MHz to 30 MHz, with good image rejection, and a minimum of unwanted products generated within its own

Table 11-5. Listing of the TI-59 Program for Intermodulation Products. Partitioning is 479.59 (Program Courtesy of Frank Ferrin, Honeywell, Inc.).

Step	Code	Key	Step	Code	Key	Step	Code	Key
000	76	LBL	054	65	×	108	02	02
001	11	A	055	43	RCL	109	85	+
002	42	STO	056	04	04	110	01	1
003	03	03	057	95	=	111	95	=
004	91	R/S	058	50	I×I	112	42	STO
005	76	LBL	059	42	STO	113	09	09
006	12	B	060	05	05	114	91	R/S
007	42	STO	061	71	SBR	115	43	RCL
008	04	04	062	01	01	116	06	06
009	91	R/S	063	15	15	117	32	X:T
010	76	LBL	064	43	RCL	118	43	RCL
011	13	C	065	01	01	119	05	05
012	42	STO	066	65	×	120	67	EQ
013	06	06	067	43	RCL	121	01	01
014	91	R/S	068	03	03	122	45	45
015	76	LBL	069	75	-	123	43	RCL
016	14	D	070	43	RCL	124	13	13
017	42	STO	071	02	02	125	32	X:T
018	07	07	072	65	×	126	43	RCL
019	91	R/S	073	43	RCL	127	05	05
020	76	LBL	074	04	04	128	77	GE
021	15	E	075	95	=	129	01	01
022	43	RCL	076	50	I×I	130	32	32
023	06	06	077	42	STO	131	92	RTN
024	75	-	078	05	05	132	43	RCL
025	43	RCL	079	71	SBR	133	14	14
026	07	07	080	01	01	134	32	X:T
027	55	÷	081	15	15	135	43	RCL
028	02	2	082	01	1	136	05	05
029	95	=	083	94	+/-	137	22	INV
030	42	STO	084	44	SUM	138	77	GE
031	13	13	085	01	01	139	01	01
032	85	+	086	97	DSZ	140	60	60
033	43	RCL	087	08	08	141	92	RTN
034	07	07	088	19	D'	142	00	0
035	95	=	089	43	RCL	143	00	0
036	42	STO	090	11	11	144	00	0
037	14	14	091	42	STO	145	71	SBR
038	68	NOP	092	01	01	146	01	01
039	68	NOP	093	85	+	147	80	80
040	68	NOP	094	01	1	148	43	RCL
041	68	NOP	095	95	=	149	19	19
042	68	NOP	096	42	STO	150	69	OP
043	68	NOP	097	08	08	151	04	04
044	76	LBL	098	01	1	152	43	RCL
045	19	D'	099	94	+/-	153	05	05
046	43	RCL	100	44	SUM	154	69	OP
047	01	01	101	02	02	155	06	06
048	65	×	102	97	DSZ	156	98	ADV
049	43	RCL	103	09	09	157	92	RTN
050	03	03	104	19	D'	158	00	0
051	85	+	105	43	RCL	159	00	0
052	43	RCL	106	12	12	160	71	SBR
053	02	02	107	42	STO	161	01	01

```
162  80  80
163  43 RCL
164  18  18
165  69 OP
166  04  04
167  43 RCL
168  05  05
169  69 OP
170  06  06
171  98 ADV
172  92 RTN
173  00  0
174  00  0
175  00  0
176  00  0
177  00  0
178  00  0
179  00  0
180  43 RCL
181  15  15
182  69 OP
183  04  04
184  43 RCL
185  01  01
186  69 OP
187  06  06
188  43 RCL
189  16  16
190  69 OP
191  04  04
192  43 RCL
193  02  02
194  69 OP
195  06  06
196  92 RTN
197  00  0
198  00  0
199  00  0
200  76 LBL
201  10 E'
202  98 ADV
203  98 ADV
204  43 RCL
205  23  23
206  69 OP
207  03  03
208  43 RCL
209  24  24
210  69 OP
211  04  04
212  69 OP
213  05  05
214  43 RCL
215  15  15
216  69 OP
217  04  04
218  43 RCL
219  01  01
220  69 OP
221  06  06
222  43 RCL
223  16  16
224  69 OP
225  04  04
226  43 RCL
227  02  02
228  69 OP
229  06  06
230  43 RCL
231  21  21
232  69 OP
233  04  04
234  43 RCL
235  03  03
236  69 OP
237  06  06
238  43 RCL
239  22  22
240  69 OP
241  04  04
242  43 RCL
243  04  04
244  69 OP
245  06  06
246  43 RCL
247  17  17
248  69 OP
249  04  04
250  43 RCL
251  06  06
252  69 OP
253  06  06
254  43 RCL
255  20  20
256  69 OP
257  04  04
258  43 RCL
259  07  07
260  69 OP
261  06  06
262  98 ADV
263  98 ADV
264  91 R/S
265  00  0
266  00  0
267  00  0
268  00  0
269  00  0
270  76 LBL
271  16 A'
272  42 STO
273  01  01
274  42 STO
275  11  11
276  85  +
277  01  1
278  95  =
279  42 STO
280  08  08
281  91 R/S
282  76 LBL
283  17 B'
284  42 STO
285  02  02
286  42 STO
287  12  12
288  85  +
289  01  1
290  95  =
291  42 STO
292  09  09
293  91 R/S

          0.  00
          6.  01
          6.  02
         90.  03
         70.  04
          0.  05
        160.  06
         60.  07
          7.  08
          7.  09
          0.  10
          6.  11
          6.  12
        130.  13
        190.  14
   50002732.  15
   50003521.  16
       2421.  17
   51001443.  18
   64002421.  19
       1443.  20
       2732.  21
       3521.  22
 1532311624.  23
 3724323136.  24
```

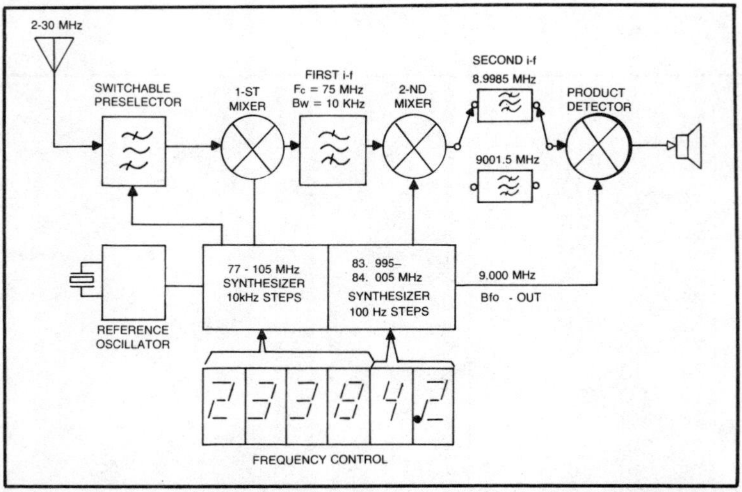

Fig. 11-11. Communications receiver block diagram. The first i-f is 75 MHz and the second i-f is 9 MHz. A two-loop synthesizer is used to provide rough and fine tuning.

Table 11-6. User Instructions for the TI-59 Intermodulation Products Program.

STEP	DESCRIPTION	ENTER	PRESS
1	Input the highest LO harmonic of interest*	m	A′
2	Input the highest rf harmonic oı interest*	n	B′
3	INPUT LO FREQUENCY	LO	A
4	INPUT rf FREQUENCY	rf	B
5	INPUT i-f FREQUENCY	i-f	C
6	INPUT BANDWIDTH Centered at i-f	BW	D
7	PRINT CONDITIONS ENTERED		E′
8	RUN PROGRAM		E

Progam finds all combinations of mxLO ± n × rf and frequencies that are equal to the i-f chosen (=i-f) and those that lie within the i-f bandwidth (*BW).

*If no input, a sixth order will be automatically analized.

USER DEFINED KEYS

A	LO
B	rf
C	i-f
D	BW
E	RUN PROGRAM
A′	m
B′	n
C′	NOT USED
D′	NOT USED
E′	CONDITIONS REPORT

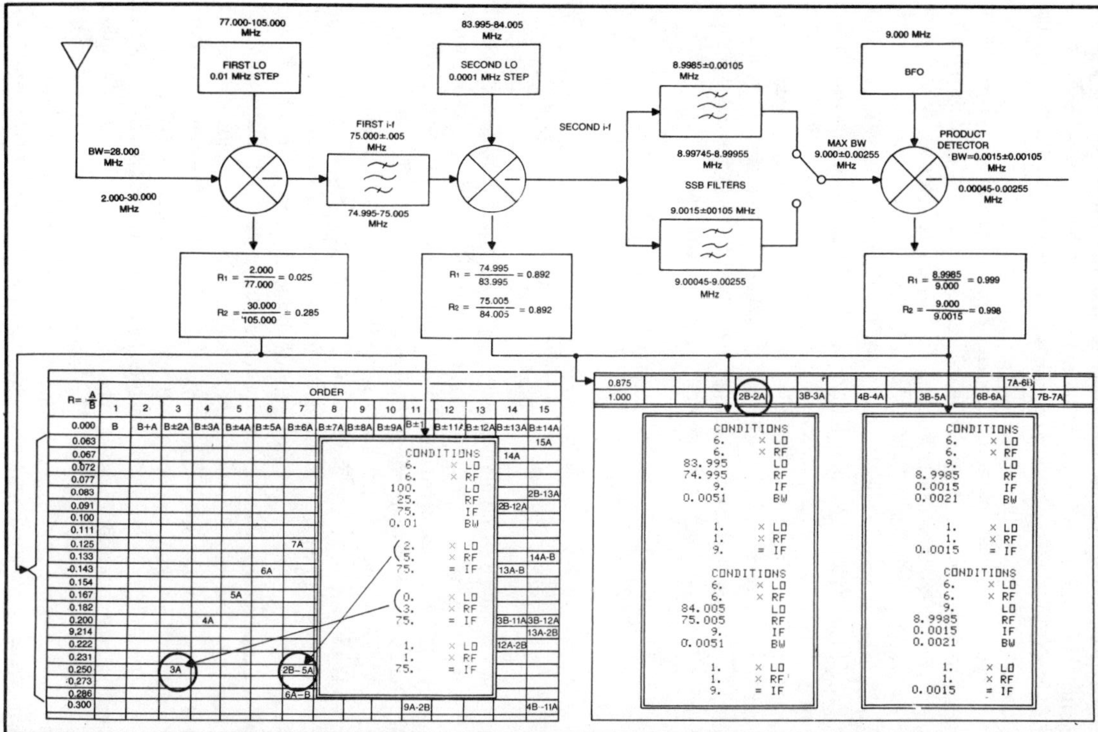

Fig. 11-12. System analysis for a double conversion, general coverage communications receiver with a first i-f at 75 MHz and a second i-f at 9 MHz.

configuration. Looking at Fig. 11-11 a double conversion approach is considered with an up-conversion first i-f compatible with commerically-available monolithic crystal filters at 75 MHz. A phase-locked synthesizer is to be used as the local oscillator for the first mixer, and will provide rough tuning in 10 kHz steps over this range. The fine tuning is achieved in the second conversion with another synthesizer which provides frequency resolution of 100 Hz within the 10 kHz rough steps.

The second i-f is at 9 MHz because of the availability of good crystal-lattice filters. Two sideband filters are used in this i-f. We will attempt to analyze the system shown with the help of the charts described earlier along with the TI-59 computer program which will perform the calculations necessary to support these charts.

Figure 11-12 shows the mathematical model for this receiver. An rf signal anywhere from 2.000 to 30.000 MHz comes from the antenna to the first mixer where it subtracts from the first local oscillator which operates so as to always produce a 75.000 MHz i-f. This local oscillator is our first synthesizer operating from 77.000 to 105.000 MHz in 0.01 MHz (10 kHz) steps. This dictates the bandwidth of the first i-f to be 10 kHz minimum, from 74.995 to 75.005 MHz in order for the second local oscillator to be able to provide fine tuning in the second i-f, as we previously discussed. The 75 MHz tandem monolithic filter (discussed earlier) is a practical example of a first i-f filter that could be used in this application. If rf = A and LO = B, two ratios $\frac{A}{B}$ can be created: R minimum and R maximum.

$$R_{MIN} = \frac{2.000}{77.000} = 0.025$$

$$R_{MAX} = \frac{30.000}{105.000} = 0.285$$

We will use the mixing product chart for subtraction since our i-f is 75 MHz, and find the entire affected band between R_{MIN} and R_{MAX} as shown in Fig. 11-12. Any product indicated within this band could be a potential problem for the corresponding received frequency.

A look at our chart indicates a series of problems (7A, 6A, 5A, 4A, 3A, 2B-5A) with the worst one at 3A. If the TI-59 program is used, we can verify this case as shown in our example. At first we can say that the third harmonic of one of the two mixing signals could be quite powerful and could indeed produce a problem, but a closer look at the system indicates that the offending frequency A, is actually a received frequency and chances are very good that a distant 25 MHz station has a level of insignificant third harmonic (75 MHz) appearing at the antenna of our receiver.

The problem is further diminished by our receiver's preselector which greatly attenuates at 75 MHz. The same conditions apply to the other products indicated by the chart. They present even a better case since they are even further removed from the received frequencies. This is a case where engineering judgement is more important than all our tools which are

```
  CONDITIONS
  7.      × LO
  7.      × RF
83.995      LO
74.995      RF
  9.        IF
0.0374      BW

  1.      × LO
  1.      × RF
  9.      = IF
```

Fig. 11-13. TI-59 program indicates no intermodulation problem in the second i-f of the double-conversion communications receiver. The −60 dB bandwidth (0.0374 MHz) of the filters was used for a worst-case analysis.

only used to warn of possible problems. The case would be different, however, if the B signal was the offender, as 3B would have been the third harmonic of the local oscillator which can be of relatively high amplitude, and can cause interference.

The filtered first i-f range of 74.995 to 75.005 MHz is further mixed with the second local oscillator operating in 0.0001 MHz (100 Hz) steps over the 10 kHz range of 83.995 to 84.005 MHz providing fine tuning for the receiver as we previously discussed. If R_{MIN} and R_{MAX} are found for the second i-f a new band of interest can be located on our chart, as shown in Fig. 11-12.

```
  CONDITIONS          CONDITIONS
  6.      × LO        6.      × LO
  6.      × RF        6.      × RF
  9.        LO        9.        LO
8.9985      RF      9.0015      RF
0.0015      IF      0.0015      IF
0.0038      BW      0.0038      BW

  2.      × LO        2.      × LO
  2.      × RF        2.      × RF
 0.003    * BW       0.003    * BW

  1.      × LO        1.      × LO
  1.      × RF        1.      × RF
0.0015    = IF      0.0015    = IF
```

Fig. 11-14. The TI-59 program indicates a 2B-2A product within the bandwidth of the 8 pole single sideband filters.

$$R_{MIN} = \frac{74.995}{83.995} = 0.89285076$$

$$R_{MAX} = \frac{75.005}{84.005} = 0.89286352$$

Since the range to be covered is only 10 kHz, this band is very narrow and in reality is expressed by the same number (0.892) since the chart only extends to three decimal points. The i-f is centered at 9.000 MHz and its bandwidth is determined by the two single-sideband filters. For simplicity, the minimum corner frequency (−3 dB) of the lower filter and the upper corner frequency for the higher filter were chosen determining a total bandwidth (−3 dB) of 0.0051 MHz (5.1 kHz, both sidebands). Figure 11-12 indicates clearly that there is no problem except for a 13th order product which can be ignored, but since the band is so narrow and so close to the 1.000 ratio which presents quite a few problems, the computer is used to completely insure safety. Unless otherwise instructed, the program will not print out the 13th product as it is programmed to only calculate products to the 12th order. If the order is increased to 14, the computer will indicate

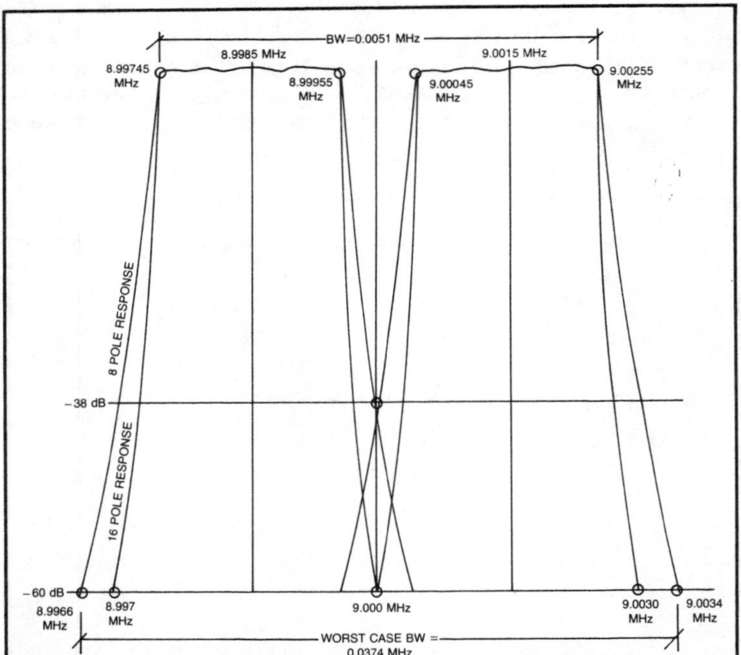

Fig. 11-15. Bandwidth characteristics of the eight pole-single sideband filters used in the communications receiver. Sixteen-pole filters could be used to eliminate a 2B-2A audio product as well as to improve the total selectivity of the receiver.

that there is really not a case for 7A-6B (case shown in Fig. 11-12) within the ratio range.

Let's now proceed to analyze the second i-f in Fig. 11-12. A total bandwidth of 0.0374 MHz (37.4 kHz) is analyzed here to completely insure freedom from intermodulation products within the slopes of the 8-pole filters used in this i-f (37.4 kHz is the –60 dB total bandwidth of the two filters) as shown in Fig. 11-13. It can be seen from this analysis that the charts can only be used to the guideline extent. For a more in-depth analysis, the TI-59 program or some other means of calculation has to be used to obtain precise answers.

We have determined that our system design provides a clean scheme to this point. The last conversion is from 9.000 MHz ± 0.00255 MHz to audio between 0.00045 to 0.00255 MHz (450 Hz to 2.550 kHz) in both single-sideband filter cases. The conversion takes place in a third mixer (product detector) as shown in our example (Fig. 11-14).

$$R_{MIN} = \frac{8.9985}{9.000} = 0.999833$$

$$R_{MAX} = \frac{9.000}{9.0015} = 0.999833$$

These identical ratios locate the intermodulation band to be analyzed very close to the 1.000 ratio in the chart and with a –3 dB bandwidth of 0.0021 MHz (2.1 kHz). No problems are found for either one of the single-sideband filters. However, if the –60 dB bandwidth (0.0038 MHz) of the 8-pole filters is used for the TI-59 computer program, the 2B-2A problem circled in Fig. 11-12 becomes evident at the 3 kHz point in the slopes of our filters. This is true for either sideband filter as shown in Fig. 11-15 and the resulting audio distortion can only be cured by improving the shape factor of the filters, in our case by doubling the number of poles to 16 for each one of the single-sideband filters.

Another way to cure this problem would be to introduce a low-pass filter in the audio portion of our receiver which will cut off all frequencies beyond the 2.55 kHz which is the highest frequency passed by the filters. A practical cut-off point would be at 2.8 kHz. The first method is, however, preferred as it also provides better adjacent-channel rejection, improving upon the overall selectivity of the receiver.

This concludes the analysis for the double-conversion general coverage communications receiver. In performing this tedious task, we have scientifically proved that we have a thoroughly analyzed sound system, with a minimum of intermodulation problems. We can now proceed with confidence to the circuit design of our receiver. In conclusion, we can safely say that all frequency schemes used in communications receivers have unwanted products within their outputs, and the fact remains that any configuration chosen is simply the best compromise in the opinion of the designer.

—•••—

Chapter 12

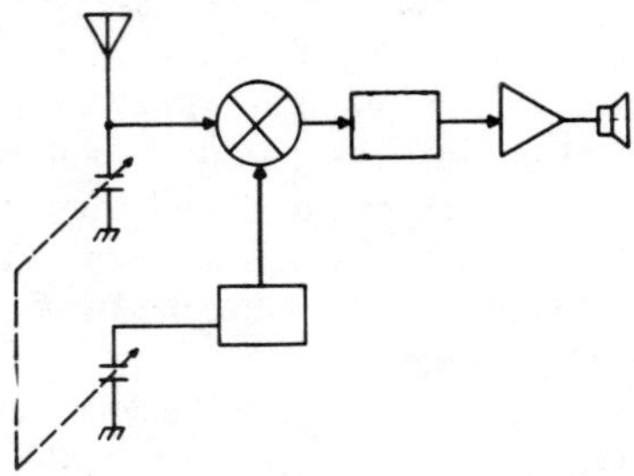

Dynamic Range

Most of the receiver design efforts of the past twenty years have been directed toward replacing already proven vacuum tube technologies with newly developed semiconductors. This was initially considered an easy task, but the results have been surprisingly poor. While such characteristics as high sensitivity and good selectivity were maintained, the new solid-state designs lacked dynamic range.

In an increasingly dense rf spectrum, the concept behind good dynamic range performance for a communication receiver extends to more than the ability to detect very weak signals in the presence of strong signals which may be greater in amplitude by as much as 100 dB. This concept should also include a high degree of rejection to spurious products produced by nonlinear interaction of many powerful signals, sometimes far removed from the receiving frequency. Let's first define dynamic range.

Dynamic range is the power range over which a device such as a radio receiver provides useful operation. The upper limit of the dynamic range (P_u) is limited by the level of two equal input signals creating a third-order intermodulation product at the output of the receiver, which is equal in amplitude to the *minimum detectable signal* (MDS) level. The *MDS* (sometimes referred to as the noise floor) is considered as the lower limit (P_L) of the dynamic range, and is defined as a signal 3 dB greater than the equivalent noise level for a specified i-f bandwidth. The minimum detectable signal can be found by using equation 12-1.

$$P_{L(dBm)} = MDS_{(dBm)} = -171\text{ dBm}^* + NF_{(dB)} + 10 \log \text{i-f} - \text{BW (Hz)}$$

Eq. 12-1

Where: MDS is lower power limit of dynamic range in dBm.
NF is System noise figure in dB
i-f-BW is i-f Bandwidth in Hz

*KTB + 3 dB = −174 + 3 = −171 dBm

P_L is lower power limit of dynamic range in dBm.

The upper limit of the dynamic range can then be expressed by using equation 12-2.

$$P_{U(dBm)} = 1/3\ (MDS + 2\ IP) \qquad \textbf{Eq. 12-2}$$

$$= 1/3\ (-171\ (dBm) + NF\ (dB) + 10\ LOG\ i\text{-}f\text{-}BW\ (Hz)\ ) + 2/3\ IP\ (dBm)$$

Where: P_U is the upper power limit of the dynamic range in dBm
IP is Receiver's third order input intercept point in dBm.

By combining the two equations, we can find equation 12-3 for the total spurious-free dynamic range:

$$SFDR\ (dBm) = P_U\ (dBm) - P_L\ (dBm) \qquad \textbf{Eq. 12-3}$$

$$= 1/3\ (MDS + 2\ IP) - MDS = 2/3\ (IP - MDS)$$

$$= 2/3\ (IP\ (dBm) - NF\ (dB) - 10\ Log\ i\text{-}f\text{-}BW\ (Hz) + 171\ (dBm)$$

Where: SFDR is the spurious free dynamic range.

It can be seen from this equation that the dynamic range is directly proportional to the intercept point (IP) and inversely proportional to the noise figure (NF), and i-f bandwidth (i-f-BW). We can then say that the dynamic range improves with the lowest noise figures, narrower i-f bandwidths and higher intercept points.

The following example shows a practical application for the dynamic range formula. Assume a typical high performance receiver with a noise figure of 8 dB, an i-f bandwidth of 2.1 kHz and an input intercept point of +20 dBm. Substituting these quantities in equation 12-3 yields;

$$SFDR = 2/3\ (+20\ dBm - 8\ dB - 10\ log\ 2100\ Hz + 171\ dBm) = 99.85\ dB.$$

$$\therefore\ SFDR = 99.85\ dB.$$

The total distribution of this number can best be understood by examining the graph in Fig. 12-1. We know that the total spurious-free dynamic range (SFDR) for our receiver is 99.85 dB, but what is not known is where does this range fit in the total picture of the receiver's sensitivity and once this is found, what does this range mean from a practical performance point of view. We had previously determined that the lower limit of the dynamic range is given by the minimum detectable signal (MDS). If using equation 12-1 for our example, we find the lower limit of the receiver's dynamic range to be −129.7 dBm.

$$MDS = -171 + 8 + 10\ Log\ 2100 = -129.77$$

We can then say that the system's noise figure for an i-f bandwidth of 2.1 kHz is 3 dB below this number, or −132.7 dBm (MDS is defined as a

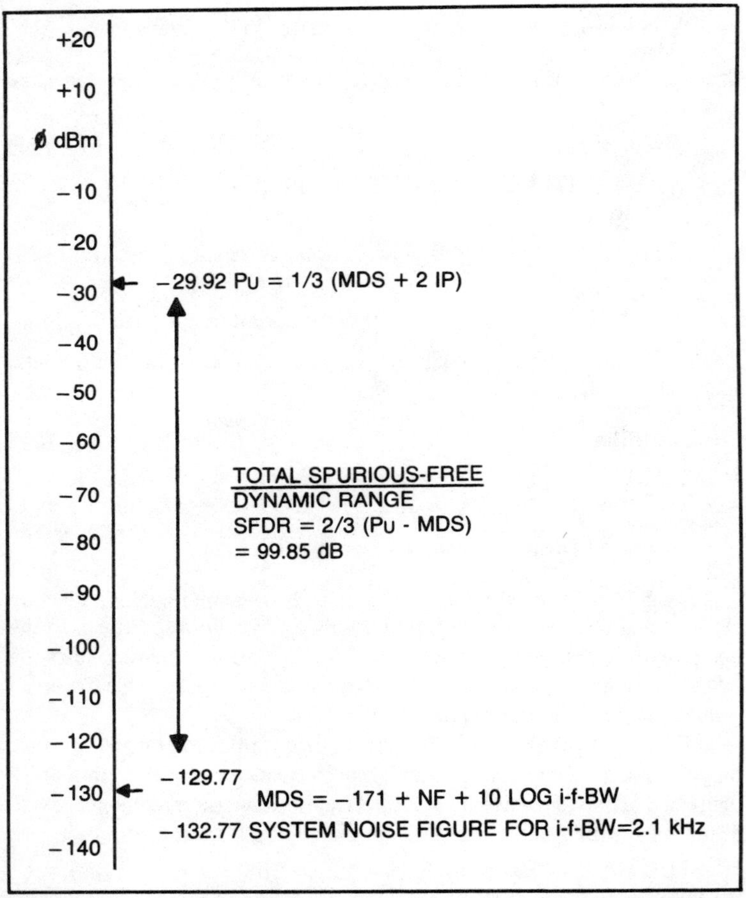

Fig. 12-1. Determining the dynamic range of a receiver with a noise figure of 8 dB, an i-f bandwidth of 2.1 kHz and an input intercept point of +20 dBm.

signal 3 dB greater than the equivalent noise level for a specified i-f bandwidth).

Knowing the MDS, the IP (20 dBm) and with the help of equation 12-2, we can determine the upper limit of our 99.85 dB dynamic range.

$$P_U = 1/3\,(-129.77 + 40)$$
$$= -29.92 \text{ dBm.}$$

The same result would be obtained if we added the total dynamic range of 99.85 dB to the MDS.

$$P_U = 99.85 + (-129.77) = -29.92 \text{ dBm.}$$

This last procedure could be used to verify the validity of equation 12-2.

If these numbers are plotted as shown in Fig. 12-1, we can conclude that the receiver in our example will perform undisturbed for all input signals varying from approximately −30 dBm to −130 dBm, with the receiver tuned to a third-order intermodulation product produced by two strong signals equal in amplitude and different in frequency from each other. The amplitude of these signals as well as their difference frequency (ΔF) were represented in our example by the +20 dBm input intercept point. In practice, this quantity is a function of the output intercept of all non-linear elements, such as mixers, amplifiers, etc., involved in the design of the receiver, as we will see below.

INTERCEPT METHOD

Figure 12-2 shows the intercept method, used as an evaluation method for the strong-signal handling capability of a radio receiver. In practice, the dynamic range of a receiver is measured with the setup shown in Figs. 12-3 and 12-4. First, the minimum detectable signal (MDS) is found as shown at 12-3.

The MDS is measured as the power necessary to generator G expressed in dB, to produce a 3 dB increase in audio output over the noise level of the receiver as indicated by the audio voltmeter. The MDS is

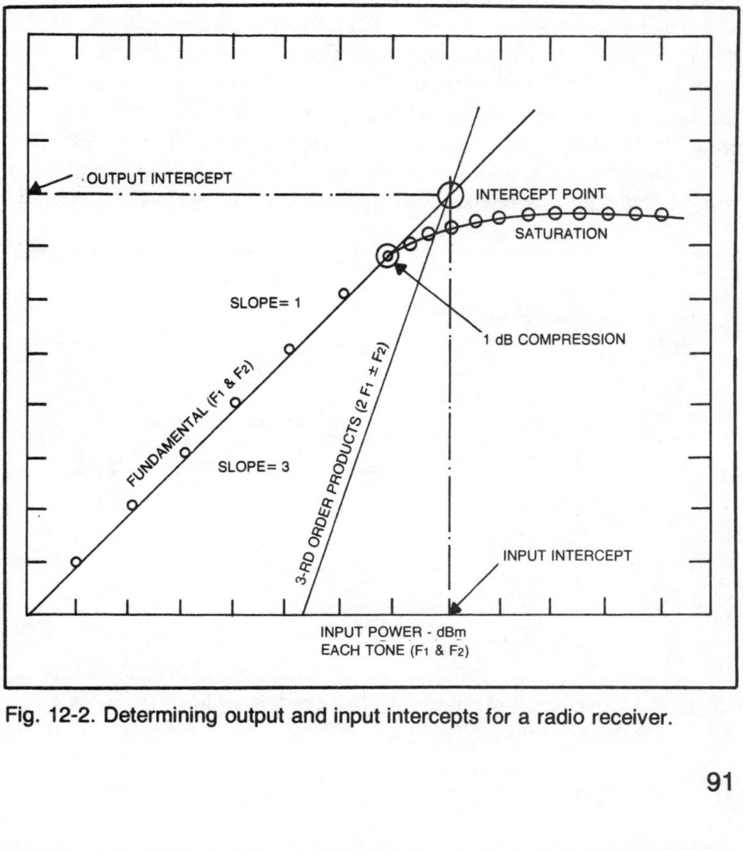

Fig. 12-2. Determining output and input intercepts for a radio receiver.

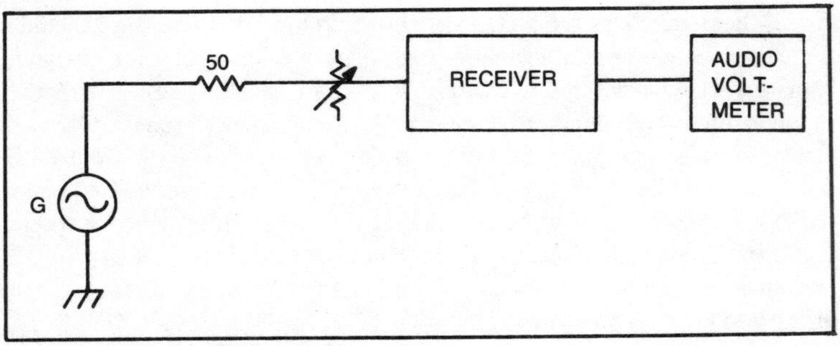

Fig. 12-3. Measurement of MDS.

specified for a given i-f bandwidth and is lowest with narrowest bandwidth available. The highest bandwidth should be used for a worst case analysis.

Knowing the MDS, the setup in Fig. 12-4 can be used to actually find the output intercept, and with this information the input intercept can be plotted as shown in Fig. 12-2.

In order to find the output intercept point the outputs of the two signal generators (G_1 and G_2) are combined in a hybrid combiner. The output of the combiner which now contains a two-tone signal is applied through a calibrated step attenuator to the receiver.

The two generators are usually 10 kHz apart with the receiver tuned to $2F_2$-F_1 or $2F_1$-F_2, a third order product. The attenuator is then varied until the response of the receiver at the frequency of the third-order product is the same as that produced by the MDS found earlier. The performance is specified by measuring and plotting the output intercept as shown.

If the receiver is well designed (Fig. 12-5), the desired output signal and the distortion product curve will intersect as high as possible, as shown

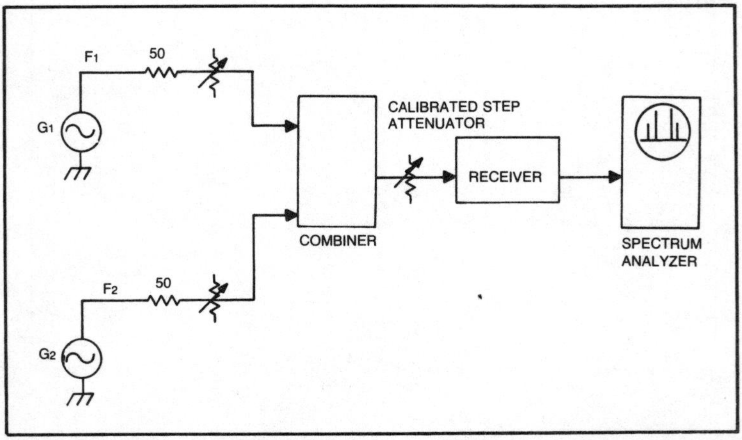

Fig. 12-4. Measurement of output intercept point. Input intercept can then be plotted and used in the dynamic range formula.

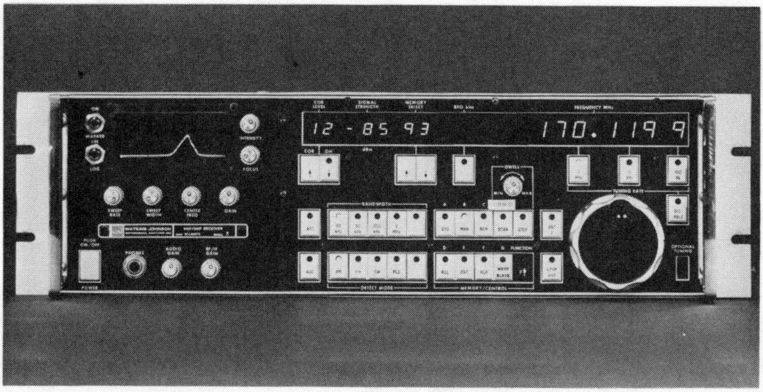

Fig. 12-5. The WJ-8617A is a high-dynamic-range vhf/uhf receiver, covering the frequency range of 20 to 1100 MHz under the control of a Motorola 6800 microprocessor which makes intelligent decisions based on the characteristics of the received signal. The receiver has a typical third order intercept point of −5 dBm, and a noise figure of 10 dB (courtesy of Watkins-Johnson Company).

in our example. This is the output intercept which describes the intermodulation response of the receiver. The input intercept can also be plotted from the intercept point as shown. This number can then be used to find the spurious free dynamic range as we previously discussed.

In conclusion, the receiver processes a weak signal in the presence of an adjacent strong signal. Because of deficiencies in the design of the first mixer and the front end, if a preamplifier is used, the receiver may not be able to copy the weak signal and it may be completely blocked out. The receiver's ability to perform under such conditions is expressed by the spurious free dynamic range. Recognizing this impediment came from military applications which required a receiver to perform properly in the presence of a number of high power transmitters broadcasting on many adjacent frequencies.

—•••—

Chapter 13

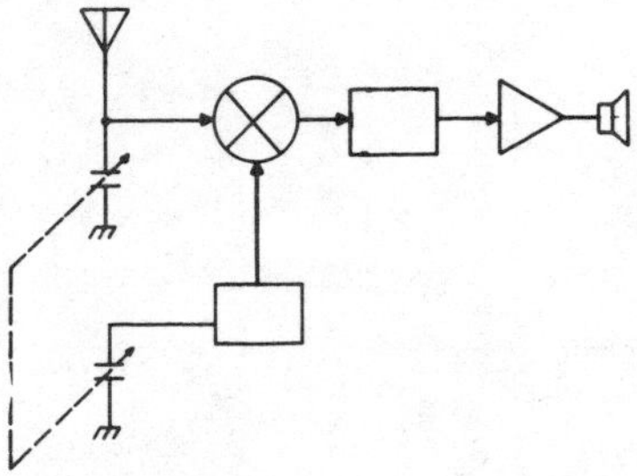

Design Considerations for the Front End

In designing a communication receiver for a particular range of frequencies, one must look at the natural noise power contribution anticipated at the antenna of such receiver for the range of interest. This tells the designer that the receiver should be quiet enough regarding its internal noise (noise figure) to hear this ambient noise. In the electromagnetic spectrum, noise contribution and its level varies considerably with frequency, type of source, location and season of the year.

DETERMINING NOISE FIGURE REQUIREMENTS

The graph in Fig. 13-1 shows the overall external noise power contribution (dB above KT_o)* for the frequency range of 10 kHz to 10 GHz. This information is based on measurements made with an omnidirectional vertical antenna over a perfectly conducting ground.**

The sky noise data and the solar noise data were obtained with a directional antenna pointed at the sources. Maximum and minimum readings were provided in the case of atmospheric and man-made noise as shown, and Equation 13-1 was used to find the available noise power F_a.

$$F_a = P_n/KT_oB = T_a/T \qquad \textbf{Eq. 13-1}$$

Where: F_a is Effective antenna noise factor

P_n is Noise power available from an equivalent loss-free antenna (WATTS)

K is Boltzman constant = 1.38×10^{-23} joules per degree Kelvin

*Where K = Boltzman's constant = 1.38×10^{-23} and t_0 = 288° Kelvin.

**CCIR Report 322. "World Distribution and Characteristics of Atmospheric Radio Noise" 10th Plenary Assembly, Geneva; 1963.

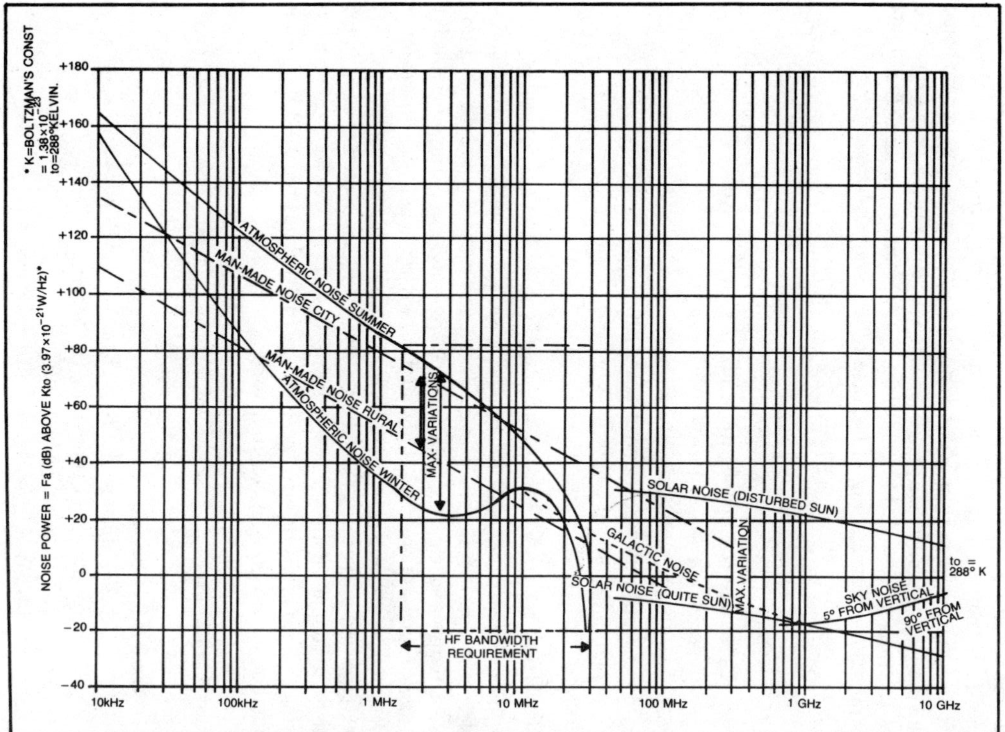

Fig. 13-1. Median radio noise-power spectral density from various sources determines the noise figure requirements for communications receivers.

T_o is Ref. temperature, taken as 288° Kelvin.
B is Effective receiver noise BW (Hz)
T_a is Effective antenna temperature in the presence of external noise (degrees Kelvin)

In HF, as seen in Fig. 13-1, things are quite noisy. Looking at the overall picture between galactic, atmospheric, and man-made noise, the external contribution to an HF receiver is such that the lowest noise level expected is about 18 dB. This is the determining factor in choosing the noise figure of our receiver. An 8 to 10 dB noise figure is typical in such a case, and anything quieter is not practical in this application. If a vhf/uhf receiver is contemplated, lower noise figures should be used. The limiting factors for such receiver would be the galactic and solar noise, as shown.

Designing for a given noise figure in a radio receiver involves all circuits, from mixers to amplifiers and local oscillators, with mixers being the predominant contributors. The noise figure of a receiver is expressed by the total noise added by this circuitry. An expression for defining this parameter in terms of signal-to-noise ratios is given by equation 13-2.

$$NF = 10 \text{ Log } \frac{PS_i/PN_i}{PS_o/PN_o} \qquad \textbf{Eq. 13-2}$$

Where: NF is Noise figure in dB
PS_i is Signal power at input
PN_i is Noise power at input
PS_o is signal power at output
PN_o is Noise power at output

SENSITIVITY

Sensitivity is a way of expressing the ability of a radio receiver to detect a signal of a certain level. There are many ways of defining how sensitivity is calculated. Included are *tangential sensitivity, signal-noise, output signal-to-noise ratio, false-alarm rate, bit-error rate, probability of detection,* etc. The type of method chosen usually depends on factors, such as type and degree of modulation, i-f bandwidths employed and type of detected output as well as the receiver's noise figure. We will deal with two simple ways of expressing receiver sensitivity. Equation 13-3 shows one method of determining sensitivity.

$$S = -174 \text{ dBm} + NF + 10 \log \text{i-f-BW} + K_{sn} + K_m \qquad \textbf{Eq. 13-3}$$

Where: S is Sensitivity in dBm
−174 dBm = KTB = Thermal noise power in a one-Hz bandwidth at room temperature
NF is Noise figure in dB
i-f-BW is Predetection i-f bandwidth in Hz

K_{sn} is Desired S+N/N in dB of the detected signal
K_m is Variable expressed in dB which is a function of type of modulation used

As it can be seen from this equation, sensitivity improves (becomes smaller) with decreased NF and i-f bandwidth.

Another method of measuring sensitivity of a receiver is expressed by the signal input expressed in (μV), necessary to produce an audio output 10 dB greater than the noise figure of the receiver. For example, if the specification reads: 1 μV for 10 dB S/N, it means 1 microvolt of signal at the antenna (30% amplitude modulated) will be heard at the speaker 10 dB over the internal noise of the receiver. This method is usually used in specifying amateur and commercial receivers.

In conclusion, sensitivity can take many quantitative forms. One fact that must be kept in mind is that sensitivity varies throughout the frequency coverage of a communication receiver because of gain variations in the front end. Good manufacturers will usually publish sensitivities at several points in the frequency coverage. See Figs. 13-2 and 13-3 and Tables 13-1 and 13-2.

PREAMPLIFIERS

In order to keep intermodulation products at a low level, good receivers use a minimum of preamplification, usually just enough to compensate for the loss in the preselector filters and the first mixer. Figure 13-4 shows a block diagram for the front end of such a receiver. This arrangement can provide a third-order intercept point of +20 dBm and a typical noise figure of 12 dB, as shown* From this example, it can be seen that in order to

*Charles E. Dexter and Robert D. Glaz. Watkins-Johnson Co., Tech-notes HF receiver design. Vol. 5 No. 2 March/April 1978.

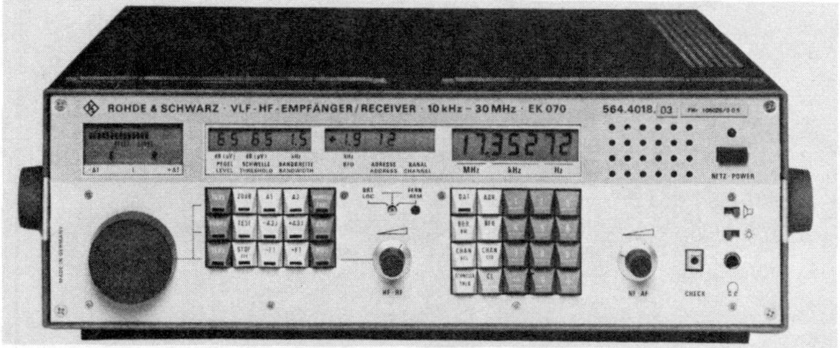

Fig. 13-2. The EK 070 communications receiver is a fully synthesized, double-conversion approach with a first i-f of 81.4 MHz and a second i-f of 1.4 MHz. This receiver covers the frequency range of 10 kHz to 30 MHz in 10 Hz steps, and includes 30 memories for computer controlled scanning. Another feature of this receiver is the logarithmic gain provided by the i-f amplifier which is a direct reading of the signal level at the antenna when the receiver is AGC'ed. This control voltage is transformed into a digital number which is displayed on the front panel to indicate the signal strength of the received signal (courtesy of Rohde & Schwarz).

Table 13-1. Specifications of the EK 070 Vlf/Hf Communications Receiver.

Parameter	Specification	
Frequency range	10 kHz to 30 MHz	
Frequency setting	1. quasi-continuously by rotary switch in increments of 10 Hz/100 Hz/1 kHz 2. digital entry via keyboard 3. remote control via data interface (setting time 50 ms)	
Readout	7-digit liquid crystal display	
Resolution	10 Hz	
Frequency drift		
after 10 minute warmup	$<3 \times 10^{-7}$ at +25°C	
within one day	$<3 \times 10^{-8}$	
caused by aging	$<1 \times 10^{-6}$/year	
in rated temperature range	$<3 \times 10^{-7}$	
Classes of emission	A1 (CW), A2 (MCW), A3 (AM) A2H, A3H (AME)} (SSB) upper A2A, A3A } and lower A2J, A3J } sideband A3B, (ISB), F1 (FSK)	
with Telegraphy Demodulator	F4 (fascimile) F6	
Antenna input	Z_{in} = 50Ω, BNC female connector	
VSWR	≤3	
Permissible input voltage	≦ 10 V EMF	
Oscillator reradiation	<10µV at antenna input with 50-Ω termination	
Sensitivity*)	for 10 dB (S+N)/N; 0.2 to 30 MHz	
with A1, B=300 Hz	<0.3µV EMF	
with A3, B=6 kHz, m=60%	<2.0µV EMF	
with A3J, B=3.1 kHz	<0.75µV EMF	
Preselection	0 to 0.5 MHz: lowpass filter 0.5 to 1.5 MHz: bandpass filter 8 suboctave filters between 1.5 and 30 MHz	
Intermediate frequencies		
1st i-f	81.4 MHz, B=12 kHz	
2nd i-f	1.4 MHz	
i-f selectivity	3 dB bandwidth: min.	60 dB bandwidth: max.
	± 75 Hz	± 225 Hz
	± 150 Hz	± 375 Hz
	± 300 Hz	± 750 Hz
	± 750 Hz	± 1875 Hz
	± 1.5 kHz	± 3.75 kHz
	± 3 kHz	± 7.5 kHz
	± 6 kHz	± 50 kHz
	+0.3 to +3.4 kHz	−0.3 to +4.0 kHz
	−0.3 to −3.4 kHz	+0.3 to −4.0 kHz
Interference immunity, non-linearities		
Intermodulation*)		
d_3 within A3J sideband	>46 dB down, wanted signals 2 × 10 mV EMF	
d_3, Δf ≧ 30 kHz	>70 dB down, unwanted signals 2 × 100 mV EMF	
d_2 (1.5 to 30 MHz), Δf ≧ 30 kHz	>70 dB down, unwanted signals 2 × 100 mV EMF	
Blocking*)	<3 dB signal attenuation, wanted signal 1 mV EMF, m=30%/1 kHz; unwanted signal 1 V EMF, Δf ≧ 30 kHz	
Crossmodulation*)	<10% modulation transfer; unwanted signal 200 mV EMF, m=30% /1 kHz; wanted signal 1 mV EMF, Δf>20 kHz	
Desensitization*)	20 dB SINAD; wanted signal 30 µV EMF, B=3.1 kHz; unwanted signal 300 mV EMF, Δf ≧ 30 kHz	
Inherent spurious signals	< 0.5µV equivalent EMF	
Spurious responses	>90 dB down at Δf > = 30 kHz	
Image frequency rejection	>80 dB	

*) Without 20 dB attenuator pad.

i-f rejection	>90 dB
Rf gain control, switchable	MGC MGC + AGC AGC
Control range	>100 dB
AGC error	<4 dB (1μV to 100 mV EMF)*)
Attack time	5 ms (level jump +60 dB)
Decay time (switchable)	0.4 s/1.8 s (level jump −60 dB)
Bfo	variable in 100 Hz steps over ± 3.1 kHz
Attenuation at i-f output	>50 dB referred to i-f level
F1 demodulator	
Limiting factor	>40 dB
Line spacing	50 to 1000 Hz
Keying speed	1 to 100 bauds
Signal distortion	<5% at 10 bauds
Single current	40 to 60 mA, variable; EMF = 60V
Double current	in compliance with CCITT V.28
Outputs	
1st oscillator 81.4 to 111.4 MHz	0 dBm, 50Ω
2nd oscillator 80 MHz	0 dBm, 50Ω
1-MHz output	50 mV into 50Ω
switchable to 1-MHz external reference input	30 to 500 mV into 50Ω
2nd i-f 1.4 MHz	50 mV into 50Ω
Recording output 12.5 kHz	0 dBm, 600Ω
Panoramic output 1.4 MHz	B=12 kHz
AF line outputs 600Ω	floating
Output level	−10 to +3 dBm, adjustable
Distortion	<1% with A3J
AF output 5Ω (headphones output 100Ω)	
Output level	1 W (12 mW, can be attenuated)
Distortion	<5%
Signal characteristics	
AF frequency response (overall)	<3 dB from 300 to 3400 Hz
AF S/N ratio	>40 dB SINAD with 1 mV EMF
Phase noise ratio with A3J	>75 dB with >300 Hz spacing and 1 Hz measuring bandwidth, 1 mV signal EMF
Remote control	interfaces in compliance iwth IEC and CCITT
IEC bus	IEC 625-1, 24-way connector (Amphenol); functions: T5, L3, SR1, RL2
or (depending on order number) RS232C	CCITT V.24, switchable to CCITT V.10 (RS 423) 110/200/300/600/1200/2400/4800/9600 bauds
Code	ASCII 7 bits

*Without 20-dB attenuator pad.

Fig. 13-3. The SR 219 A/AFC receiving system can cover the frequency range of 20 MHz to 4 GHz, with proper rf plug-in tuners. The receiver is used for reception of AM, FM, cw and pulse signals and can be operated as a fixed or portable unit (courtesy of Norlin Communications).

Table 13-2. Specifications of the SR 219 A/AFC Receiving System.

SENSITIVITY LEVELS (dBm)

Tuning Head Noise Figure	I-F BANDWIDTH													
	10 kHz	20 kHz	30 kHz	50 kHz	60 kHz	75 kHz	100 kHz	200 kHz	300 kHz	500 kHz	1 MHz	2 MHz	3 MHz	4 MHz	8 MHz
4 dB	−111	−108	−106	−104	−103	−102	−101	−98	−96	−94	−91	−88	−86	−85	−82
5 dB	−110	−107	−105	−103	−102	−101	−100	−97	−95	−93	−90	−87	−85	−84	−81
6 dB	−109	−106	−104	−102	−101	−100	−99	−96	−94	−92	−89	−86	−84	−83	−80
7 dB	−108	−105	−103	−101	−100	−99	−98	−95	−93	−91	−88	−85	−83	−82	−79
8 dB	−107	−104	−102	−100	−99	−98	−97	−94	−92	−90	−87	−84	−82	−81	−78
9 dB	−106	−103	−101	−99	−98	−97	−96	−93	−91	−89	−86	−83	−81	−80	−77
10 dB	−105	−102	−100	−98	−97	−96	−95	−92	−90	−88	−85	−82	−80	−79	−76
11 dB	−104	−101	−99	−97	−96	−95	−94	−91	−89	−87	−84	−81	−79	−78	−75
12 dB	−103	−100	−98	−96	−95	−94	−93	−90	−88	−86	−83	−80	−78	−77	−74
13 dB	−102	−99	−97	−95	−94	−93	−92	−89	−87	−85	−82	−79	−77	−76	−73
14 dB	−101	−98	−96	−94	−93	−92	−91	−88	−86	−84	−81	−78	−76	−75	−72
15 dB	−100	−97	−95	−93	−92	−91	−90	−87	−85	−83	−80	−77	−75	−74	−71

RF PLUG-IN TUNERS

Model	Tuning Range	Noise Figure	I-F Rejection	Image Rejection
SH-230P-2/AFC	20-100 MHz	5.0 dB	60 dB	60 dB
SH-200P-1/AFC	20-45 MHz	4.5 dB	90 dB	65 dB
SH-201P-1/AFC	30-100 MHz	5.5 dB	55 dB	55 dB
SH-202P-1/AFC	90-300 MHz	6.5 dB	80 dB	50 dB
SH-203P-1/AFC	250-500 MHz	10.0 dB	90 dB	60 dB
SH-204P-1/AFC	490-1000 MHz	12.0 dB	90 dB	80 dB
SH-205P-1/AFC	1-2 GHz	13.0 dB	90 dB	60 dB
SH-206AP-1/AFC	2-4 GHz	15.0 dB	90 dB	60 dB

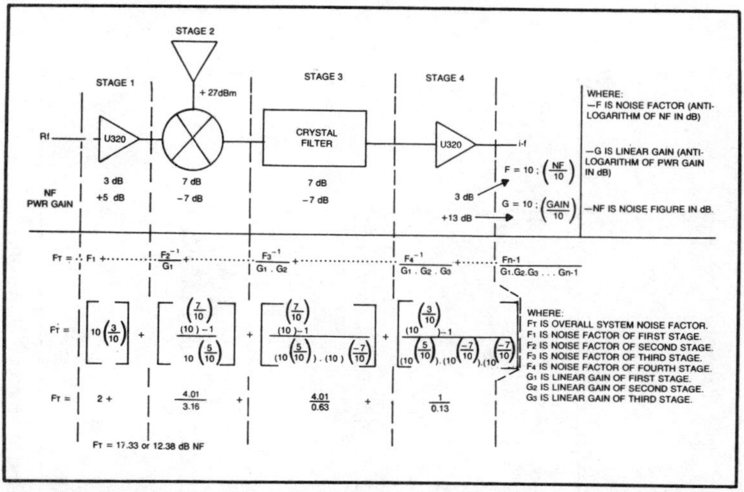

Fig. 13-4. Communications receiver front end, has a high intercept point and a low noise figure, due to the careful choice of devices.

achieve these specifications, a very high level mixer was chosen requiring +27 dBm of local oscillator drive. The Watkins-Johnson M9E can provide a third-order intercept point of greater than +28 dBm. (The typical doubly-balanced mixer LO drive requirement is +7 dBm).

A better and more economical approach is shown in Fig. 13-5. By eliminating the preamplifier, and by reconsidering each stage for noise

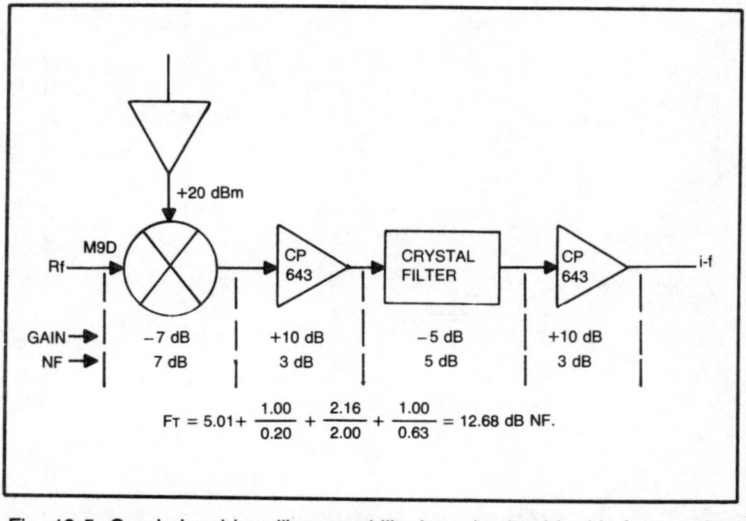

Fig. 13-5. Good signal handling capability is maintained in this front end arrangement which has no preamplifier.

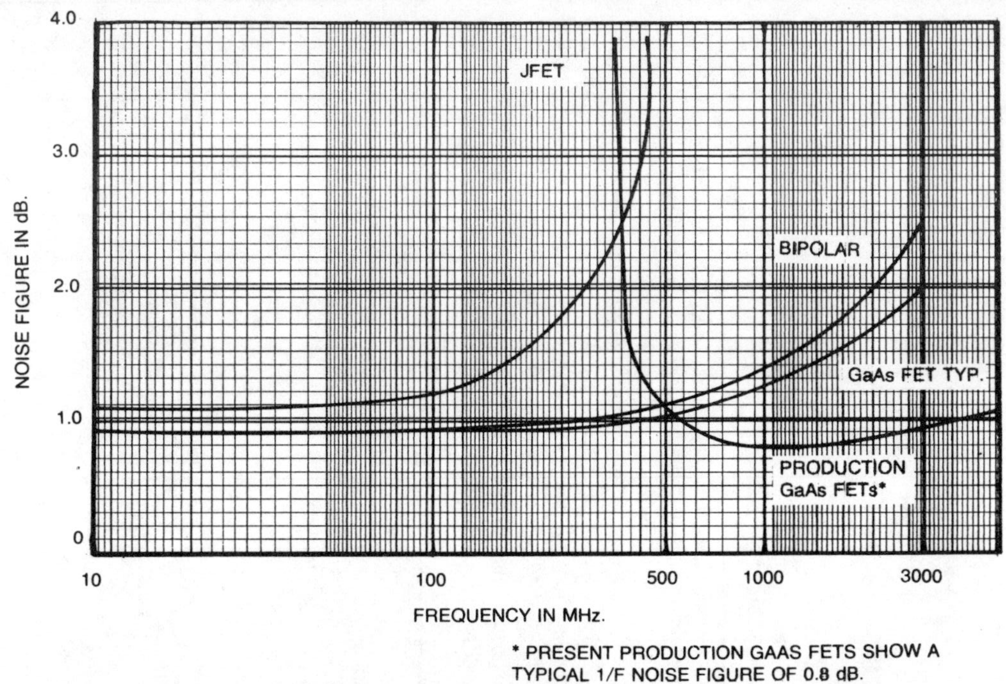

Fig. 13-6. Overall noise figure for JFET, bipolar transistor, and GaAs FET technologies used in preamplifier design.

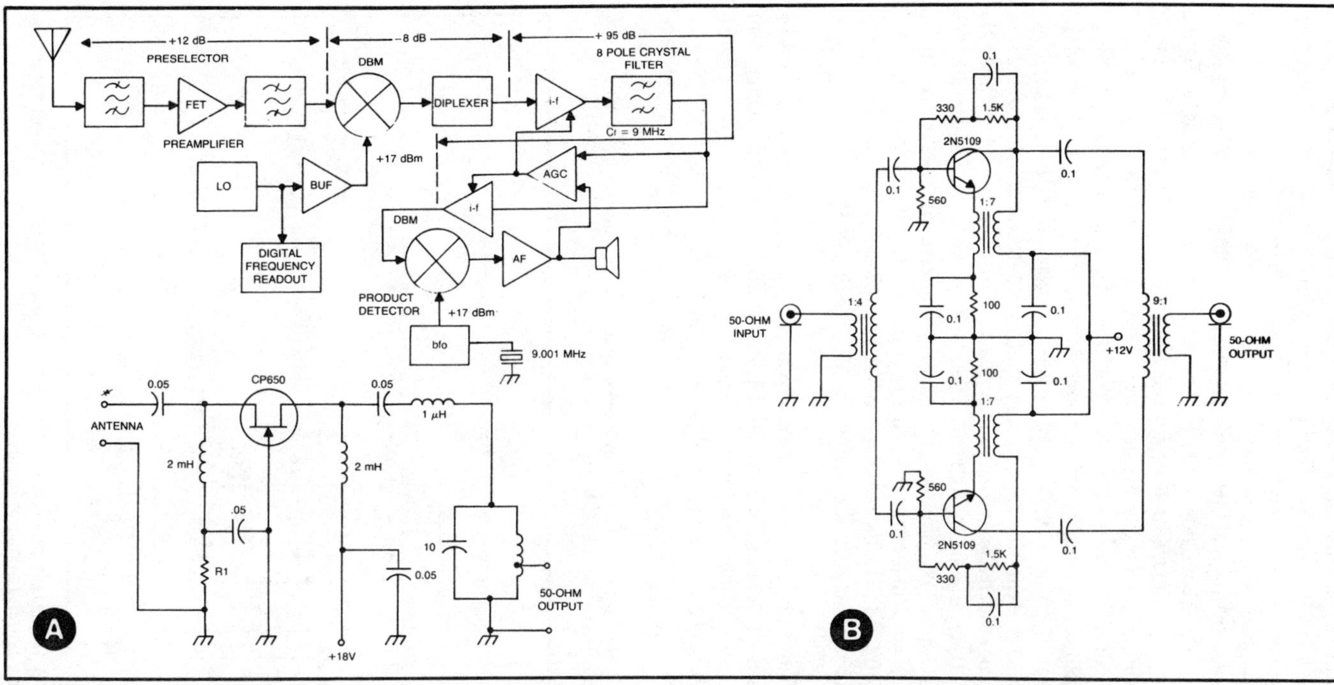

Fig. 13-7. Two high-dynamic range preamplifiers are shown in this example of a 9 MHz i-f communications receiver. Both examples exhibit a noise figure of 2 dB.

figure, the signal-handling capability of the entire front end is maintained with less local oscillator drive. This last approach is used in the WJ-8718 HF communication receiver manufactured by Watkins-Johnson.

If a preamplifier must be used, typically a JFET circuit is chosen for frequencies below 300 MHz. Bipolar transistors can yield low noise figures up to 2 GHz. Above this, GaAs FETs are used to achieve low noise figure performance. Figure 13-6 shows the noise figure performance for these types of devices for the frequency range of 10 MHz to 3000 MHz.

Figure 13-7 shows two different approaches to high-dynamic range preamplifiers as used in a single-conversion communication receiver with a 9 MHz i-f. They provide a typical 2 dB noise figure. The CP650 is manufactured by Teledyne-Crystalonics and consists of fifteen FETs in parallel on a single chip which allows for a transconductance (gm) in the order of 150,000 μmhos (typical 100,000 μmhos). The input impedance of the circuit at A (1/gm) is about 20 ohms, resulting in a mismatch to the 50 ohm antenna (1 dB loss) which in turn improves the noise figure over the range by approximately 0.5 dB. The output is a combination low-pass filter, and matching autotransformer with the output set at 50 ohms for the entire frequency range.

Figure 13-7 shows a push-pull arrangement which uses two uhf power transistors (2N5109) manufactured by RCA. This low-noise arrangement uses emitter feedback and provides about 12 dB of gain with a noise figure of 2 dB, and an input intercept of +22 dBm*.

---

*Ulrich L. Rohde - Optimum design for high frequency communications receivers. HR, Oct. 1976. Ulrich L. Rohde, Eight Ways to Better Radio Receiver Design-Electronics, Feb. 1975.

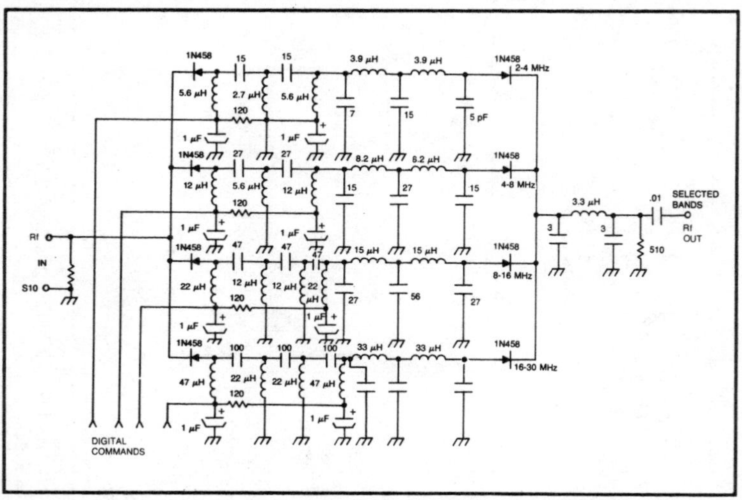

Fig. 13-8. Bandpass preselector filters can be switched at the input of a general coverage communications receiver.

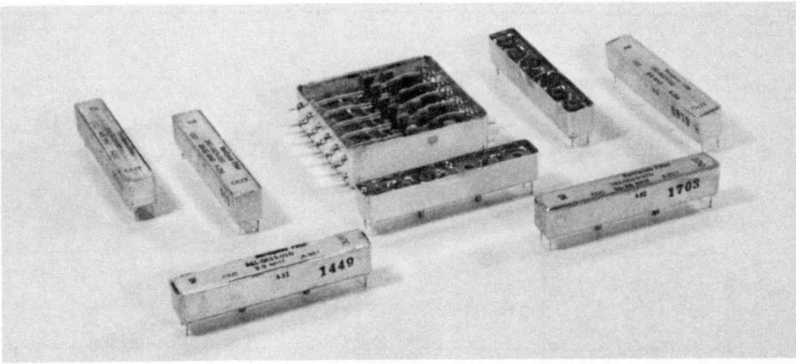

Fig. 13-9. Half-octave filters designed to be used in the preselector of modern communications receivers (courtesy of Communication Coil Co.).

MODERN PRESELECTORS AND THEIR SWITCHING

General coverage communication receivers today depart from the traditional preselectors described earlier by using banks of switchable bandpass filters. The switching is usually accomplished with the use of low leakage diodes (1N458) or miniature relays. Signal diodes such as the IN914 or the IN4148 are usually avoided.

Because diodes are not perfect switches, and because of their nonlinear nature, they can contribute to the deterioration of the intercept point of the receiver. Another factor that can contribute to the degradation of the dynamic range in a communication receiver is the size of the iron-powder toroid cores usually used in the construction of the preselector filters. It has

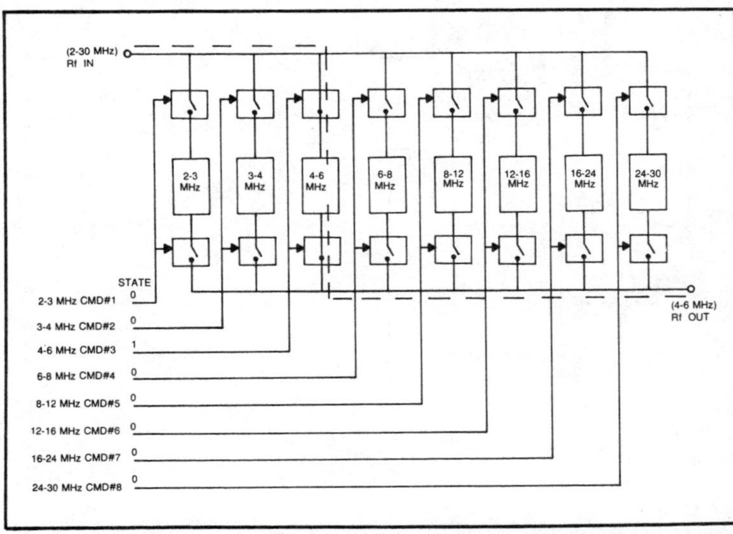

Fig. 13-10. Block diagram for a typical preselector assembly using half-octave filters, showing signal flow with command line # 3 activated (4-6 MHz).

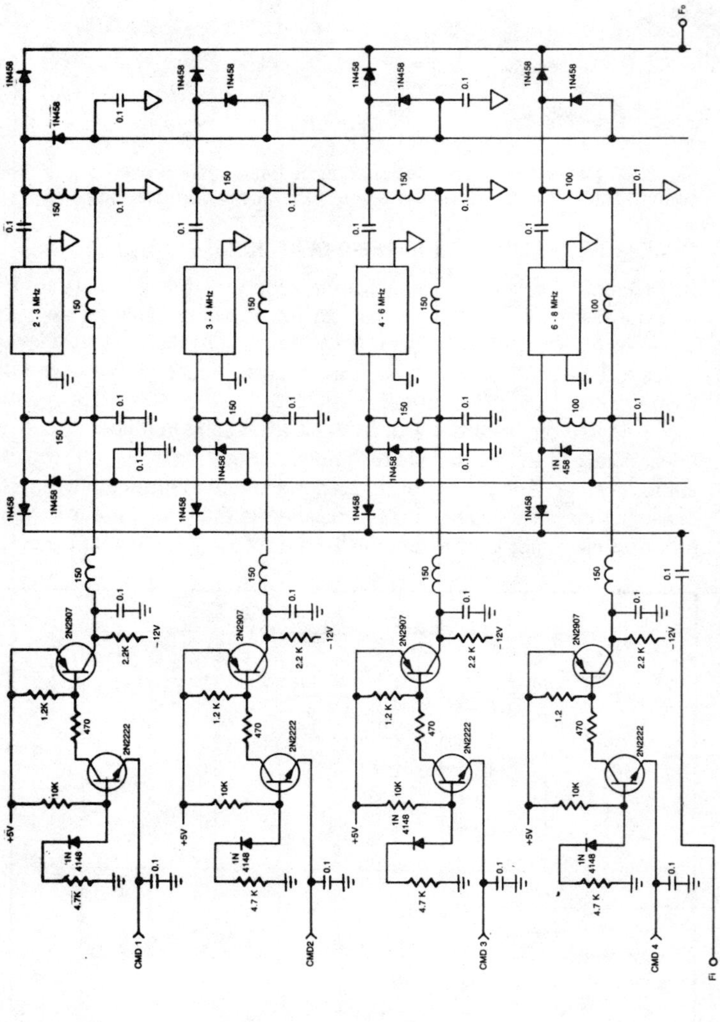

2 - 3 MHz
3 - 4 MHz
4 - 6 MHz
6 - 8 MHz
CMD 1
CMD2
CMD 3
CMD 4
Fi
Fo
2N2907
2N2222
1N458
1N 4148
+5V
−12V

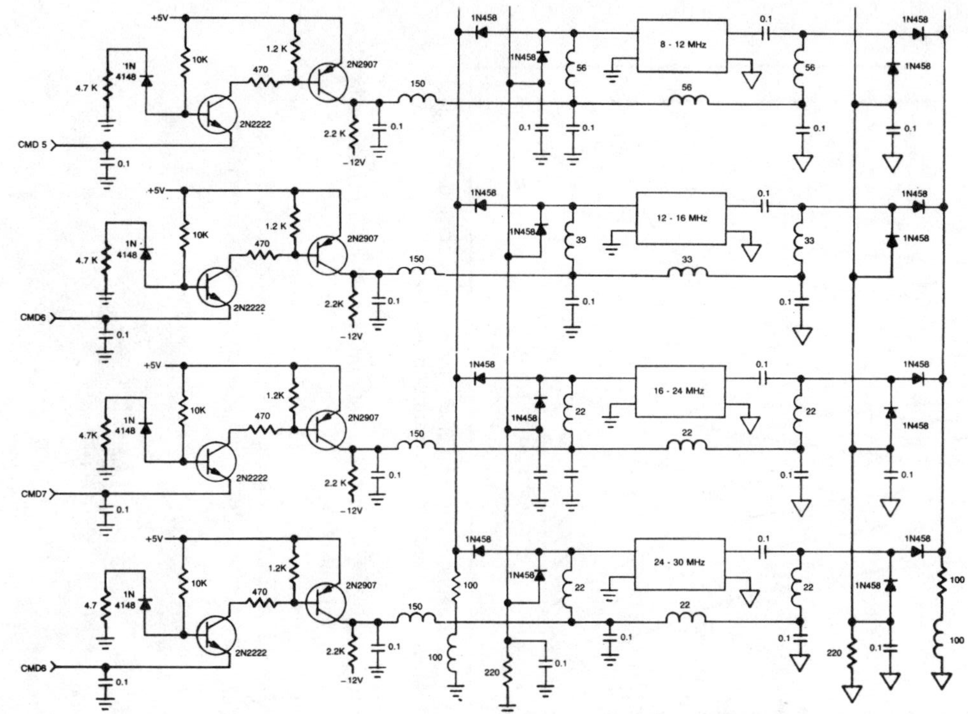

Fig. 13-11. Circuit diagram for a typical implementation of a half-octave preselector assembly.

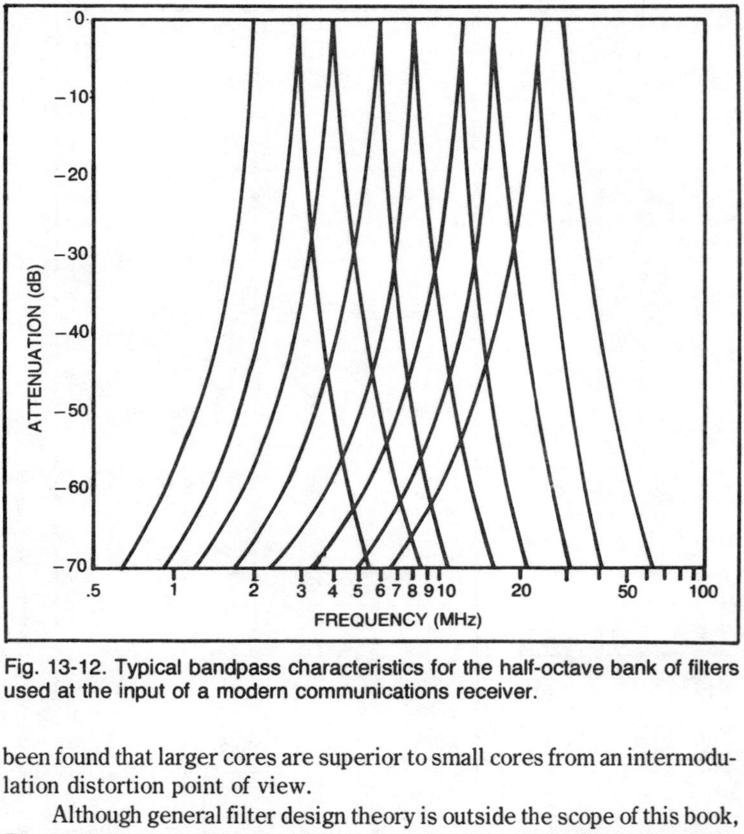

Fig. 13-12. Typical bandpass characteristics for the half-octave bank of filters used at the input of a modern communications receiver.

been found that larger cores are superior to small cores from an intermodulation distortion point of view.

Although general filter design theory is outside the scope of this book, Fig. 13-8 shows a schematic diagram for several bandpass filters that could be used at the input of a general coverage communication receiver.

Fig. 13-13. The Rockwell International 651S-1B is a fully-synthesized communications receiver equipped with an automatically switched preselector (courtesy of Rockwell International).

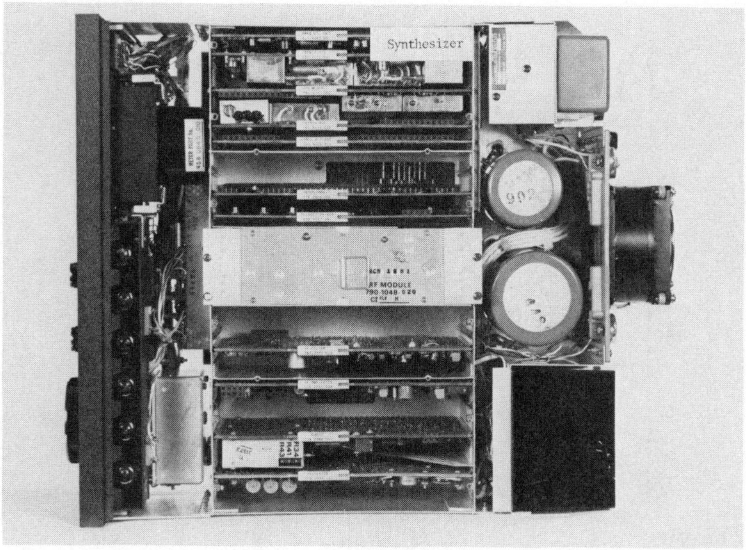

Fig. 13-14. Internal view of the 651S-1 communications receiver (courtesy of Rockwell International).

Preselector filters in modern receivers are usually of the half octave*, Bessel design type, in order to keep second-or higher-order products out of the pass-band characteristics of the receiver's front end as well as to provide steep attenuation beyond the cut-off points. Figure 13-9 shows several half-octave preselector filters manufactured by Communication Coil Co. for this purpose.

The block diagram in Fig. 13-10 shows how such filters can be incorporated in the design of a front end for a communication receiver, and Fig. 13-11B shows a typical circuit design for this stage. This design shows how the half-octave filters can easily be switched in by digital logic levels that can be provided by the receiver's synthesizer or digital frequency display. Figure 13-12 shows the total composite frequency characteristic for the half-octave filters from 2 to 30 MHz.

These examples show that the old manual "switch and peak" method of preselecting has been surpassed by this "hands off" method. While until recently this method only applied to expensive communication receivers, new lower-cost versions of such preselectors have been introduced by receiver manufacturers especially in Japan. As technology evolves this method of preselection will become the rule rather than the exception in the design of communication receivers of all classes. See Figs. 13-13 and 13-14.

*Musical term applied to the frequency domain describing the interval between eight notes, or a basic frequency ratio of two. Example: one octave = 2 to 4 MHz. ½ octave = 2 to 3 MHz, etc.

—•••—

Chapter 14

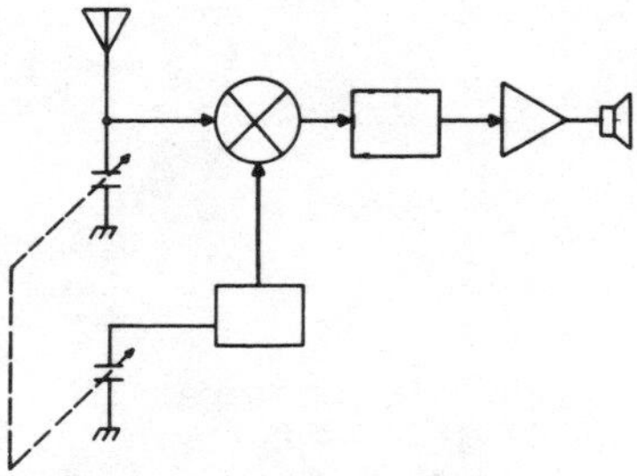

Mixers

One of the most important parts in a radio receiver is the mixer. Figure 14-1 A and B show an actual implementation as applied to our previous 9 MHz i-f communications receiver example. The level of the intermodulation performance of this device plays a crucial role in the design of such instrument, as we previously discussed. Today, the most commonly used mixer is the hot carrier diode type, although new technologies have recently evolved, as we will analyze later.

THE DIODE MIXER AND DIPLEXER

In using the diode mixer, one must pay particular attention to impedance matching particularly of the i-f output. Proper termination of this port helps achieve the predicted intermodulation distortion of the entire system. The use of a bandpass filter, called a diplexer is recommended for achieving this goal. Figure 14-1C shows an actual implementation for the 9 MHz i-f communications receiver example. The diplexer terminates the unwanted products coming out of the mixer, sometimes up to the sixth order. Several diplexers can be used at the same time following a mixer in order to terminate different mixer products. Today there are other ways of matching the output of a mixer to the i-f amplifier as we will see later.

THE DOUBLY-BALANCED MIXER AND ITS PERFORMANCE CHARACTERISTICS

A doubly-balanced mixer works in the following manner: referring to Figs. 14-2 and 14-3 it can be seen that the voltage at the secondary of the local oscillator (LO) transformer forms currents that flow through the diode pair D1, D2, or D3, D4 according to polarity, causing alternate conduction, and making the ends of the secondary of the rf transformer points B and C to appear alternately at ground potential, at the frequency of the signal which is applied to the LO port.

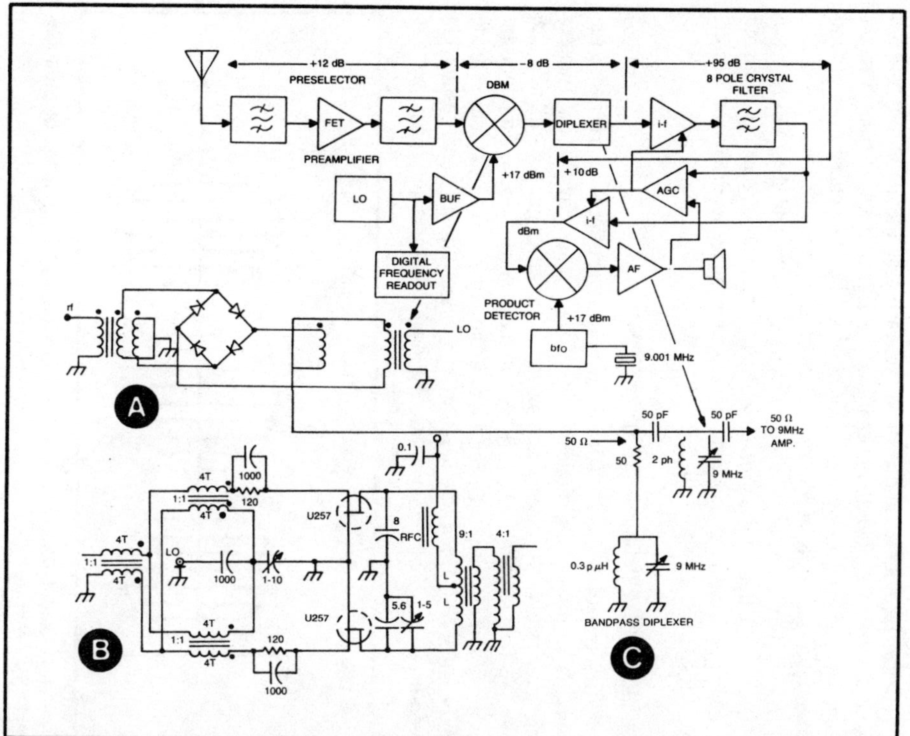

Fig. 14-1. First mixer and diplexer.

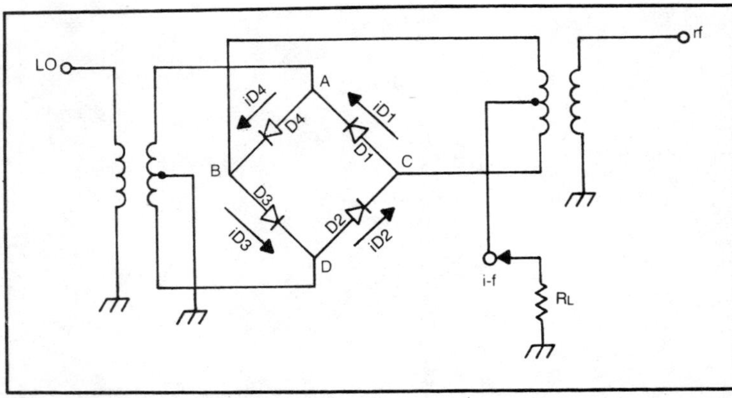

Fig. 14-2. Electrical operation of a doubly-balanced mixer.

It can be seen that the signal appearing at the i-f output will be influenced by the level and polarity of the signal at the rf transformer's secondary, as well as by which terminal of that secondary is at ground potential at that time as shown in Fig. 14-2.

The output at the i-f port contains the sum and difference of the frequencies of the signals present at the LO and rf ports, plus other mixing products created by the same harmonic frequencies as discussed earlier. See Figs. 14-3, 14-4 and 14-5.

HARMONIC INTERMODULATION OF MIXERS

This type of intermodulation distortion is sometimes referred to as the spurious response of a diode mixer, and is created by the i-V transfer characteristics of the diode junctions involved, which generate all $nF_{(LO)} \pm mF_{(rf)}$ (where n and m = 0, 1, 2, 3 ... ∞) harmonic products. An important property of the doubly-balanced mixer is the suppression by common-mode cancellation of all internally-generated harmonic outputs which result from even-order harmonic components (Although odd harmonic components of

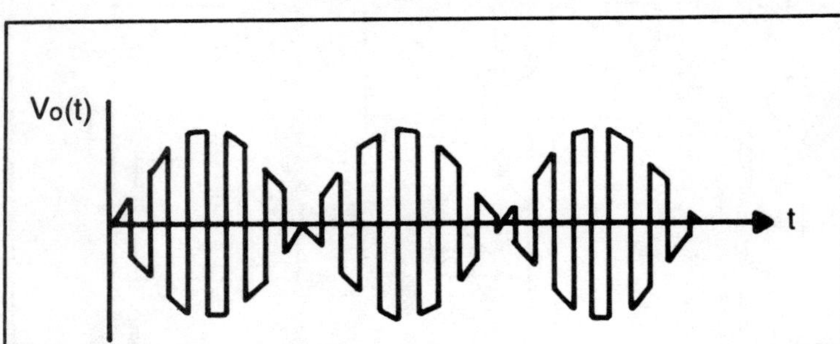

Fig. 14-3. Output waveform resulting from the switching action of a doubly-balanced mixer.

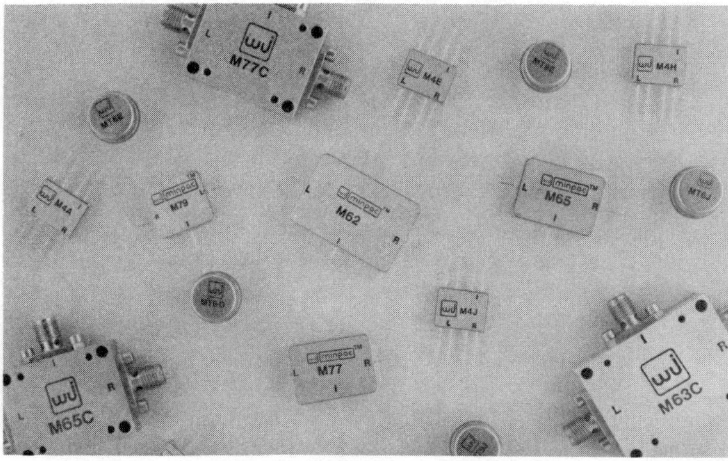

Fig. 14-4. Several doubly-balanced mixers (courtesy of Watkins-Johnson Company).

the 3±3, 5±5 order cannot be suppressed by any mixer cancellation, they can be reduced by higher LO drives in Class II and Class III devices, as will be discussed later in the two-tone intermodulation ratio section.) of the LO and/or rf input signals, such as $2F_{(LO)}$, $4F_{(LO)}$, $2F_{(rf)}$, $2F_{(LO)} \pm F_{(rf)}$, $F_{(LO} \pm 2F_{(rf)}$ AND $2F_{(LO)} \pm 2F_{(rf)}$.

This cancellation is, however, not a perfect matter due to the imbalance of the elements involved, even though great care is usually exercised in matching all diodes. For those interested, Fig. 14-6 shows suppression performance that can be expected from a Class I doubly-balanced mixer. There are 3 classes of doubly-balanced mixers, defined by the LO drive requirements: Class I is + 7 dBm to + 9 dBm, Class II is + 17 dBm to + 20 dBm, and Class III + 23 dBm to + 27 dBm.

Fig. 14-5. The WJ-M6E is an example of a Class I doubly-balanced mixer (courtesy of Watkins-Johnson Company).

Conditions:

1. Rf = −10 dBm at 100 MHz
   LO = +7 dBm
   Broadband resistive terminations at all ports.
2. The mixing products are referenced to the desired output $f_{LO} + f_{rf}$ and are expressed in dB below that level.
3. Rf harmonics are referenced to the rf input.
4. LO harmonics are referenced to the LO input.

HARMONICS OF F_{rf}		F_{LO}	$2F_{LO}$	$3F_{LO}$	$4F_{LO}$	$5F_{LO}$	$6F_{LO}$	$7F_{LO}$	$8F_{LO}$
$7F_{rf}$	99	95	100	90	95	99	100	85	95
$6F_{rf}$	97	90	95	90	97	95	100	90	100
$5F_{rf}$	85	80	85	70	90	67	92	65	85
$4F_{rf}$	85	80	95	85	95	85	95	85	90
$3F_{rf}$	70	60	65	70	70	55	75	55	75
$2F_{rf}$	60	70	70	65	80	60	73	70	98
F_{rf}	30	0	33	13	35	35	42	35	45
F_{rf}	$\pm F_{LO}$	35	40	46	54	54	55	56	55

Fig. 14-6. Typical spurious response chart for Class I doubly-balanced mixer.

AMPLITUDE MODULATION NOISE SUPPRESSION

Another property of the doubly-balanced mixer is the suppression of AM noise associated with the local oscillator. AM noise components on the other side of a local oscillator carrier will mix with the rf signal and produce noise sidebands within the i-f pass-band as we'll discuss later. While this phenomenon is common to all mixers, the balanced mixer tends to cancel this effect so that the noise sidebands in the i-f are typically down from the converted carrier level by 40 to 50 dB at frequencies up to 100 MHz. (The spurious levels will typically increase at higher frequencies.) It can be seen that the amount of AM noise suppression can actually determine the mixer's noise figure with a particularly noisy local oscillator. Put another way, it can determine the maximum local oscillator drive level allowed for the particular mixer for a desired noise figure. In spite of this, not all AM noise is cancelled in a DBM. Uncancelled AM noise is sometimes caused at the i-f port, by the FM noise of the local oscillator moving through the slopes of the following i-f filter. AM noise suppression plays an important part in choosing a good doubly-balanced mixer and implementing it properly in a particular receiver design. The mechanization of such a test is, however, beyond the scope of this book. The reader is again directed to the specialized material provided by mixer manufacturers. See Figs. 14-7 and 14-8 and Table 14-1.

Fig. 14-7. The Watkins-Johnson WJ-8888 B is a highly versatile hf receiver which covers the 0.5 to 30 MHz frequency range in 10 Hz increments. Low intermodulation distortion and wide dynamic range performance are achieved in this design (courtesy of Watkins-Johnson Company).

CONVERSION LOSS IN DIODE MIXERS

Conversion loss in a diode mixer is usually referred to as the *single sideband conversion loss,* and is simply the power difference (expressed in dB) between the input rf level (expressed in dBm) and the output i-f level

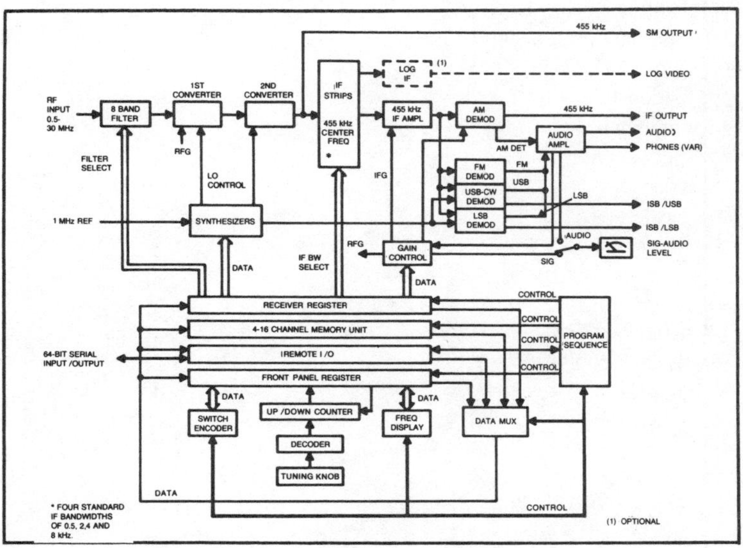

Fig. 14-8. Simplified block diagram of the WJ-8888 receiver.

Table 14-1. Specifications of The WJ-8888 Receiver.

Tuning Range	0.5 - 30 MHz
Tuning Resolution	10 Hz, 100 Hz, 1 kHz, 10 kHz; switch-selectable
Preselection	Sub-octave filters automatically switched
Input Impedance	50 ohms, unbalanced
Oscillator Radiation	10 μV or less at receiver input
Antenna Input Protection	The antenna input will withstand the effects of rf power to +15 dBm and static build-up. The protection circuit automatically resets.
i-f Bandwidths (3 dB)	4 Bandwidths supplied, 6 positions available
Normally Supplied	0.5-, 2-, 4-, and 8-kHz
Optional	0.2-, 1-, 3-, 6-, 12-, or 16-kHz
i-f Shape Factor (Typical)	i-f BW — 50 dB:3 dB 0.2 kHz — 10:1 0.5 kHz — 7:1 1 kHz — 5:1 2 kHz — 3:1 16 kHz — 1.9:1 Filters with higher selectivity may be special ordered.
Detection Modes	AM Noise Limiter, AM, CW Fixed, CW Variable, USB, LSB, ISB, FM
Gain Control Modes	Manual, Normal AGC, Hold AGC
AGC and Manual Range	100 dB, minimum, for input signals above 2 μV
AGC Threshold	2.0 μV, minimum
AGC Attack Time	20 ms, nominal
AGC Release Time	Normal AGC, 0.1 seconds. Hold AGC, 2 seconds; both nominal.
Internally Generated Spurious	1 μV, maximum (equivalent input signal)
Tuning Speed (Remote)	5 ms, typical; 15 ms, maximum
Frequency Stability	6 x 10^{-8} per day, 2 x 10^{-6} per year
Standard Frequency Output	50 mV, minimum at 1 MHz into 50 ohms. Provision is made for an external 1 MHz standard.

Frequency Display	7 Digit, LED (dot matrix) display
Remote Control	Via Input/Output TTL serial synchronous differential pair, 64-bit word; formats as shown below:

DATA WORD FORMAT

MSB

Bits	1	1	26	2
	Spare	AGC Dump	Frequency	Gain Mode

LSB

Bits	3	3	1	11	1	7	1	7
	I-f BW	Det. Mode	Spare	BFO	Spare	Gain	Fault	Signal Level

Remote Control Interface:

Input Trigger	Differential pair, TTL strobe pulse, positive logic, 5 ms minimum pulse width; from controlling device to command receiver to supply clock signal for synchronous transfer of data.
Clock Input/Output	Two differential pairs, TTL clock train, 25 kHz, 10% duty cycle, square wave; effects serial data transfer to/from controller; may be OR wired.
Receiver Address	One differential pair, discrete, TTL level Logic 1 = receiver addressed Logic 0 = receiver not addressed

Memory Mode	Four channel memory capacity supplied. The entire 64-bit word is stored. Complete operating mode of the receiver may be internally stored and recalled later.
Optional Memory	Up to a total of sixteen channels in groups of four channels.
Ac Power Interrupt	While the receiver is in the Local Mode, all the above functions are placed in a separate memory and reappear when power returns.
Control Modes	Local, Memory and Remote
Sensitivity:	
AM Sensitivity	The input signal levels specified * 50% AM modulated at a 400-Hz rate will produce a 10 dB (S+N)/N ratio at the audio output (1 kHz and greater i-f bandwidths).
Cw Sensitivity	The CW input signal levels specified * will produce a 16 dB (S+N)/N ratio at the audio output.
FM Sensitivity	The input signal levels specified FM modulated at a deviation equal to 30% of the IF bandwidth at a 400 Hz rate, will produce a 17 dB (S+N)/N ratio at the audio output (6 kHz and greater i-f bandwidths).
LSB, USB, ISB	0.56 μV for 10 dB (S+N)/N or greater in 2.8 kHz bandwidth (8 kHz i-f bandwidth also in use).
Audio Outputs:	
Line Audio	1 mW, minimum, transformer coupled, balanced into 600 ohms at 2.0 μV input level or greater.
Audio Distortion	Less than 5%
Audio Amplifier Frequency Response	Within 3 dB from 100 Hz to 15 kHz
Phones	10 mW, minimum into 600 ohms, front panel adjusted
ISB (LSB, USB)	Two, each provides 1 mW, minimum, transformer coupled, balanced, into 600 ohms at 0.56 μV input level
I-f Output	455 kHz, 50 mV, minimum, at 2 μV input level or greater
Signal Monitor Output	455 kHz center frequency, bandwidth limited by first i-f filter

I-f rejection	Greater than 100 dB
Image Rejection	Greater than 100 dB
Unwanted Sideband Rejection	50 dB at 350 Hz into unwanted sideband
Intermodulation:	
Third Order Input Intercept Point	+20 dBm, minimum, for the undesired signals separated by more than 50 kHz
Second Order Input Intercept Point	+60 dBm, minimum
Reciprocal Mixing	With a desired signal of 25 microvolts, in the 2 kHz i-f bandwidth, the desired signal to noise ratio will be greater than 20 dB, when an undesired signal 70 dB higher in amplitude and removed 30 kHz in frequency is present.
Cross Modulation	With desired signal at 50 μV, an undesired signal at 50 mV greater than 50 kHz away, AM modulated 50% produces an output at least 20 dB below the output level of desired signal in the 2 kHz i-f bandwidth.
Non-Remote Control Functions	Phone level, squelch, memory channel select and rf / Audio Meter
Signal Meter	Indicates rf input signal level or line audio output level
Size	19 inches wide, 5.25 inches high, and 19.5 inches deep
Weight	Approximately 40 pounds
Operating Temperature Range*	0°C to 50°C
Power Consumption	Approximately 0.8 amps at 115 Vac
Input Power Requirements	115/220 Vac ±10%, 48-62 Hz

*Operation within published specifications guaranteed at 25°C ±5°C.

I-f Bandwidth kHz	Input Level		I-f Bandwidth kHz	Input Level	
	Microvolts	dBm		Microvolts	dBm
0.2	0.40	−115.5	4.0	1.3	−105
0.5	0.45	−114	6.0	1.7	−102.5
1.0	0.64	−111	8.0	1.8	−102
2.0	0.89	−108	12.0	2.4	−99.5
3.0	1.2	−105.5	16.0	2.5	−99

(expressed in dBm) of the desired product (sideband) as shown in Equation 14-1, where L_c is the conversion loss.

$$L_{c(dB)} = P_{(rf)dBm} - P_{(i\text{-}f)dBm} \qquad \textbf{Eq. 14-1}$$

If a −10 dBm number is used for the rf input, and the desired sideband measures −16 dBm*, then:

$$L_c = -10 - (-16) = 6 \text{ dB}$$

Although single sideband conversion loss is usually specified by the mixer manufacturer, (typically 6 dB for a doubly-balanced mixer) its importance is crucial to the design of a radio receiver. Knowledge of this value dictates not only the amount of amplification needed to make up for it, but more importantly, it determines the noise figure of the mixer which impacts the total performance of a radio receiver as we will see in the next paragraph.

NOISE FIGURE OF MIXERS

Referred to as the single sideband noise figure, this value is a direct function of the conversion loss plus the estimated contribution in electrical noise caused by the diode junctions, a value which is typically 0.5 to 1 dB as shown in Equation 14-2.

$$\text{SSB} \quad \text{NF} \leq \text{LC} + 1 \text{ dB} \qquad \textbf{Eq. 14-2}$$

As we discussed earlier in this book, this number can determine the noise figure of the system, (if predominant) as well as the bottom level of the dynamic range of a particular stage and enters the total dynamic range picture in a multiconversion receiver.

If a receiver were to be designed that has an expected noise contribution at the antenna of less than this number, such as in a space communication receiver, a lower noise figure mixer of a different type (or a lower noise figure preamplifier used ahead of the doubly-balanced mixer) should be used in order to bring down the bottom level of the dynamic range.

TWO-TONE INTERMODULATION RATIO

The intercept method described earlier for the overall performance of a radio receiver is a form of a *two-tone intermodulation ratio* test which can be applied to all or any parts that make up the system, including amplifiers (rf, i-f and af) and mixers. The total performance depends on the individual performance of these parts, and the mixer is often the most important contributing element to this performance. Shown in Fig. 14-9 is a typical two-tone test setup.

When two $F_{(rf)}$ signals are simultaneously applied to a mixer they will combine with the $F_{(LO)}$ producing the sums and the differences as well as the undesired intermodulation products, as discussed earlier. The largest

*This experiment requires a spectrum analyzer, or a power meter.

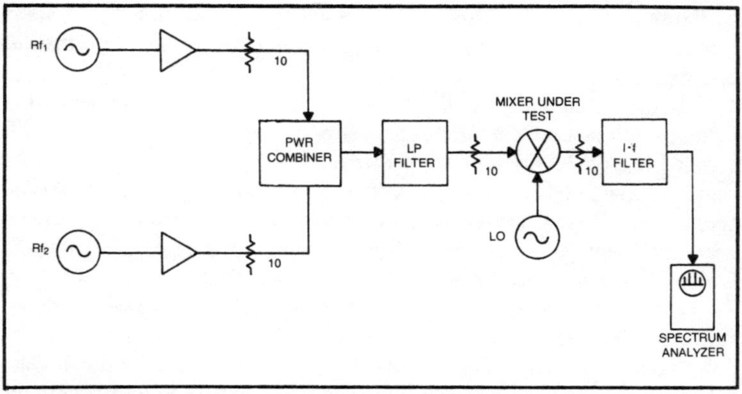

Fig. 14-9. Typical two-tone test set-up.

magnitude of undesired intermodulation product level is the third-order product and it was chosen as a means of performance specification for mixers as well. The Two-Tone Intermodulation ration is the ratio of the third order IM product to an i-f output level, at a specified power level for the two rf inputs.

If the output products versus rf input power are plotted on a log-log graph, the third order IM products present a 3:1 slope while the i-f outputs present a 1:1 slope. The two slopes meet at the third order intercept point

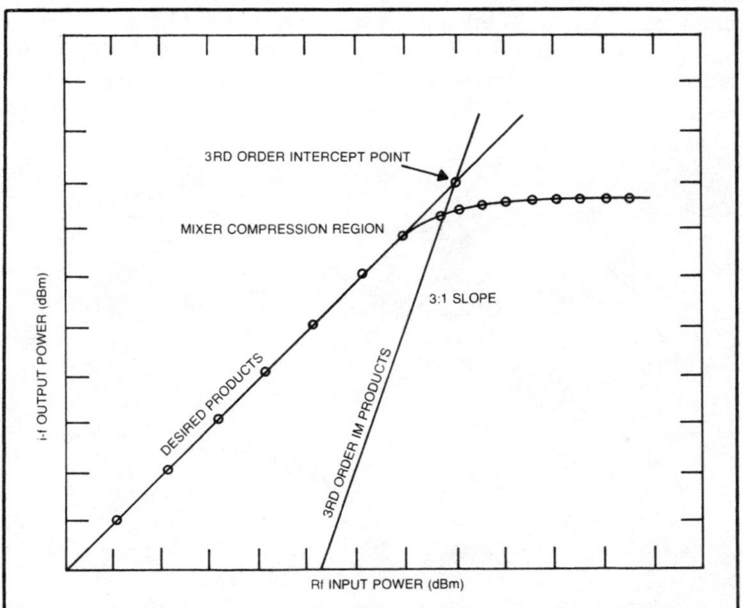

Fig. 14-10. Two-tone intermodulation distortion for doubly-balanced mixers.

as shown in Fig. 14-10. The intercept point is specified in terms of its rf input power level and is the means of expressing the mixer's performance. See Fig. 14-11.

A rule of thumb is that the higher the intercept point, the better the mixer performance. Generally, improved performance can be obtained with higher LO drive levels which will switch the diodes further into the linear regions at both ends of their I-V curves while reducing the amount of time spent in the nonlinear region centered around V=0. While the forward current through each diode pair is almost unlimited, the reverse voltage is limited by the forward voltage drop of the unconducting diode pair.

The solution for this problem is adding more series diodes, adding series resistance to each diode, or choosing diodes having a higher breakdown voltage. This in turn increases the requirement for LO drive levels, which increases power consumption. The improvement in intercept point as effected by the local oscillator power is exemplified in the test results shown in Fig. 14-12. In this experiment, performed by Watkins-Johnson Co., three typical classes of mixers were tested in a two-tone intermodulation setup. The results show clearly that the higher the class is, (and consequently the higher LO drive requirement) the better the intercept point is. For example: the class I, WJ-M62 mixer requires + 9 dBm of LO drive power and has an intercept point of + 13 dBm, while the class II, WJ-M40 mixer with a requirement of + 20 dBm LO drive power, exhibits a superior intercept point of + 20 dBm. The best intercept point in this experiment was obtained with a class III mixer, the WJ-M9E which has an intercept point of + 33 dBm, using LO power of + 23 dBm.

Spectrum analyzer photographs for the intermodulation performance of 3 classes of mixers are shown in Fig. 14-13 A, B, and C. They also show reduced intermodulation distortion with increased LO drive for the typical

Fig. 14-11. Internal view of the WJ-M9G mixer (courtesy of Watkins-Johnson Company).

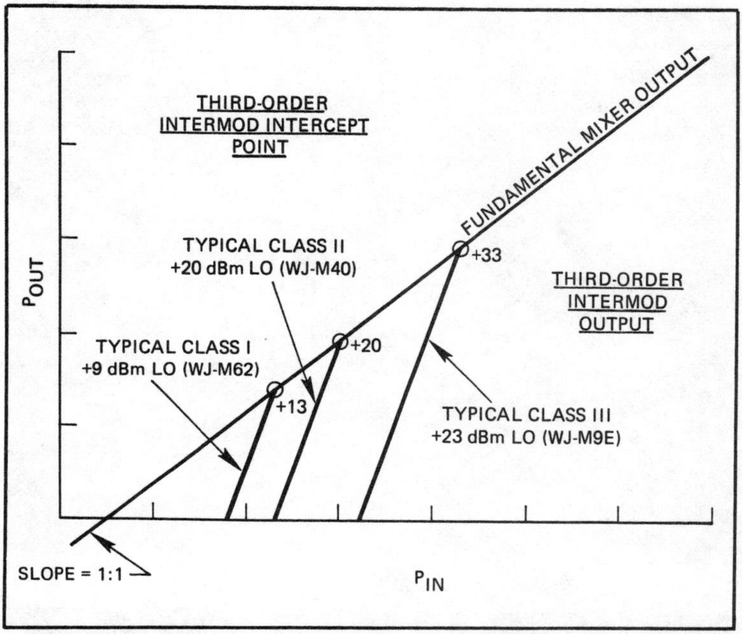

Fig. 14-12. Typical two tone intermodulation performance for 3 classes of mixers, showing improvement in performance with increased LO drive (courtesy of Watkins-Johnson Company).

3 classes of mixers, in this case a WJ-M1 for class I, a WJ-M9BC for the class II, and the WJ-M9E for the class III.

COMPRESSION POINT (– 1 dB)

When a single rf input is applied to the rf port of a mixer, along with a large LO signal, the conduction transfer characteristic of the diodes changes as the rf input level is increased beyond a point, and the output will eventually start saturating until no increase is obtained no matter how much the rf input level is increased.

The point at which the output level drops 1 dB from following the input is the – 1 dB compression point (Fig. 14-14). This point plays an important role in determining the upper limit of the dynamic range of a radio receiver.

DESENSITIZATION LEVEL

Another receiver parameter that depends on the mixer performance is the 1 dB desensitization created by a mixer. This level is the rf input power of an interfering signal that causes the small-signal conversion loss to increase by 1 dB. This level is typically 2 to 5 dB above the compression point level and should not be confused with it. Figure 14-15 shows a typical setup for measuring this parameter.

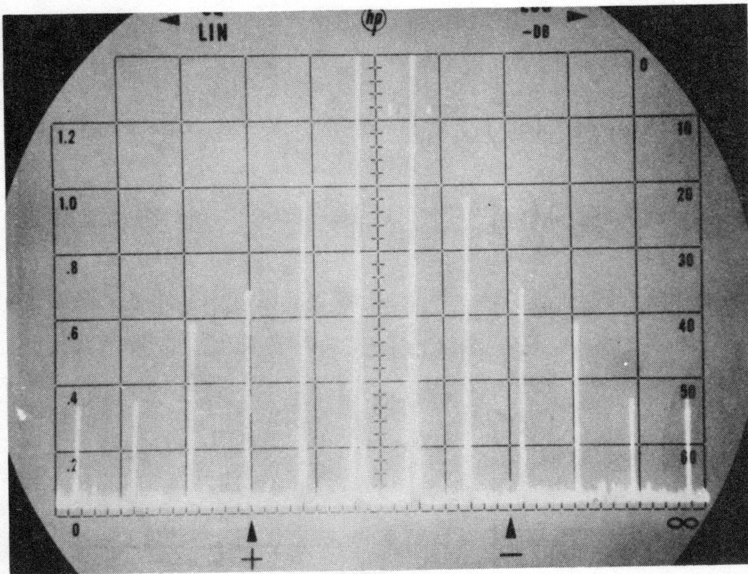

A

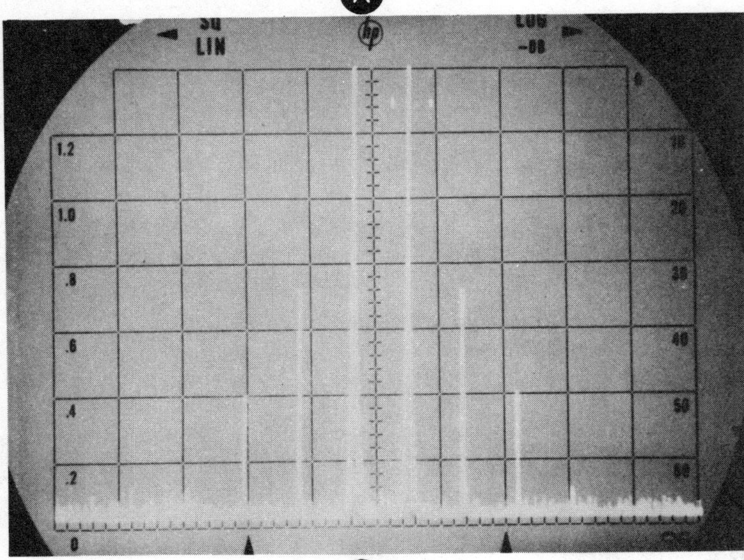

B

Fig. 14-13. Spectrum analyzer representation for the intermodulation performance of 3 Classes of mixers (courtesy of Watkins-Johnson Company). Good performance is achieved with a Class I mixer, the WJ-M1. LO= +7 dBm (A). Better performance is achieved with a Class II mixer, the WJ-M9BC. LO= +17 dBm (B). Best performance is achieved with a Class III mixer, the WJ-M9E. LO= +27 dBm (C).

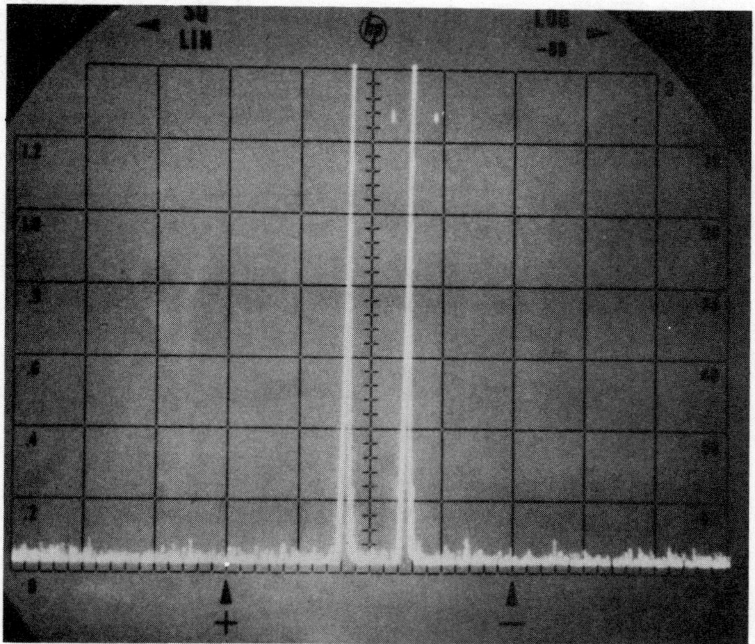

C

Fig. 14-13. Continued from page 124.

ISOLATION IN BALANCED DIODE MIXERS

Isolation between two mixer ports is expressed by the amount of attenuation that an input signal at one port experiences when measured at the other port. This parameter is usually important in preventing unwanted signals from getting back into the system.

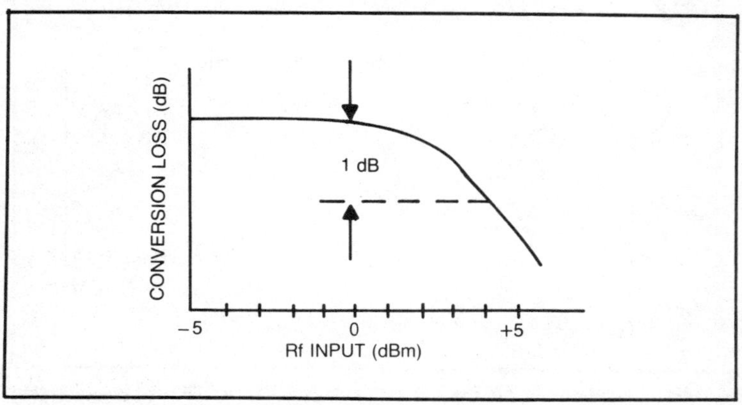

Fig. 14-14. Finding the compression point.

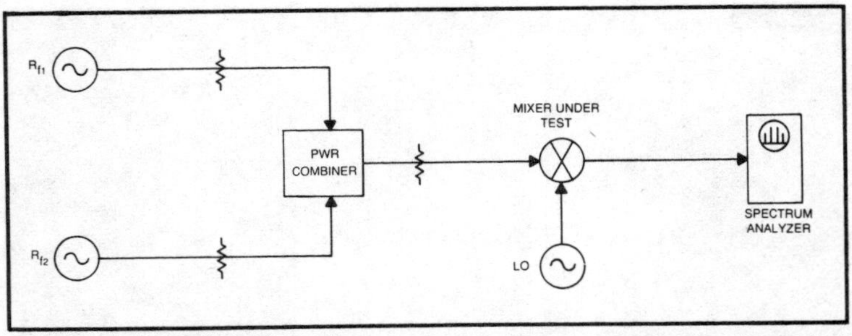

Fig. 14-15. Typical measurement set-up for the desensitization level of a doubly-balanced mixer.

Although most doubly-balanced mixers have good isolation (typ. 50 dB) some systems might require additional filters and amplifiers for improving this characteristic. The setups shown in Fig. 14-16 A and B are used to measure this isolation. See Fig. 14-17.

OTHER TYPES OF DIODE MIXERS

While the doubly-balanced diode mixer remains the most popular type, there are other types in use today.

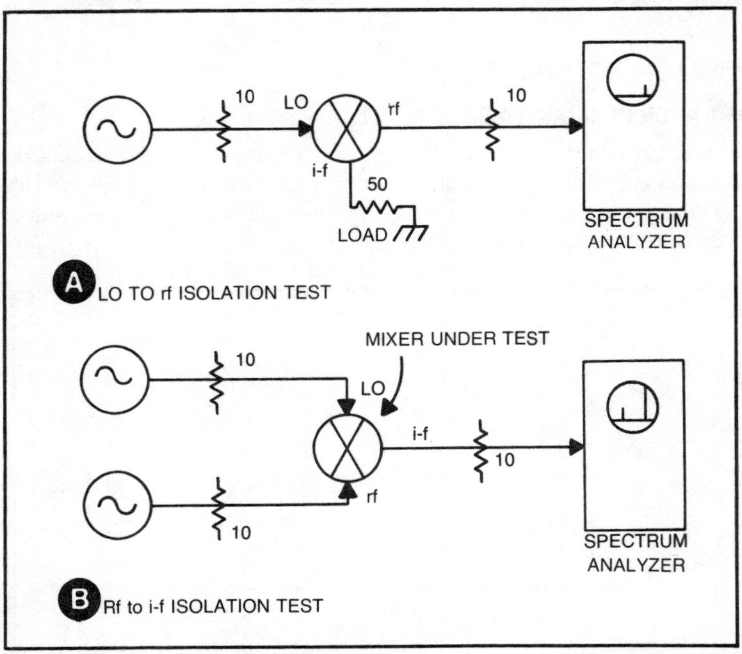

Fig. 14-16. Isolation test set-up. LO to rf isolation test (A). Rf to i-f isolation test (B).

Fig. 14-17. Internal construction of the WJ-M9D high level doubly-balanced mixer (courtesy of Watkins-Johnson Company).

THE SINGLY-BALANCED MIXER

Figure 14-18 shows a singly-balanced mixer which uses a balanced transformer to feed the LO signal to the two matched mixing diodes, while the rf signal is coupled via a high-pass LC network. If the circuit is truly balanced and symmetric, the voltage dropped in each diode is the same and consequently, no LO signal will appear at the rf port and vice-versa. Filtering is generally used in the i-f port in order to increase the isolation between the rf and the i-f port as shown.

IMAGE-REJECTION MIXERS

The image problem discussed earlier in receivers with relatively low i-fs can be improved with the help of the so-called image rejection mixers.

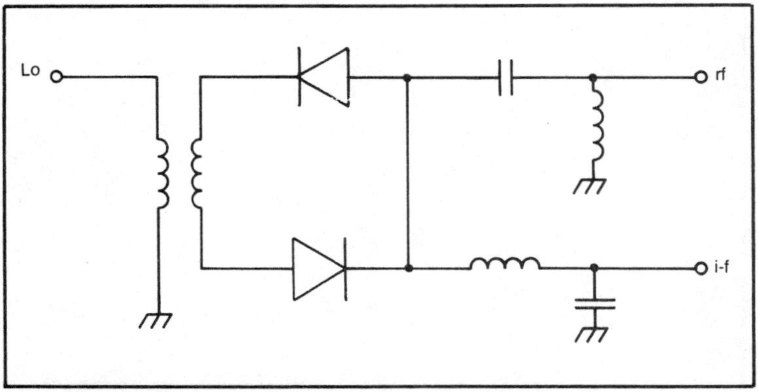

Fig. 14-18. Representative schematic of a singly-balanced mixer.

This is particularly true in guidance receivers which, because of their tight packaging requirements cannot afford double-conversion approaches or preselectors. Doubly-balanced mixers are combined with other wideband devices to accomplish this type of mixer. Figure 14-19 shows a pair of gain-and phase-matched mixers arranged to provide image rejection by cancellation in the hybrids. A low-i-f radio receiver employing such a mixer will exhibit a typical image rejection of 20 to 30 dB without any preselection.

IMAGE-RECOVERY MIXERS

We previously discussed how mixers respond to signals at the image frequency, however mixers can also generate energy at this frequency. This phenomenon is accomplished through two mechanisms. First, the second harmonic of the LO can mix with the incoming rf signal creating energy at the image frequency as shown in Equation 14-3

$$F_{\text{image (internally generated)}} = 2F_{LO} \pm F_{rf} \qquad \textbf{Eq. 14-3}$$

Secondly image energy can also be created by the i-f signal being reflected back into the mixer (do to mismatching at the i-f port) and remixing with the LO energy as shown in Equation 14-4.

$$F_{\text{image (internally generated)}} = F_{LO} \pm F_{i\text{-}f} \qquad \textbf{Eq. 14-4}$$

To verify these mechanisms, the reader is encouraged to use the image model previously presented in Fig. 4-7. If using the values from the example in Equations 14-3 and 14-4 we can easily prove the internal generation of image energy in mixers. In the case of Equation 14-3:

$$F_{\text{image (internally generated)}} = (2 \times 14.545) - 14.090 = 15.000 \text{ MHz.}$$

In the case of Equation 14-4:

$$F_{\text{image (internally generated)}} = 14.545 + 455 = 15.000 \text{ MHz.}$$

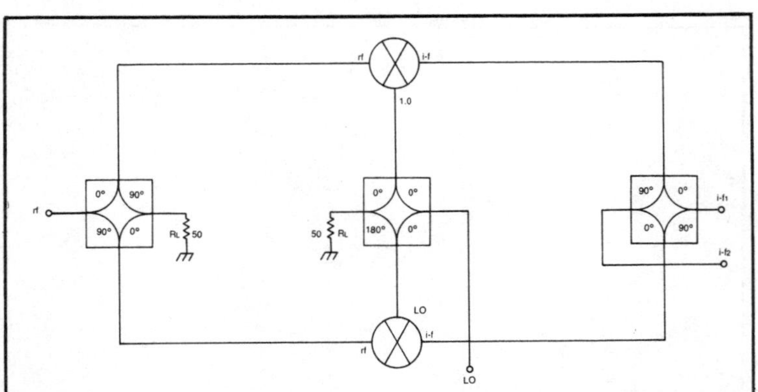

Fig. 14-19. Internal configuration for an image-rejection mixer.

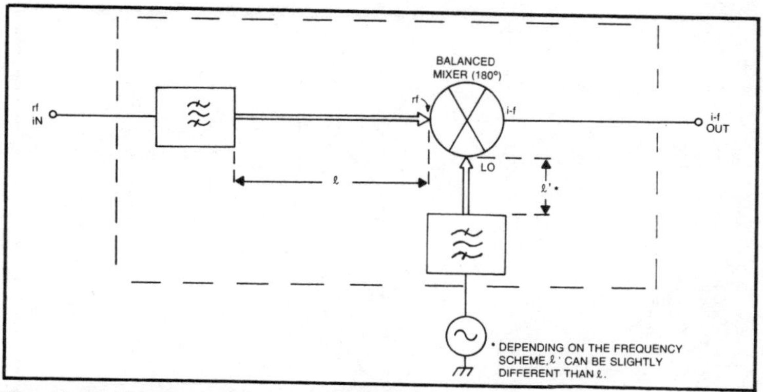

Fig. 14-20. The image-recovery mixer principle.

In both cases, the image created energy can be recovered and used so as to create additional power at the i-f frequency, reducing the conversion loss of the mixer. Figure 14-20 shows the implementation of an image recovery mixer.

In this arrangement the filters at the rf and LO ports of the mixer serve the function of reflecting the internally generated image energy, back into the mixer, while providing a path for the rf and the LO signals. Proper adjustment of the electrical lengths e and e' between the filters and the mixer provides the right phasing between the two resultant i-f contributions due to signal and image. This in turn results in a decrease in the conversion loss of the mixer. Typical improvements of 2 dB have been reported with this type of mixer.

Although not widely used, the image recovery mixer finds increased application in systems where gain cost is at a premium. It is important to note, however, that if not properly designed, the image recovery mixer can be lossier than a typical balanced mixer, presenting a problem rather than a solution to a communications system.

JFET MIXERS

JFETs are also used and are gaining a good intermodulation reputation. Figure 14-21 shows a favorite scheme, with a carefully balanced and matched set of FETs which provides a +30 dBm typical intercept point. Other FET mixers use VMOS devices. Shown in Fig. 14-22 is a suggested singly-balanced mixer using the Siliconix VMP 4 FET. With proper balance of the dc bias, a typical 43 dBm intercept point has been reported.*

INTEGRATED CIRCUIT MIXERS

Active mixers in IC form are also common. They are inexpensive and provide gain along with mixing, an important factor when designing

*Modern Receiver Mixers for High Dynamic Range, Doug DeMaw and George Collins, QST, January, 1981.

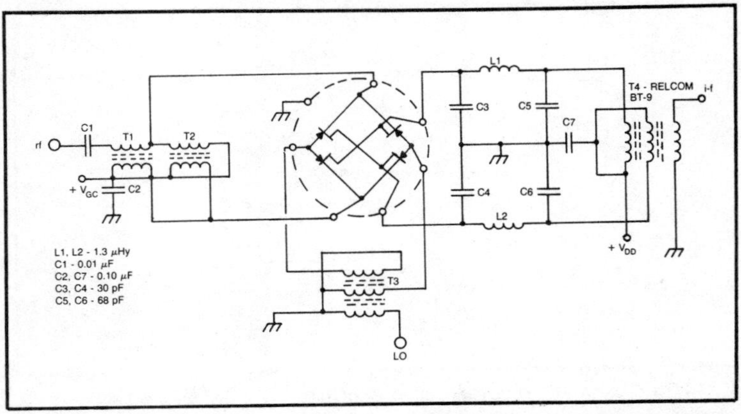

Fig. 14-21. Doubly-balanced mixer using four closely-matched junction FETS.

economically. Figure 14-23 shows the internal configuration of a motorola MC1596. Also shown is the surrounding circuitry required to achieve an active doubly-balanced mixer with this IC.

IMPLEMENTATION OF IC MIXERS IN BILATERAL SYSTEMS

Active mixers are widely used in systems like the transceiver shown in Fig. 14-24. This scheme is economical to build, and is ingenious for

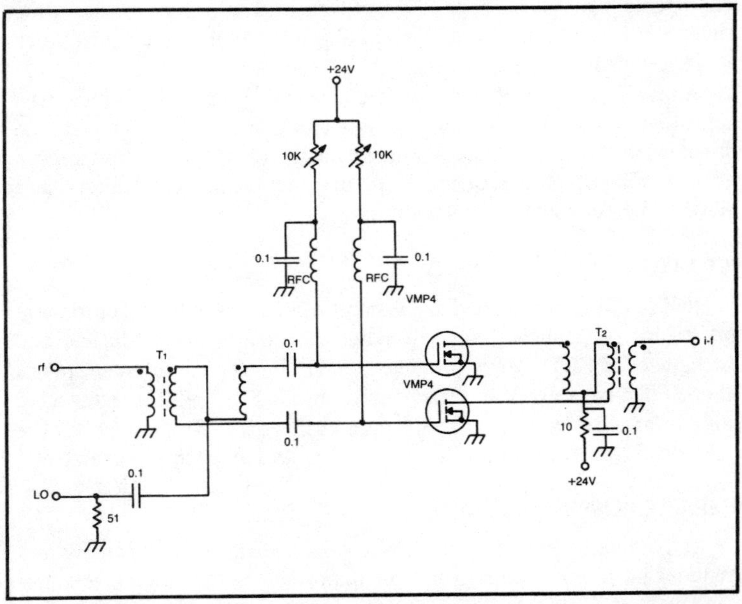

Fig. 14-22. Suggested singly-balanced mixer using the Siliconix VMP 4 V-MOS FET.

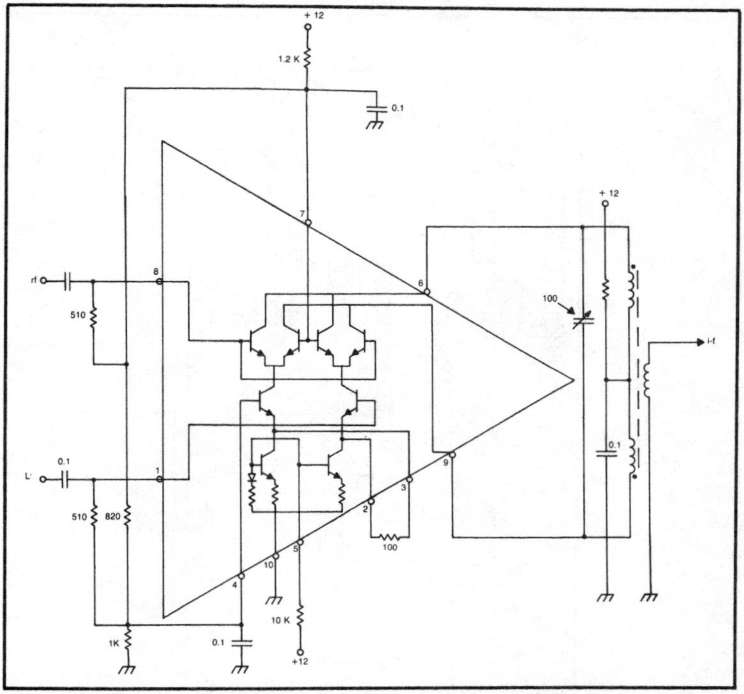

Fig. 14-23. Internal configuration and application circuit for an MC1596 active doubly-balanced mixer.

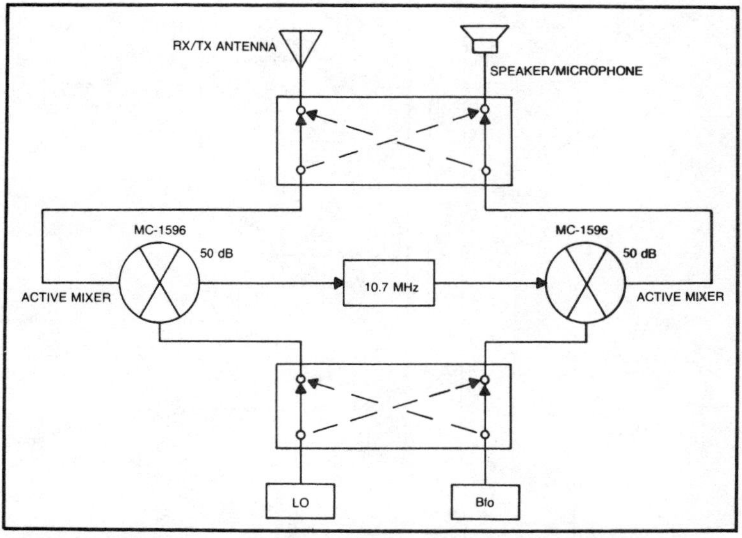

Fig. 14-24. Implementation of active mixers in a bilateral transceiver.

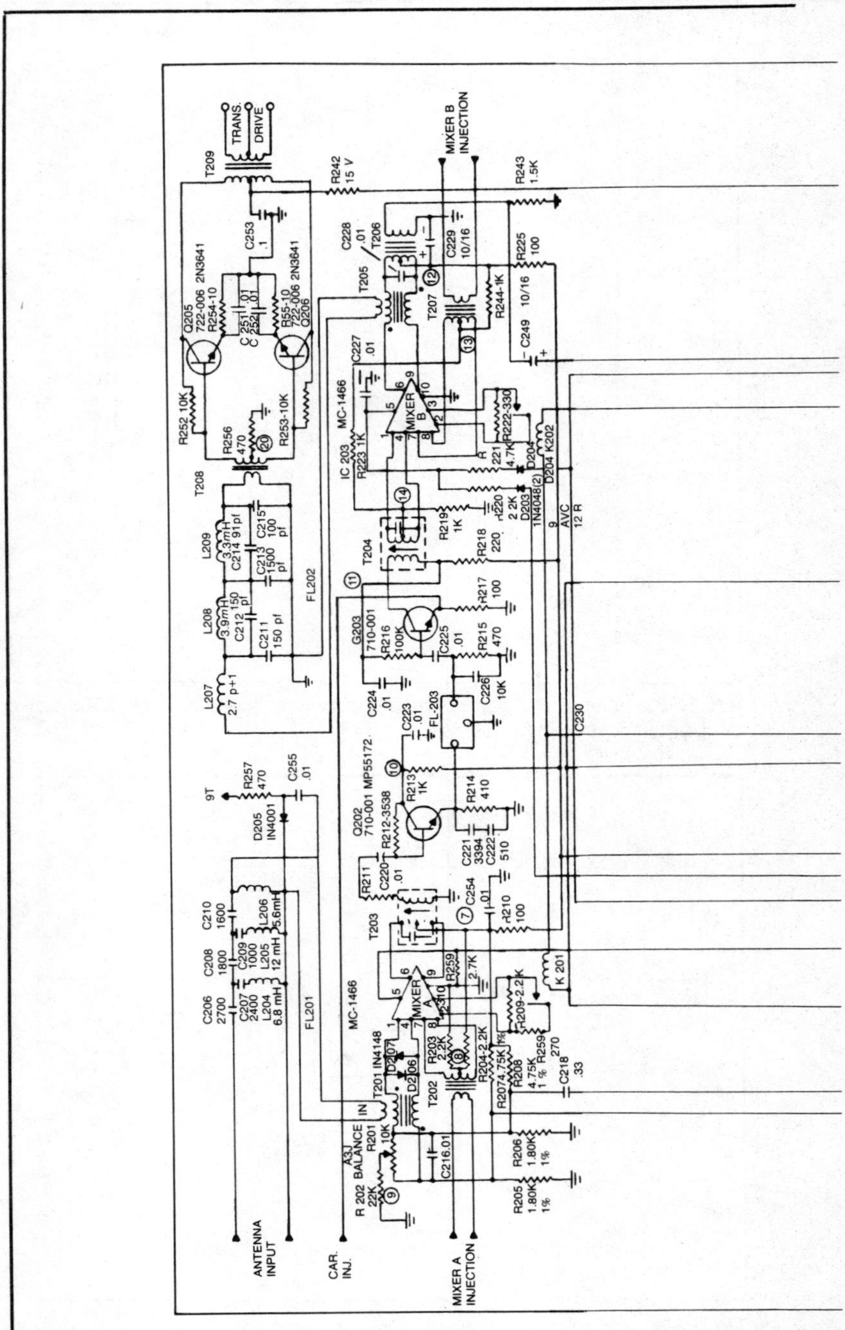

ANTENNA INPUT
CAR. INJ.
MIXER A INJECTION
MIXER B INJECTION
TRANS. DRIVE
T209
R242 15 V
R243 1.5K
Q205 722-006 2N3641
Q206 722-006 2N3641
R254-10
R55-10
C251 .01
C252 .01
C253 .1
R252 10K
R253-10K
R256 470
T208
L207 2.7 p+1
L208 3.9mH
L209 3.3mH
C211 150 pf
C212 150 pf
C213 1500 pf
C214 91pf
C215 100 pf
FL202
FL201
C206 2700
C207 2400
C208 1800
C209 1000
C210 1600
L204 6.8 mH
L205 12 mH
L206 5.6mH
9T
R257 470
D205 IN4001
C255 .01
MC-1466
MIXER A
MIXER B
IC 203
R223 1K
C227 .01
C228 .01
T205
T206
T207
C229 10/16
R225 100
R244-1K
C249 10/16
R222-330
K202
D204
D203
1N4048(2)
AVC
12 R
R219 1K
R220 2.2K
R218 220
R221 4.7K
T204
T203
T202
T201
IN
Q203 710-001
R216 100K
C225 .01
R217 100
R215 470
C224 .01
C226 10K
FL-203
C223 .01
Q202 710-001 MP55172
R212-3538
R213 1K
R214 410
C221 3394
C222 510
C220 .01
R211
C254 .01
R210 100
R259 2.7K
K 201
R209-2.2K
R259 270
C218 .33
R203 2.2K
R204-2.2K
R207 4.75K
R208 4.75K 1%
R206 1.80K 1%
R205 1.80K 1%
C216 .01
R201 10K
A3J BALANCE
R 202 22K
D207
D206
IN4148
C230

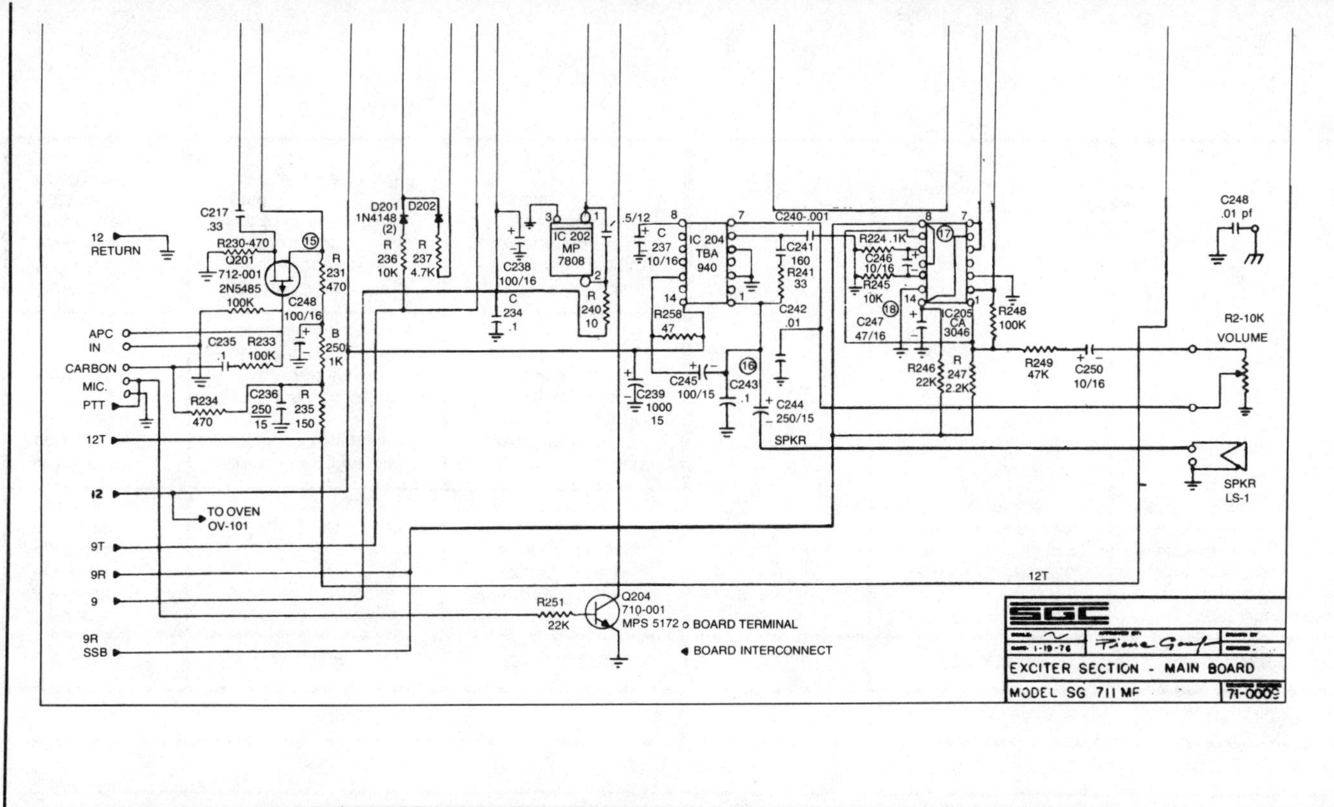

Fig. 14-25. The heart of a bilateral transceiver using MC 1596 active mixers, Stoner-Goral model 711 (courtesy of SGC Inc.).

Table 14-2. Trade-Offs in Mixer Technologies.

MIXER TYPE					
ACTIVE			PASSIVE		
DEVICE	ADVANTAGES	DISADVANTAGES	DEVICE	ADVANTAGES	DISADVANTAGES
Bipolar transistor in a doubly-balanced configuration.	-low noise figure that provides high gain.	-easily overloaded. -Intermodulation Distortion (IMD) performance poor.	Diode mixers in a doubly-balanced configuration.	-low noise figures that provides excellent dynamic range. -excellent overload characteristics.	-high Local Oscillator drive required. -conversion loss. -needs proper termination to source-load.
JFET in a doubly-balanced configuration.	-low noise figure that provides conversion gain. -good IMD, blocking, cross-modulation performance. -not easily overloaded.	-high Local Oscillator power required. -operates as square-law detector for large-input a-m signals.			
Dual-gate MOS FET and VMOS devices in a doubly-balanced configuration.	-good IMD, blocking, cross-modulation performance. -provides conversion gain. -not easily overloaded.	-higher noise figure. -operates as square-law detector for large input a-m signals.			

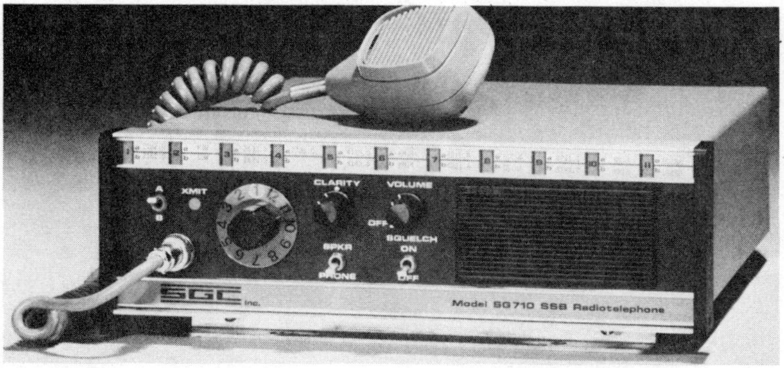

Fig. 14-26. The Stoner-Goral model SG-710 uses an improved bilateral system (courtesy of SGC Inc.).

achieving a transceiver with a minimum number of parts. There are two MC1596s in this system. The antenna is directly coupled to the first one, and the first oscillator provides conversion with gain (about 50 dB) to a 10.7 MHz i-f. The second oscillator (bfo) injects into a second MC1596 and provides detection with another 50 dB of gain. The MC1596 can be used either as a first mixer, a product detector, or a sideband generator, with only minor changes in values. This system takes advantage of this fact by using a switching mechanism involving relays and diodes transforming the receiver into a transmitter. The actual implementation of the heart of this transceiver is shown in Fig. 14-25. An improved version of this approach is used in the transceiver shown in Fig. 14-26 from Stoner-Goral Corporation.

MIXER TECHNOLOGY TRADE-OFFS

It has been seen that linearization of mixer performance plays an important role in the design of a communication receiver. Several types of mixers were described and a few technologies were analyzed. Today, the doubly-balanced diode mixer is the most widely used, and despite its relatively high conversion loss (6 dB typical) it provides a typical dynamic range approaching 100 dB with high output intercepts (>+23 dBm). Its disadvantages are that it is susceptible to odd-order harmonic mixing and it requires a proper termination.

New recent developments in doubly-balanced mixer design indicate third-order intercept point achievements of + 38 dBm (VAY-1, from Mini-Circuits Labs.) with LO drives in excess of 28 dBm. The JFET mixer, while relatively more expensive, provides superior noise figures (4 dB typical), combined with gain and high intercept (> + 25 dBm). The bipolar transistor in an active doubly-balanced form will continue to provide a low-cost approach, while experiments with MOS and VMOS FETS look promising from a performance standpoint. Shown in Table 14-2 is a list of trade-offs between some of the rf mixer technologies.

—•••—

Chapter 15

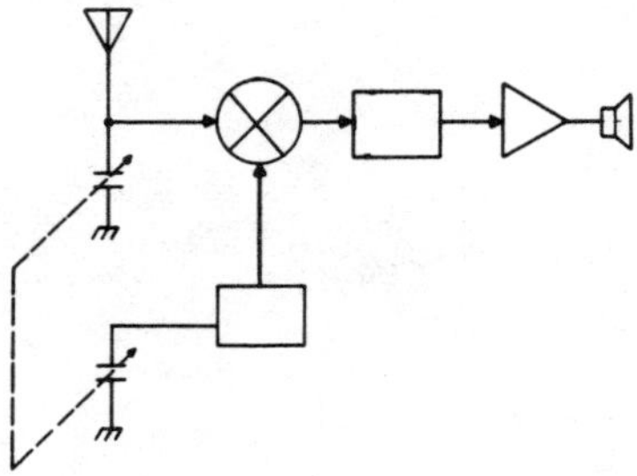

Variable Frequency Oscillators

We will discuss now the first local oscillator (LO) which usually is a variable frequency oscillator working directly into the first mixer of a communication receiver. Figure 15-1 shows a 5 MHz Vfo operating as an integral part of the 9 MHz i-f receiver we have previously chosen for our example. This Vfo can provide 500 kHz of coverage (5.0-5. MHz) allowing our receiver to tune the 14 MHz amateur band, if the first mixer operates in a subtracting mode (14−5=9). If the first mixer operates in the adding mode (4+5=9), the 3.5 MHz amateur band can also be covered with the same Vfo (dial tuning would be reversed). Switching between the two bands would be accomplished by changing the preselector from one band to the other. Although this technique has been used for many years, the drawback remains the inconvenience of the reversed tuning for one of the bands if a mechanical dial is used. In the case of a digital dial, this problem is greatly diminished due to the instantaneous frequency readout.

If the same direction in tuning was desired, and several other bands were required for our receiver, the Vfo would have to switch ranges as shown in Fig. 15-1 in order to tune all bands of interest. This technique has been widely used by many manufacturers, however, switched tuned circuits in Vfos proved to be a poor choice of design in some receivers due to dirty contacts in switches resulting in frequency instability especially at higher frequencies. Improved performance is obtained if the Vfo is mixed with fixed crystal oscillators which inherently have good stability.

MIXING VFO'S WITH CRYSTAL OSCILLATORS

In most receivers, a mixing scheme such as the one in Fig. 15-2, is used to provide injection to the first mixer. The Vfo operating at 5.0 to 5.5 MHz is mixed with several crystal oscillators as shown in our example. The

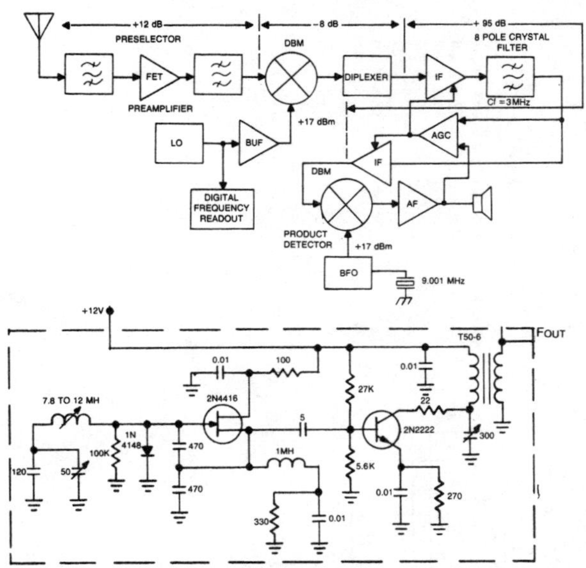

RECEIVER FREQ. BAND MHz.	Vfo FREQ. MHz.
1.8	10.8 - 10.9
3.5	12.5 - 13.0
7.0	16.0 - 16.5
14.0	5.0 - 5.5
21.0	12.0 - 12.5
28.0	19.0 - 19.5
28.5	19.5 - 20.0
29.0	20.0 - 20.5
29.5	20.5 - 21.0

Fig. 15-1. Implementing a switched Vfo in a 9 MHz i-f communications receiver.

mixing products are carefully filtered out at the output of the mixer which in our case is an active MC1596. The corresponding capacitors are automatically introduced in the transformer/filter circuit T_1 by the band switch wafers SC and SD as shown. This scheme provides good results over the switched Vfo by basing most of its stability on the crystal oscillators. If a different frequency scheme is contemplated, even a lower frequency Vfo could be considered, putting most of the stability burden on the crystal oscillators. (A Vfo should be kept as low as possible in frequency, because stability can be achieved easier there).

The mixing scheme should be chosen with extreme care since this additional mixing process generates a multitude of products which could interfere with the receiver's performance. A full system analysis should be performed prior to the circuit design by using the method previously discussed. It can be seen from this discussion that the plain Vfo would be preferred from a spurious response point of view, however stability problems introduced by switching, yield mixing schemes which in turn introduce unwanted spurious products which could carry through the entire

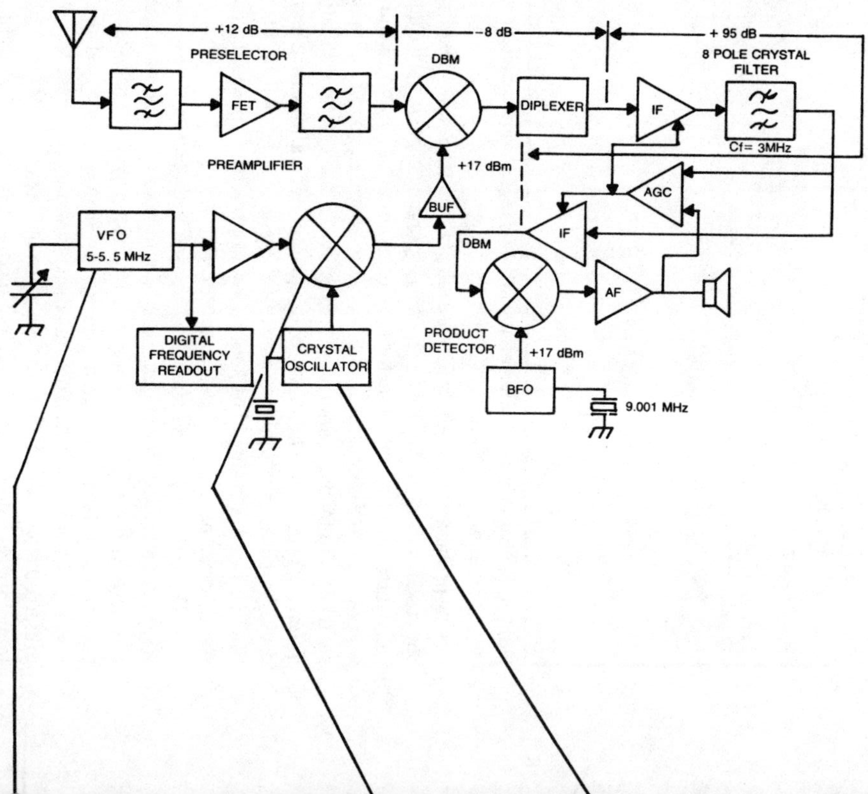

ALL FREQUENCIES ARE IN MHz.

BAND F	CRYSTAL F	Vfo F	TOTAL INJ F
1.8	5.8	5-5.5	10.8
3.5	7.5	5-5.5	12.5
7.0	11.0	5-5.5	16,0
14.0	--	5-5.5	5.0
21.0	6.990	5-5.5	12.0
28.0	13.990	5-5.5	19.0
28.5	14.5	5-5.5	19.5
29.0	15.0	5-5.5	20.0
29.5	15.5	5-5.5	20.5

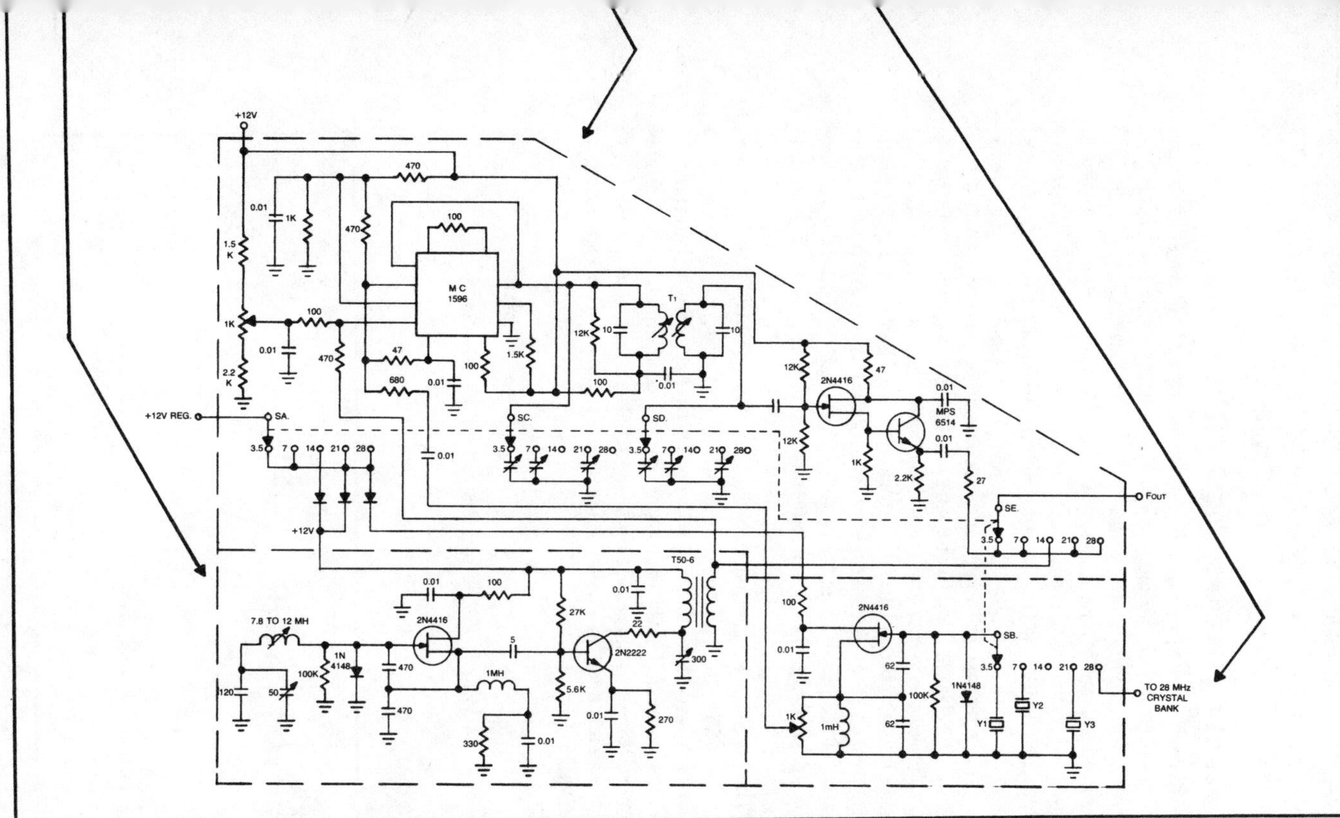

Fig. 15-2. Requirements for a mixed Vfo in a 9 MHz i-f communications receiver.

receiver. The real solution for solving the stability problem is the frequency synthesizer, but the design becomes much more complex, contrary to popular belief.

INTRODUCING THE FREQUENCY SYNTHESIZER AS A VFO

Although there are many types of frequency synthesizers, the phase-lock type is probably the most widely used approach in communication receivers. It is very simple to understand in its basic form. Figure 15-3 shows a block diagram for such a synthesizer. This scheme could be used to generate the entire range of frequencies of 10.8 to 21 MHz necessary for the local oscillator injection in our 9 MHz i-f communication receiver example.

In this synthesizer, a reference frequency is derived from a crystal oscillator (100 Hz) and fed to one side of a phase comparator/detector. In order to make the vco operate at the required frequency, the entire loop searches until the second line to the phase detector becomes equal in phase and frequency to the reference frequency (100 Hz). It is at this point that the dc output of the phase detector (which is the control line to the voltage-controlled oscillator) assures that the desired frequency is developed at its output.

The loop tracks continuously so that the stability of the output is always based on the stability of the reference frequency. If a change in frequency is desired, all that has to be done is to change the division number N and the synthesizer will search for whatever vco frequency divided by N will give a 100 Hz signal at the second input to the phase detector, (in Phase with the reference) as shown. This relationship can be expressed by Equation 15-1.

$$F_{out} = N \times F_{ref} \qquad \textbf{Eq. 15-1.}$$

Where: F_{out} is the desired output frequency
N is the division number
F_{ref} is the reference frequency

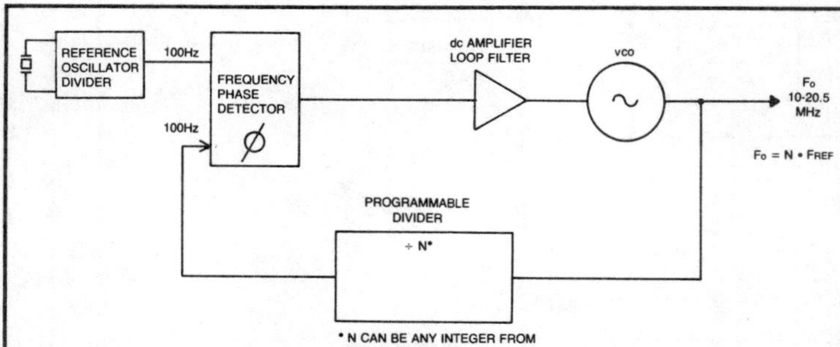

Fig. 15-3. A typical phase-locked loop synthesizer. Resolution is 100 Hz.

It can be seen from this equation that the phase-locked loop approach can provide injection frequencies to the first mixer of a communication receiver in fractional steps, unlike the Vfo which provides continuous coverage within the specified frequency range.

To approach this continuous resolution, the synthesizer bases itself on the reference frequency, and can not tune continuously. The reference frequency determines the switching step, in our case 100 Hz. Every time the division number is switched digitally, a 100 Hz step is obtained which will sound just like stepping through a musical staircase generator when tuning an outside signal with the receiver (if the bfo is on). In reality, this synthesizer is not practical.

The 100 Hz reference which is needed to create the tuning resolution, would make the switching extremely slow, so that it may take approximately half a second to achieve locking, which would be a long time to retune from an operator standpoint. Secondly, the 100 Hz reference would introduce sidebands to the local oscillator which would intermix with the incoming signal and create intermodulation distortion. Thirdly, such a system would be affected by hum since the reference frequency is kept low as in our case.

PRACTICAL PHASE-LOCKED SYNTHESIZERS

The solution for fast switching is increased reference frequency. This would also put the sidebands further away from the signal, making filtering more possible. However, increasing the reference frequency would change the resolution step. In order to achieve 100 Hz resolution with higher reference frequencies, the synthesizer becomes much more complex as we will see.

MULTI-LOOP SYNTHESIZERS, A REALITY

By using two loops, or two synthesizers with slightly different reference frequencies, and by stepping them in a specific mathematical relationship, and finally subtracting them in a mixer, fast switching and proper resolution can be achieved at the same time. By using this approach, a radio receiver can be made to scan over the entire range of interest. The cost is medium digital complexity. Figure 15-4 shows a block diagram approach for a dual synthesizer as applied to the 9 MHz i-f communication receiver. The two loops use separate reference frequencies which differ from each other by the resolution (derived from same source) frequency (10.000 Hz - 9.900 Hz = 100 Hz). Also shown in Fig. 15-4 is a small part of the tedious algebraic manipulation of the two numbers (N_1 and N_2) necessary to tune the receiver over a 30 kHz range from 1.000 to 1.030 MHz. Note that there are 300 individual combinations of N_1 and N_2 alone to cover this range. A specially designed interface would have to be used to accommodate all possible combinations of N_1 and N_2. The long term stability of this type of Vfo is determined entirely by the reference oscillator stability, and it can be as good as few Hz when using an oven-controlled crystal oscillator.

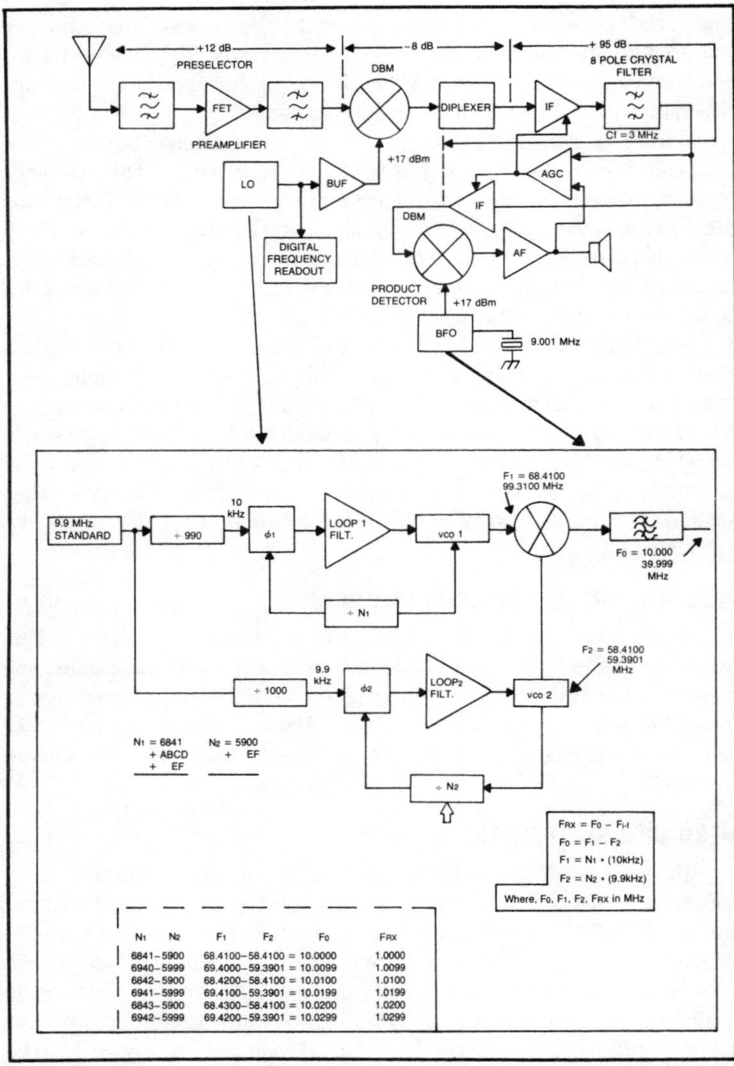

Fig. 15-4. Using the dual loop synthesizer in the 9 MHz i-f communications receiver provides fast switching and high resolution.

In order for this stability to be maintained throughout the entire receiver, a similar dual synthesizer would have to be used in the place of the bfo. In this synthesizer, N_1 and N_2 would have such values that a 9 MHz signal would be generated at the output. Offsetting the bfo would also be possible by changing these numbers. It is entirely possible by using this method to slave the entire receiver to the reference frequency, and this receiver would become an ideal candidate for remote controlling.

Figure 15-5 shows a block diagram of the reference generator which is a combination of two forms of synthesizers in itself, a brute force and a fixed phase-locked approach. By using a reference frequency of its own, the 9.9 kHz and the 10 kHz reference frequencies necessary for all loops can be derived from a typical 4 MHz, oven controlled crystal oscillator.

Figure 15-6 shows the schematic diagram for the digital portion of a dual synthesizer. To simplify the circuit design, the Hughes HCTR-0320 LSI chip was used for both loops. This *large-scale integration* (LSI) CMOS device is equipped with a programmable divide-by-N counter, a phase/frequency detector and an adder/decoder which provides proper interface with the digital inputs. Figure 15-7 shows the internal configuration of this versatile integrated circuit.

Multi-loop synthesizers are a fact of life in today's technology. As many as four or five loops are sometimes used in multi-conversion receivers. Complicated steering mechanisms provide fast switching with minimum noise problems over wide frequency ranges, while lock-up times have been reduced to milliseconds and sometimes to microseconds. See Fig. 15-8.

Figure 15-9 shows the block diagram of the Rockwell-Collins 651S receiver. This triple-conversion approach uses a frequency synthesizer which provides injection signals for all mixers involved. The block diagram of the synthesizer used in this receiver is shown in Fig. 15-10.

In this approach, all loops are slaved to a single reference frequency (9.9 MHz). The Rockwell-Collins 851S-1 shown in Fig. 15-11 is an even

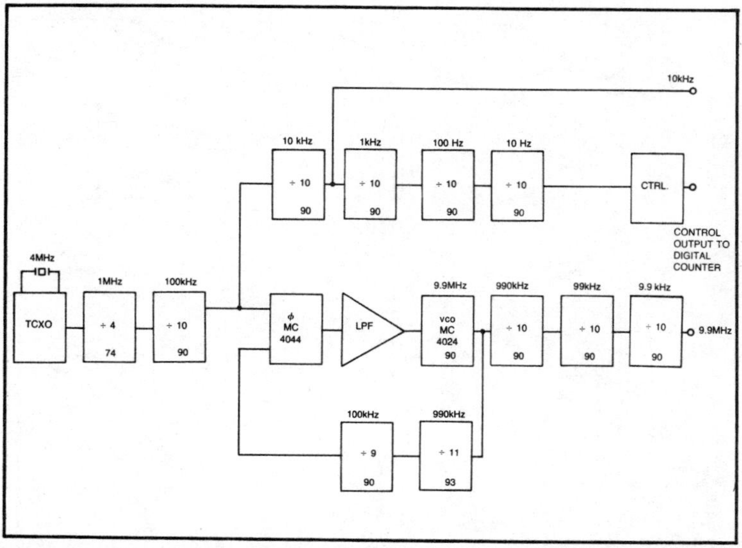

Fig. 15-5. Deriving the 9.9 kHz and the 10 kHz references from a 4 MHz oven-controlled crystal oscillator. This synthesizer uses a brute-force approach combined with a fixed phase-locked approach.

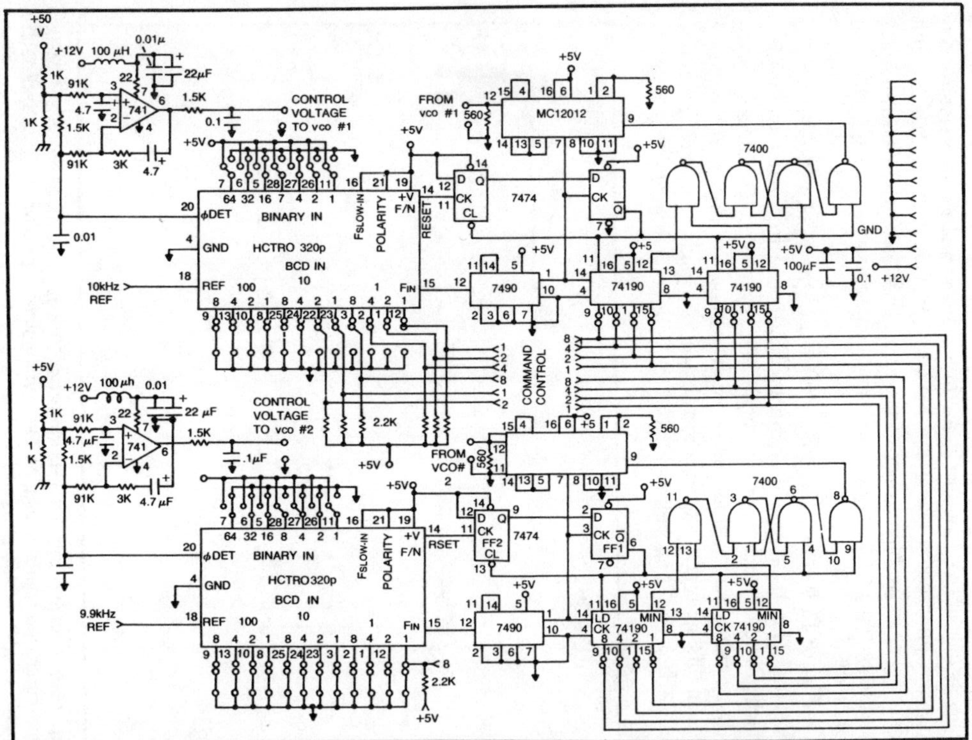

Fig. 15-6. Circuit diagram for a dual synthesizer using the Hughes HCTR-0320 LSI chip. Circuit includes loop filters, and variable modulus prescalers.

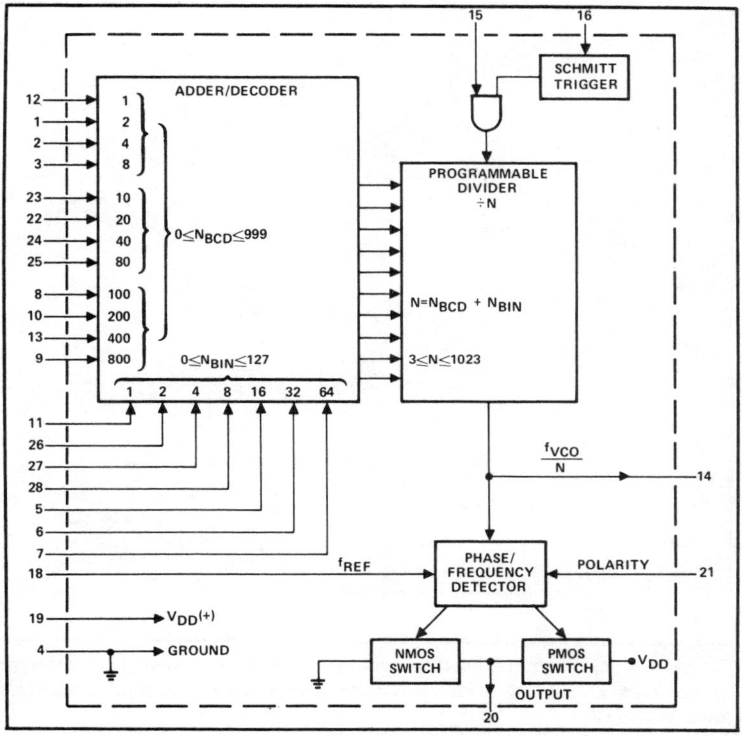

Fig. 5-7. Internal configuration of the Hughest HCTR-0320 LSI synthesizer chip.

Fig. 15-8. Special ICs are being manufactured to provide all functions necessary in a phase-locked loop synthesizer. The Plessey NJ 8811/12 contain a reference divider, a programmable divider and a phase/frequency detector (courtesy of Plessey Semiconductors).

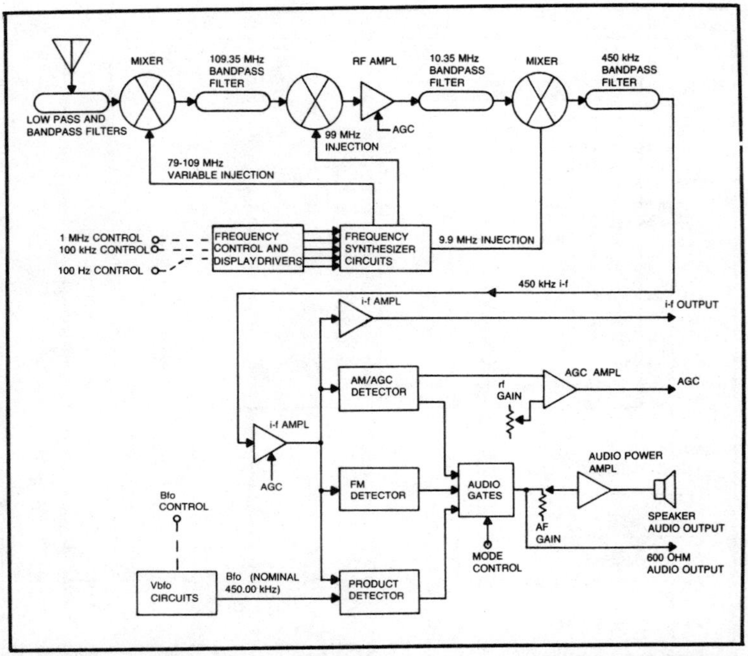

Fig. 15-9. Block diagram of the Rockwell-Collins 651S communications receiver. Frequency synthesizer provides all injections necessary in this triple conversion receiver.

more advanced form of a multi-conversion synthesized communication receiver. Figure 15-12 shows the internal configuration of this receiver with the synthesizer and the digital interface shown.

The frequency stability of this receiver is 5×10^{-7} Hz over the 0° to 50° C temperature range, and the drift rate will not exceed 3×10^{-8} Hz per week. The lock-up time is on the order of 2 milliseconds. Typical specifications for the Rockwell-Collins 851S-1 are listed in Table 15-1.

The multi-loop phase-locked approach just described is only one form of implementation of frequency synthesis, but many other forms are being used today either alone or in combination with the above, depending on the requirements of the system design for a particular receiver.

To better understand what can be done in receiver design we will now discuss the several forms of synthesizers that can be used to provide local oscillator injection in communication receivers.

SYNTHESIZER FORMS

While a brute-force or a harmonic approach can be used for rough tuning, a phase-locked or a frequency-lock combination can be used to fine tune the receiver. The cost/performance trade-offs are usually the deter-

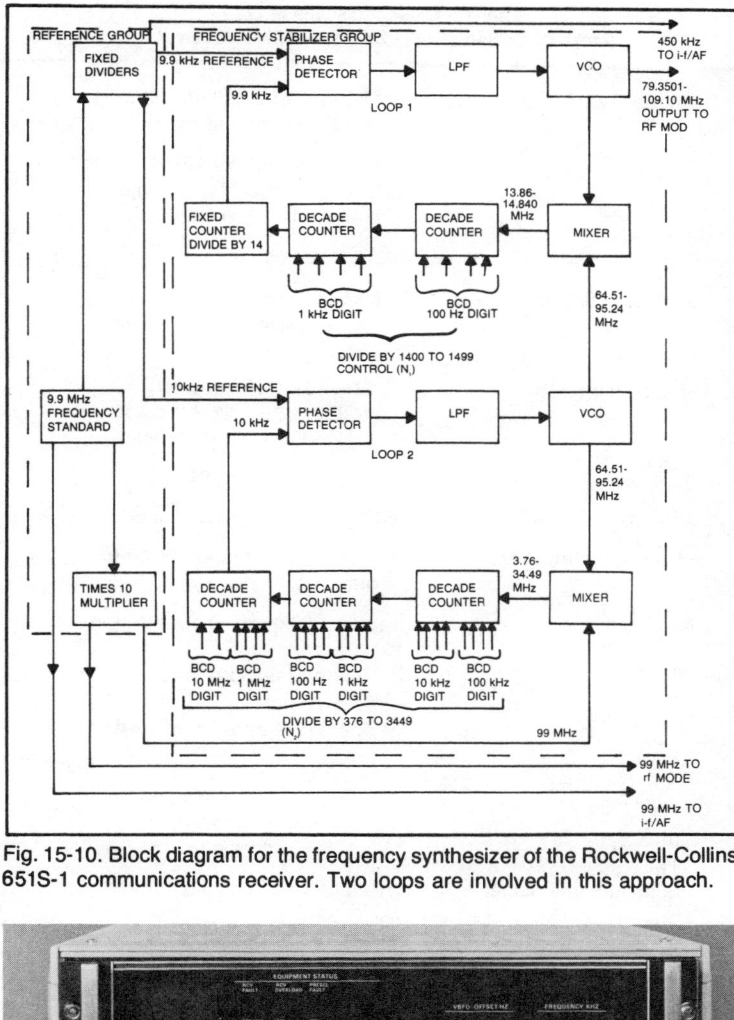

Fig. 15-10. Block diagram for the frequency synthesizer of the Rockwell-Collins 651S-1 communications receiver. Two loops are involved in this approach.

Fig. 15-11. The Rockwell-Collins 851S-1 communications receiver uses an advanced synthesizer to provide fast switching for tuning (courtesy of Rockwell International).

Table 15-1. Specifications of the Rockwell-Collins 851S-1 Communications Receiver.

Frequency coverage	250 kHz to 29.99999 MHz (down to 12 kHz with vlf option).
Frequency stability	Not more than 5×10^{-7} from 0 to +50 °C. Drift rate not more than 3×10^{-8} per week.
Channeling speed	2 ms nom; 5 ms max.
Sensitivity (SSB)	0.25 to 1.6 MHz: 2 μV "hard"; 1.6 to 29.9 MHz: 0.7 μV "hard" (10 dB (s+n)/n).
Modes of operation	A3A/A9B (ISB optional); A3J (USB or LSB); A3/A3H (AM); A1 (cw); and F3 (FM optional).
Bandwidths	16 kHz; 2.7 kHz USB/LSB.
Antenna input impedance	50 ohms nom, unbalanced.
Rf overload protection	Up to 100-V rf input power on or off.
Audio outputs	Line: 600 ohms, 0 dB mW nom, balanced. Headphone: 600 ohms, +10 dB mW max. Speaker: 8 ohms nom, 2 watts peak.
Squelch	Operates on audio signal-to-noise ratio. Applicable to speaker output only.
Audio hum and noise	-40 dB on line audio output.
AGC threshold	3 μV max.
AGC control	6 dB max rise in audio output for signal increase from 3 μV to 0.3 V.
Primary power	100/115/215/230 V ac ±10%, single phase, 47 to 63 Hz; 100 W max.
Physical	Size: 483 mm (19 in) wide x 178 mm (7 in) high x 483 mm (19 in) deep (531 mm deep with handles).
Mounting	EIA 483-mm (19-in) equipment rack, or optional case.

Weight	17.2 kg (38 lb), nominal.
Environmental	Temperature: 0 to 50 °C. Humidity: 0 to 95% relative humidity. Altitude: 3,658 m (12,000 ft) operating; 12,190 m (40,000 ft) nonoperating.

OPTIONAL FEATURES

- 10 Hz
- Internal oven standard or external standard
- CW or FSK offset filters (qty 5)
- ISB
- 1/2-octave rf input filters (7 bands)
- 200-V input protection with external preselector
- Digitally synthesized variable bfo ±9.990 kHz
- Remote manual or processor control
- LED status indicator (mode and bandwidth) with remote control option
- Wiring for external preselector (HF-8060)

NOTE: Includes space for future options of:

- FM
- Vlf rf module with performance specified down to 12 kHz
- Noise blanker
- AFC (automatic frequency control)
- Preset frequency control

Fig. 15-12. Internal configuration of the Rockwell-Collins 851S-1 receiver showing the synthesizer portion at right (courtesy of Rockwell International).

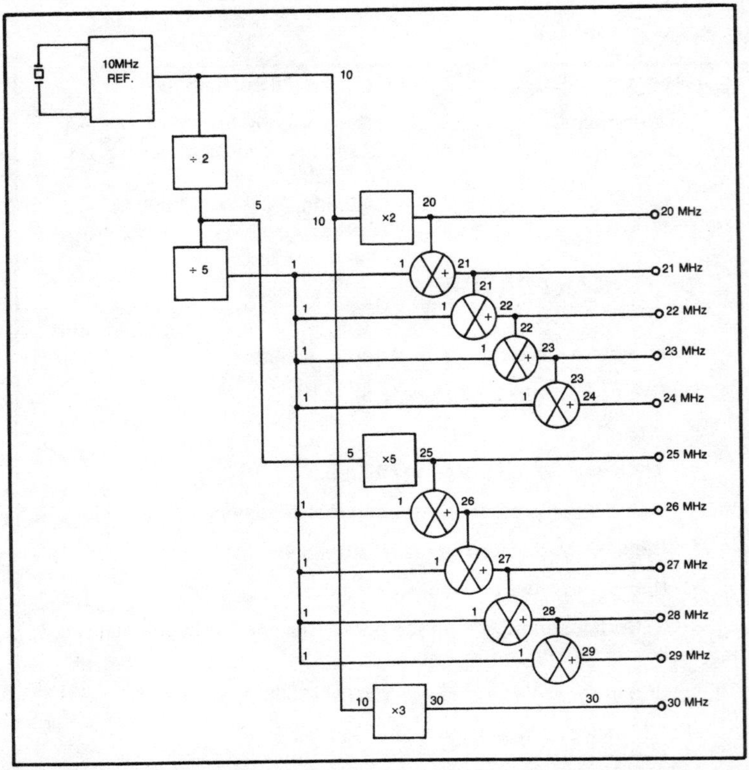

Fig. 15-13. An example of a direct synthesizer using dividers, multipliers and mixers.

mining factors in choosing the right combination. Proper understanding of the advantages and disadvantages offered by the different types will help the designer to make the right decisions before implementing a system, and not after.

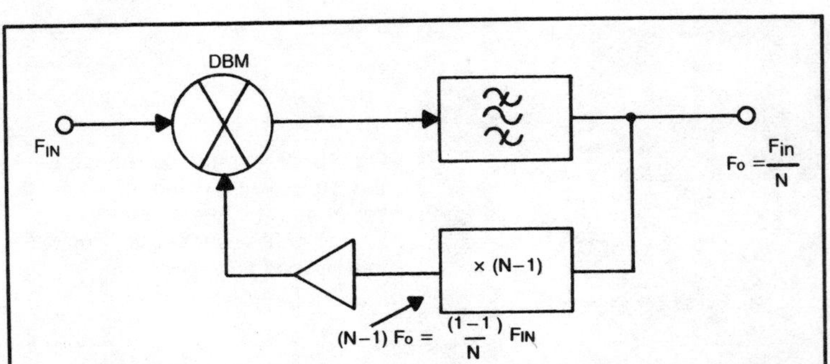

Fig. 15-14. Wideband analog divider using a mixer.

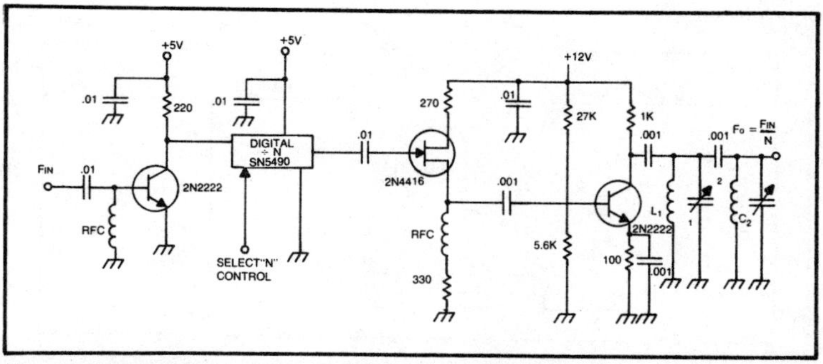

Fig. 15-15. Wideband digital divider using an SN 5490 integrated circuit. (FIN MAX ≤ 20 MHz).

DIRECT (BRUTE FORCE) SYNTHESIZER

The *direct synthesizer* uses multiplication, division, and mixing to produce desired frequencies from a *reference frequency source*. This type of synthesizer was first discovered in the 1940s and it usually applies to cases that require the generation of a relatively small number (up to ten) of frequencies such as in a limited coverage dual-conversion receiver in which the first local oscillator selects the rough bands to be fine tuned later in the second local oscillator. Figure 15-13 shows an example of a direct synthesizer that generates the frequency range of 20 to 30 MHz in 1 MHz steps. For reasons of simplicity, the extensive filtering required in this type of synthesizer is not shown in this example, however the system should be designed following the rules presented earlier because of the many products that can be generated by the multipliers, dividers, and mixers used in this approach.

Digital and/or analog techniques can be used to accomplish the functions necessary for this synthesizer. Figures 15-14 and 15-15 show typical implementation of a ÷ N circuit using analog and digital techniques. Extreme care in filtering the output of the digital approach should be observed because of the many products contained in the resulting square wave, as shown in Fig. 15-16.

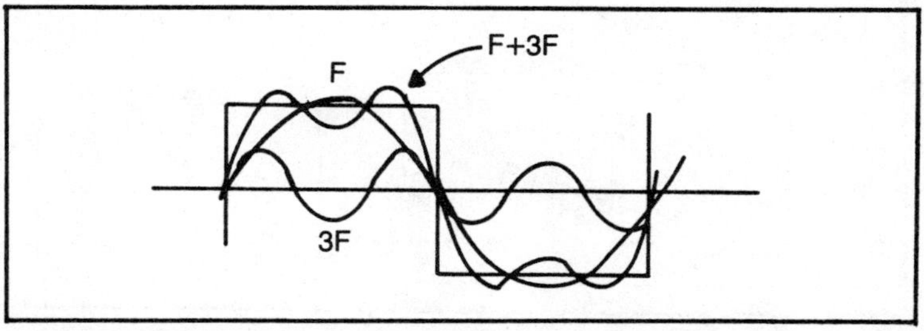

Fig. 15-16. The output of the wide-band digital divider contains many odd harmonics which have to be carefully filtered out.

÷	CONFIGURATION	REMARKS
2	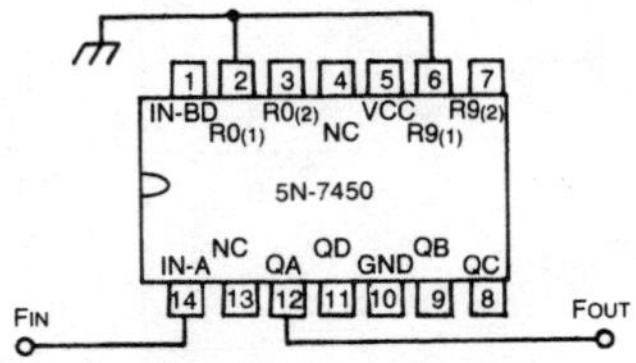	÷5 SECTION UNUSED
3	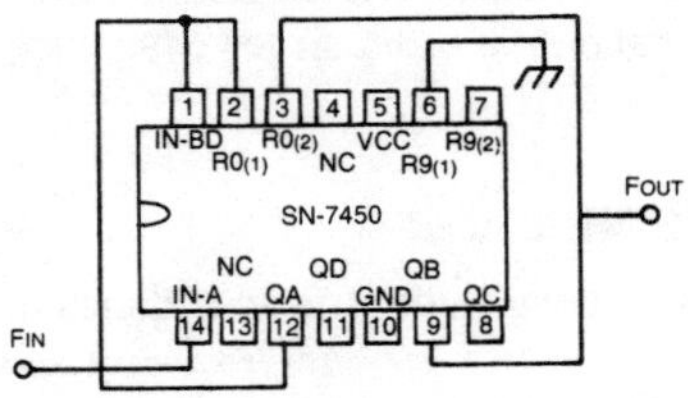	BCD COUNT IS OBTAINED ON QA OR QB
4	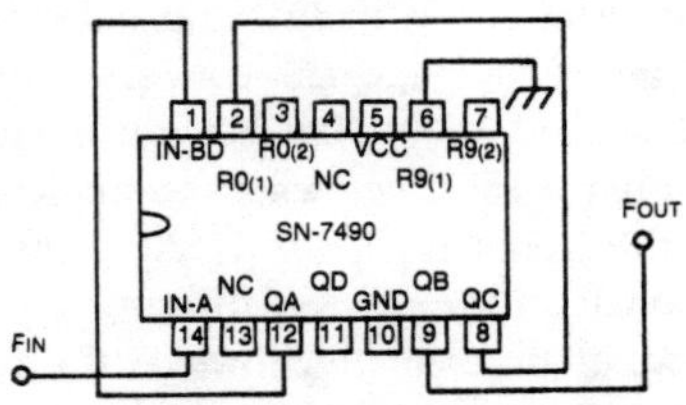	BCD COUNT IS OBTAINED ON QA OR QB
5	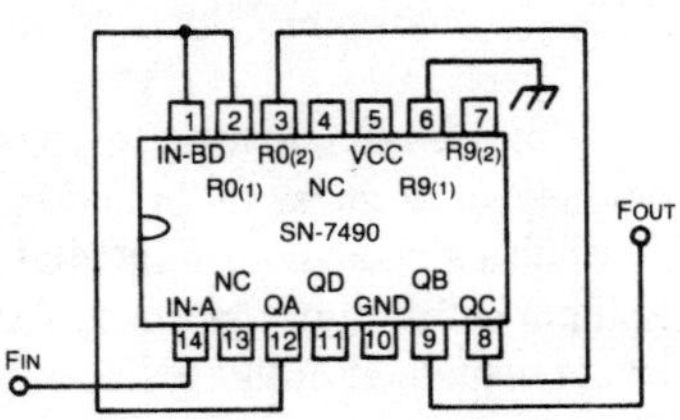	BCD COUNT IS OBTAINED ON QA OR QB OR QC
5	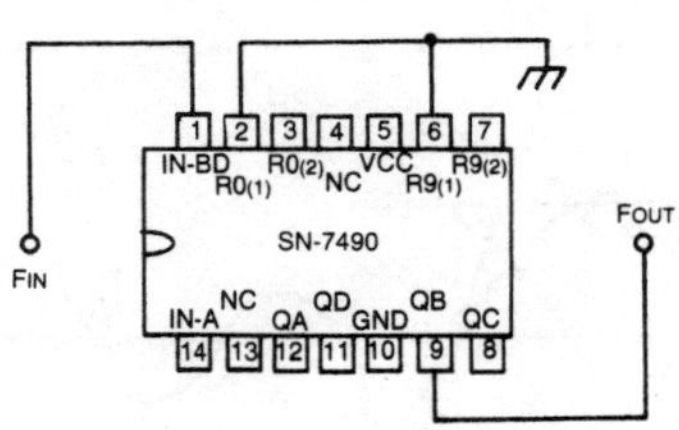	BCD COUNT IS OBTAINED ON QB, QC, OR QD.

Fig. 15-17. Frequency division using the SN 5490 (7490). Any division between 2-10 can be obtained without any additional parts.

6	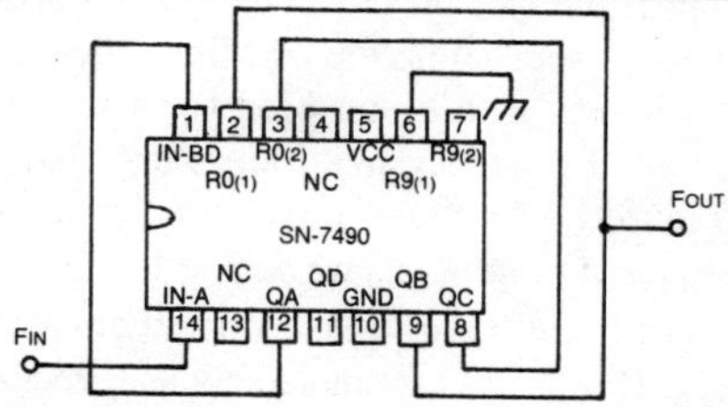	BCD COUNT IS OBTAINED ON QA OR QB OR QC
7	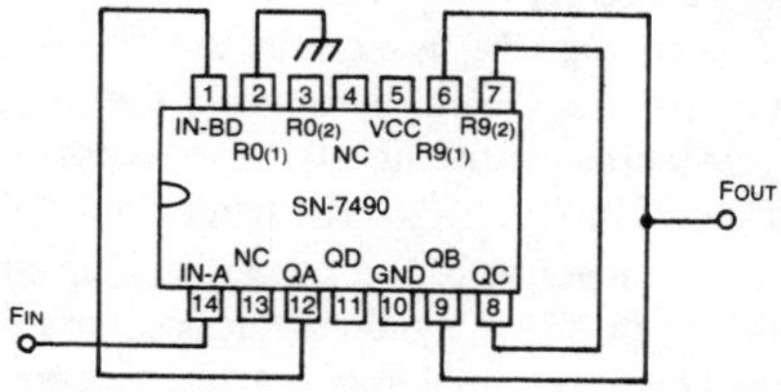	BCD COUNT IS NOT AVAILABLE Rgi IS USED.
8	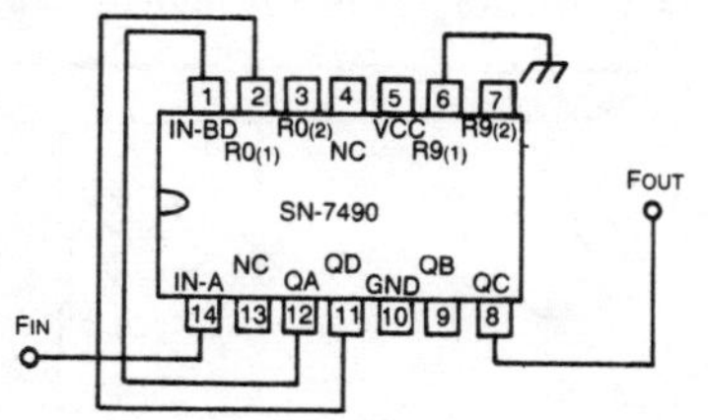	BCD COUNT IS OBTAINED ON QA, OR QB, OR QC, OR QD.
9	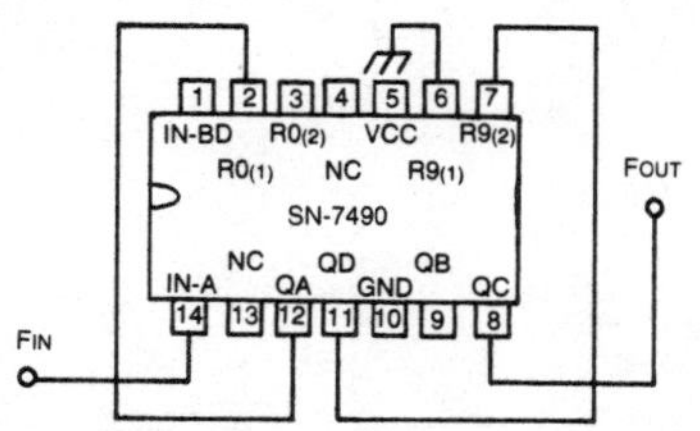	11
10	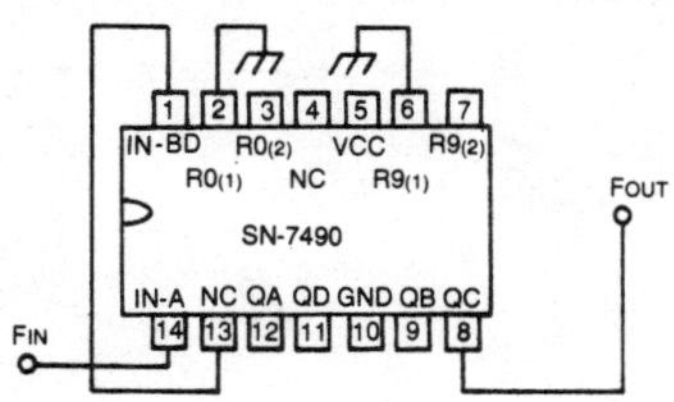	11

Figure 15-17 shows the pin configuration for implementing any division number between 2 to 10 for the SN5490 (7490) in the above circuit. Although faster digital devices could be used, the limitation of the input frequency for this device is approximately 20 MHz. Other devices such as flip-flops (SN-5474-7474) can also be adapted to this circuit in order to perform simple divisions. Frequency multipliers are also used in brute-force synthesizers. Figure 15-18 shows circuit implementation for a ×2, ×3, and ×4, narrow-band multiplier using analog techniques.

Another narrow-band multiplier can be obtained with digital techniques as shown in Fig. 15-19. This example acts as a doubler of the input frequency, by properly selecting the time constants for the dual one-shot configuration. The same analog interface can be used here, although filtering is not as important as previously, because of the inherent harmonic rejection caused by the one-shot timing elements (approx. – 35 dB). A wide-band doubler is shown in Fig. 15-20. The bandwidth of this approach is limited by the frequency response of the power divider and mixer used.

In conclusion the brute-force synthesizer presents several advantages. It is relatively simple to implement if the system application is correct, and it presents fast switching which is unparalleled by any other

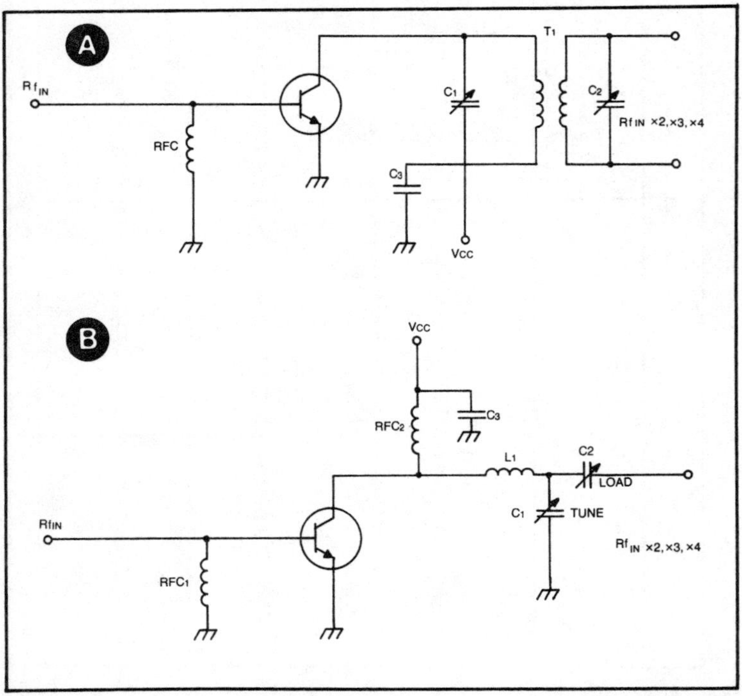

Fig. 15-18. Narrow band multipliers using tuned circuits in the output of a typical amplifier configuration. Several stages can be cascaded to maintain signal level or provide amplification.

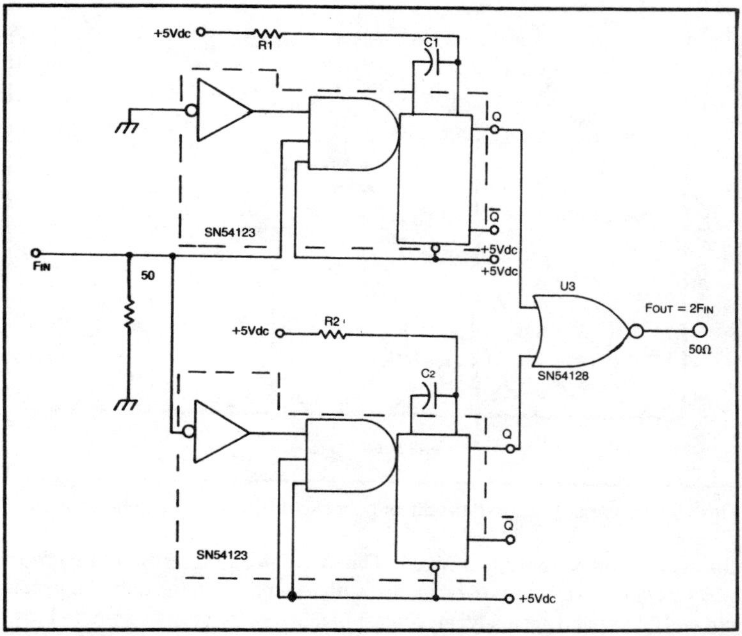

Fig. 15-19. This narrow-band frequency doubler has built-in rejection to unwanted products.

method. The problems associated with this type of synthesizer are usually the unwanted products generated in the mixing, multiplying, and dividing process, as well as the relatively large phase noise generated by this process as shown in Fig. 15-21. This type of synthesizer is relatively expensive to build because of the extensive shielding and filtering required. In spite of this, it is widely used in many receivers, from CB sets to guidance systems.

THE HARMONIC SYNTHESIZER

The *harmonic approach* is used whenever channel spacing between adjacent frequencies is the same, such as in a communication receiver. This approach uses the harmonic content of a reference frequency which equals

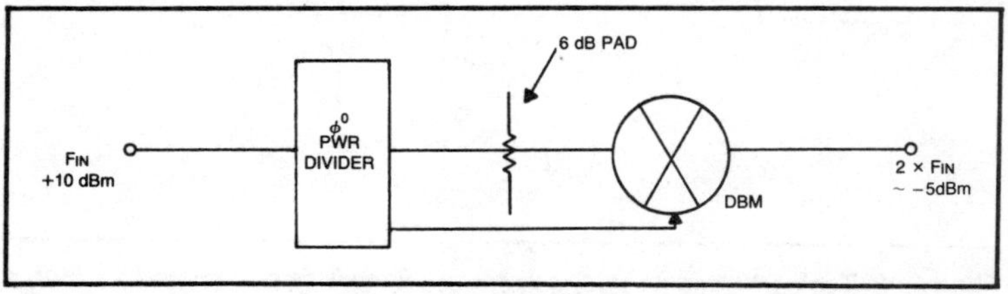

Fig. 15-20. Wide-band analog doubler using a doubly-balanced mixer.

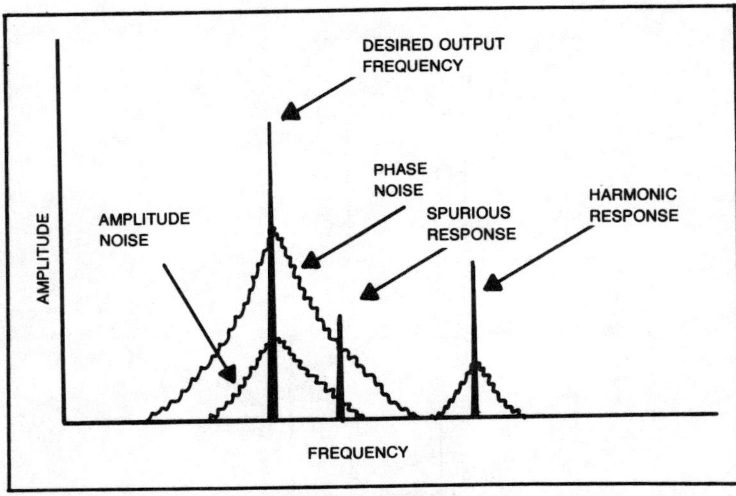

Fig. 15-21. Some unwanted products generated by a direct synthesizer.

the step required (usually 1 MHz). This oscillator is referred to as a *comb generator* because of the resemblance with a comb when the output signal is viewed on a spectrum analyzer. Each individual output can be selected with the help of a variable bandpass filter as shown in Fig. 15-22.

Another approach is presented in Fig. 15-23. In this example, there is no variable bandpass filter, as selectivity is provided by the 9 MHz crystal filter. Selection of the proper harmonic is obtained through the process of double mixing provided by the tuned oscillator.

This system can offer superior spurious performance over the previous one because of the i-f type processing offered by the crystal filter however, an image problem develops because of the heterodyne approach.

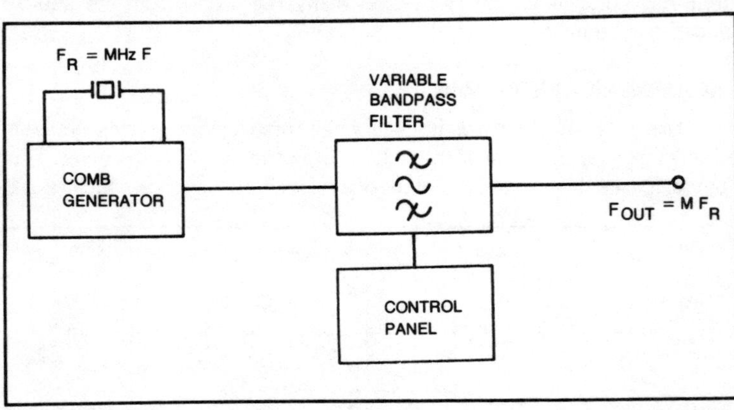

Fig. 15-22. Typical harmonic synthesizer uses a comb generator and a variable band-pass filter to generate multiples of F_R.

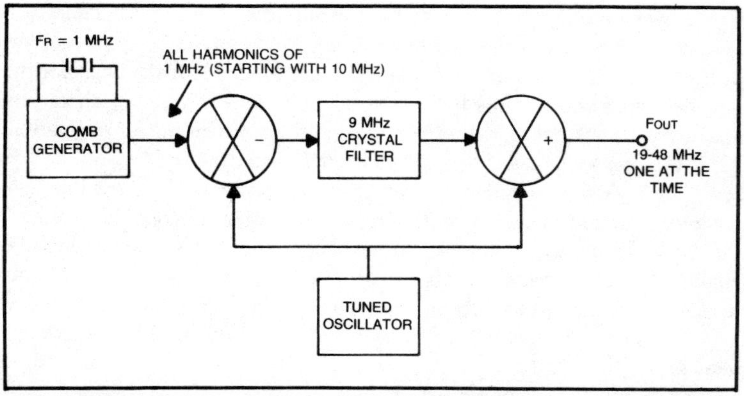

Fig. 15-23. Harmonic synthesizer uses tuned oscillator and crystal filter to select harmonic frequencies.

The methods presented here can be used in communication receivers either alone or in combination, depending on system requirements. A brute-force approach combined with a phase-locked approach was previously shown in Fig. 15-5. This combination provided two reference frequencies that were derived from one source for the dual-loop synthesizer discussed earlier. Many other combinations are also possible by using the harmonic approach.

THE PHASE-LOCKED LOOP (PLL) AND ITS ANATOMY

The phase-lock principle was discussed earlier. This approach is nothing more than a modern implementation of a closed loop motor control system as shown in Fig. 15-24.

A control signal of a certain frequency which is proportional to the desired speed of the electric motor is fed to one side of the phase comparator and since the phase difference between this signal and that produced by the tachometer is at the same value, the motor will be forced into syncronism at the point of coincidence between the two phases.

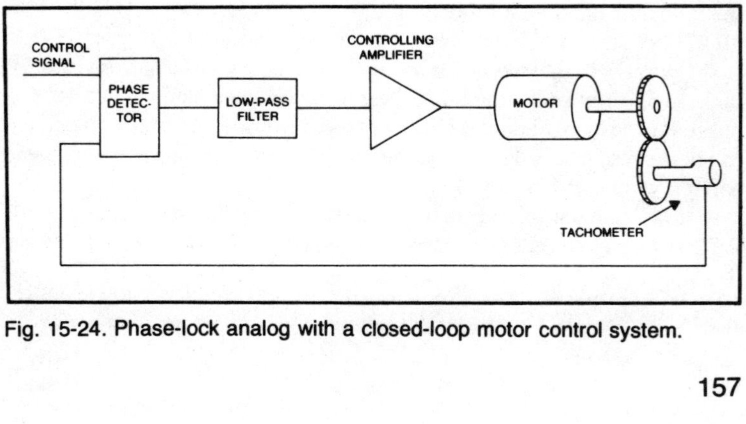

Fig. 15-24. Phase-lock analog with a closed-loop motor control system.

However simple this may seem, the application of this principle to a fast-switching phase-locked synthesizer used in a communication receiver presents a rather complex mathematical case, which is usually analyzed by laplace transformation methods. These methods allow the designer to analyze the transient and steady state conditions, providing for evaluation of all the operating cases in a phase-locked system.

Although such detailed mathematical discussion is a must in understanding how such a system works, it is nevertheless beyond the scope of this book. The reader is directed to special publications which will provide ample information in this direction. The bibliography at the end of this book lists some of the works recommended for this purpose.

The basic PLL synthesizer shown in Fig. 15-3 contains five functional blocks.

- Reference oscillator/divider
- Frequency/phase detector
- Dc amplifier/loop filter
- Voltage controlled oscillator (vco)
- Programmable divider

We will now analyze each of the five blocks.

REFERENCE OSCILLATOR/DIVIDER AND REQUIREMENTS FOR CALIBRATORS

The *reference oscillator* can be called the heart of the PLL because it establishes the stability performance characteristic of the synthesizer's output. It is sometimes not more than a simple crystal oscillator. It can take the form of a highly stable, temperature controlled, frequency source which in turn could be locked to atomic standards or time stations such as WWV or WWVB. It can provide frequency accuracies as good as 1×10^{-12} Hz/sec.*

Several forms of reference oscillators are shown in Fig. 15-25. Commercially available reference oscillators usually operate in the frequency range of 1 to 10 MHz. The 4 and 5 MHz models are probably the most widely used today. If stability is not of extreme importance, the designer can build his own. Figure 15-26 show some digital approaches to implementing this function. For higher frequencies, and sinewave outputs, Fig. 15-27 show how a reference oscillator can be built.

In practice, the reference oscillator frequency is further divided by digital means in order to obtain the reference frequency needed for the phase-locked loop. Figure 15-28 shows how a 100 kHz reference frequency can be derived from a 4 MHz commercially available temperature compensated crystal oscillator (TCXO).

In a communication receiver, a reference oscillator can also serve another function if the receiver is not synthesized. It can be used as a time

*Typical short term stability specification, over the temperature range of −55°C to +125°C. Long-term stability of $\pm 5 \times 10^{-10}$ Hz/day have been achieved for the same temperature range with proportional ovens.

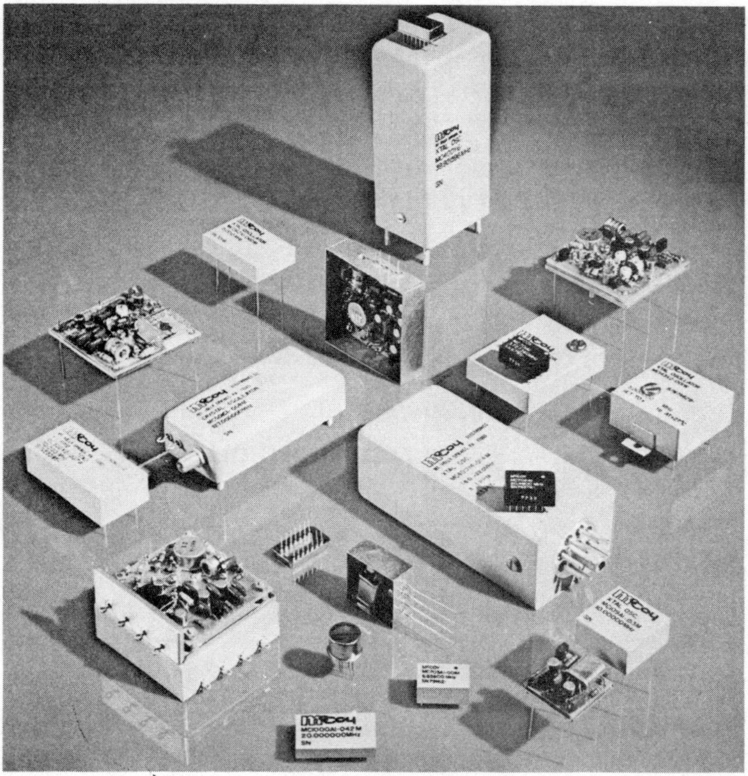

Fig. 15-25. Several types of crystal oscillators (courtesy of McCoy Electronics).

base for a digital counter used for measuring the tuning frequency, or with some slight modifications it can serve as a frequency calibrator for a receiver with an analog dial.

The calibrator is usually equipped with a 100 kHz crystal oscillator. The output of this oscillator is coupled with the receiver's antenna and the harmonics generated by the oscillator appear every 100 kHz within the receiver's tuning range at different points of the dial, allowing for calibration. Figure 15-29 shows a simple implementation of a frequency calibrator using the Motorola MC4024. This chip contains two independent vco's. Our design uses the spare one to provide pulsed 100 kHz bursts which can be easier identified than a continuous carrier at the output of the receiver.

Figure 15-30 shows a more elaborate calibrator which can be pulsed or continuous and it can also provide 10, 20, and 50 kHz calibration signals as well as the 100 kHz signal. Selection of these frequencies can be obtained with the use of the S_1 A, B, C switch. This design offers considerable flexibility as only one IC is used to accomplish the different frequency divisions.

TTL OSCILLATOR - 1 MHz to 10 MHz

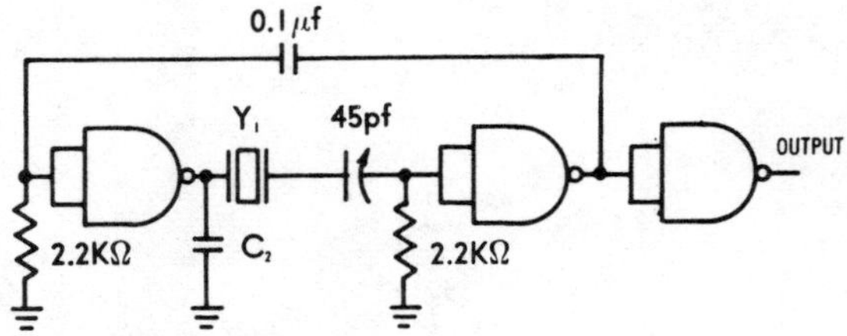

NOTES:

1. $C_2 = \frac{1}{f} \times 10^{-4}$; (f is in Hz) - prevents spurious frequency.
2. IC's are 7400/7404

VOLTAGE CONTROLLED CRYSTAL OSCILLATOR

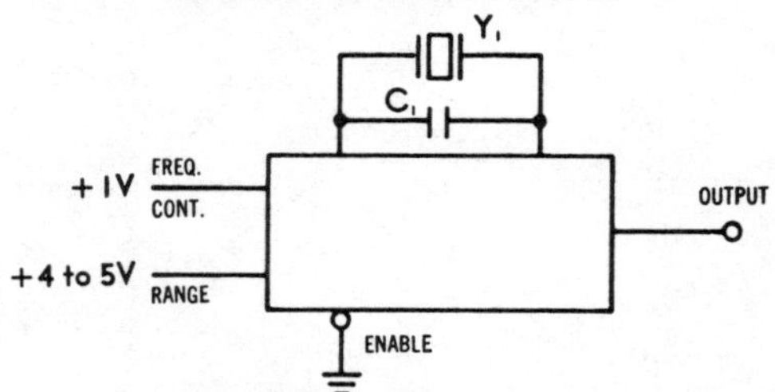

1. For frequencies ≤ 1MHz, C_1 = 5 to 15 pf. For frequencies ≥ 1 MHz, C_1 may be eliminated.
2. IC is SN74S124 for f max. of 60 MHz. IC is SN74LS124 for f max. of 35MHz.

CMOS OSCILLATOR - 1 MHz to 4 MHz

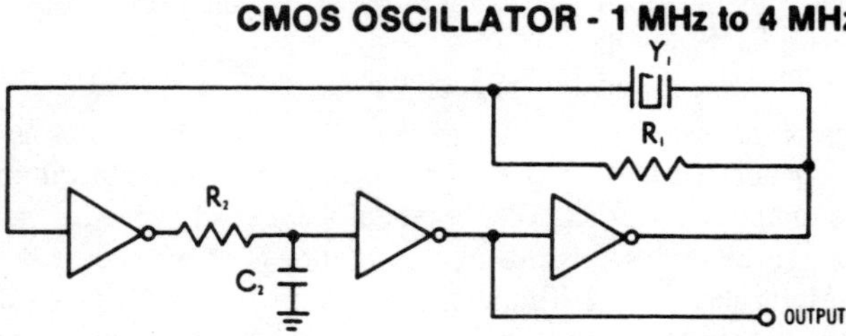

NOTES:

1. $1M\Omega < R_1 < 5M\Omega$
2. Select R_2 and C_2 to prevent spurious frequency.
3. IC's are 74C04 or equivalent.

Fig. 15-26. Digital implementation of reference oscillators.

HIGH FREQUENCY
50 MHz to 150 MHz-Overtone

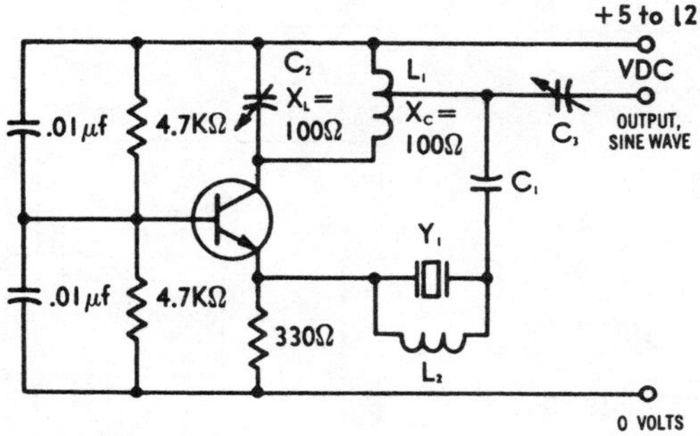

NOTES:

1. Y_1 is "AT" cut, overtone crystal.
2. Tune L_1 and C_2 to operating frequency.
3. L_2 and shunt capacitance (C_0) of crystal (approximately 6 pf) should resonate to the output frequency of the oscillator. (L_2 = .5 uh at 90 MHz). This is necessary in order to tune out effect of C_0 of the crystal.
4. C_3 is varied to match output.

HIGH FREQUENCY
10 MHz to 80 MHz
Fundamental or Overtone

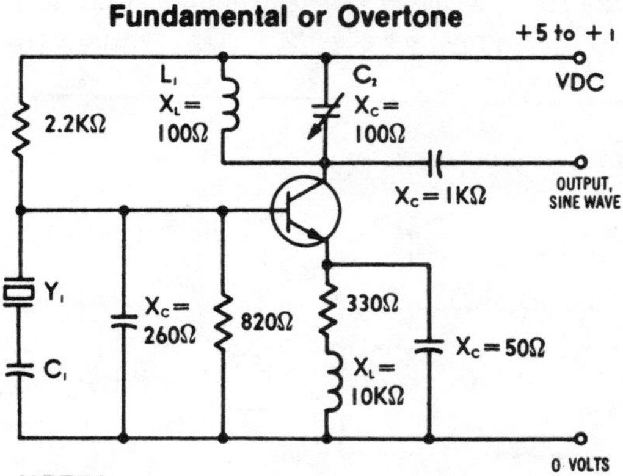

NOTES:

1. Y_1 is "AT" cut, fundamental or overtone crystal. (Finish at 24 pf.)
2. Tune L_1 and C_2 to operating frequency.

Fig. 15-27. Analog implementation of reference oscillators for the frequency range of 10 to 150 MHz.

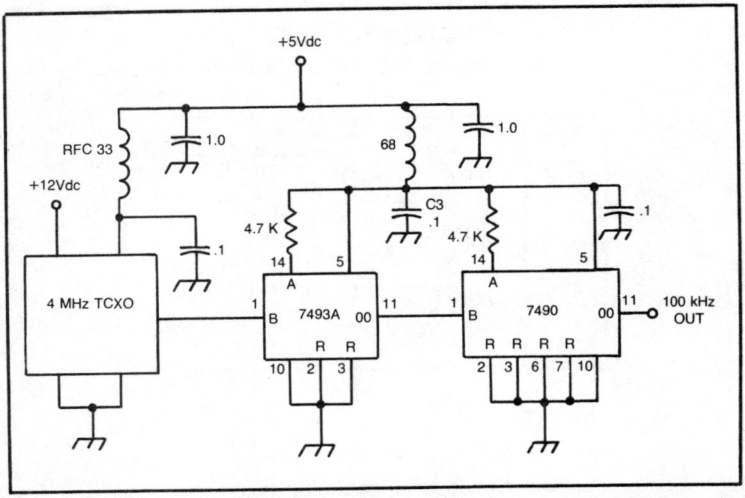

Fig. 15-28. Obtaining 100 kHz reference frequency from a 4 MHz TCXO.

PHASE DETECTOR

An element necessary in the design of a phase-locked loop is the *phase detector*. It is considered the brain of the system because it senses errors and outputs proper commands to the vco in order to lock to the reference frequency. The phase detector, or as it is sometimes called, the *frequency/ phase comparator* uses the reference frequency to compare against the signal from the vco, as modified by the programmable dividers, and outputs a dc signal to the vco proportional with the correction necessary to bring it

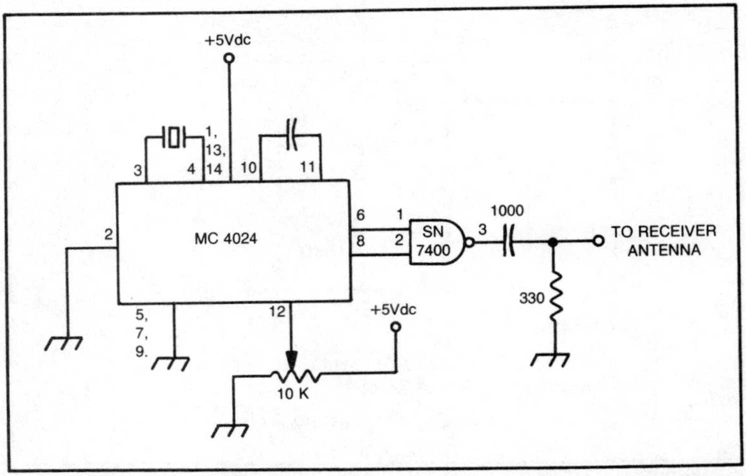

Fig. 15-29. Pulsed 100 kHz signals can be used to calibrate a communications receiver.

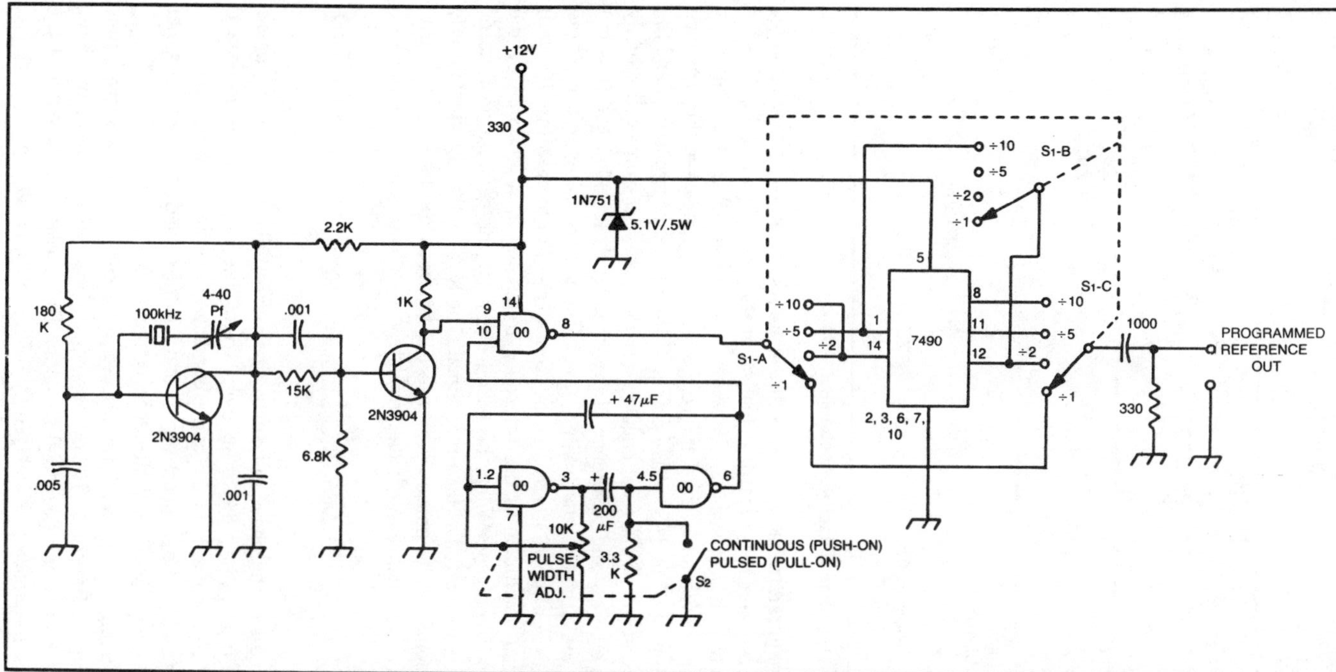

Fig. 15-30. Pulse programmable frequency calibrator has four positions, 10, 20, 50, and 100 kHz selectable from S_1. Pulse frequency can be adjusted and turned off for continuous operation from pulse adjust potentiometer and S_2 which are on the same shaft.

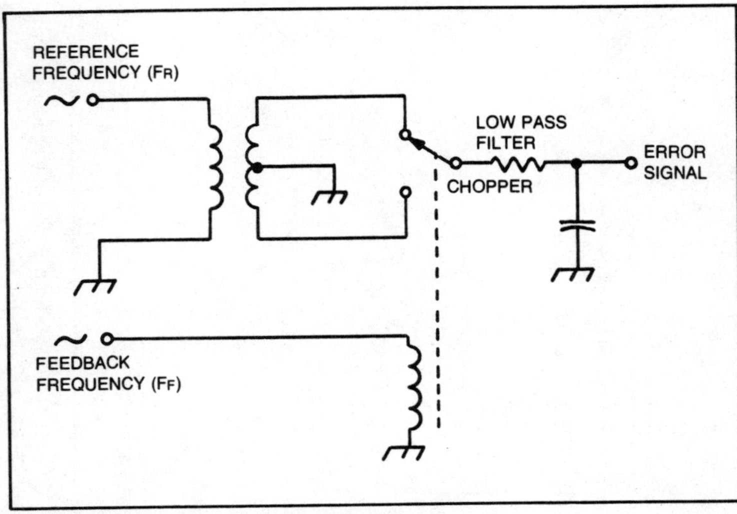

Fig. 15-31. Phase detector mechanical analogy.

to the required frequency. To better understand how the phase detector works, consider the electromechanical analogy illustrated in Fig. 15-31.

As the chopper switches between the two sides of the transformer at the feedback frequency rate, the reference signal appears at the armature alternately at 0° or 180° phase points. If the reference signal (F_R) and the feedback signal (F_F) are not equal in frequency, the armature will actually see the sum and the difference of the two signals (F_R+F_F) and (F_R-F_F) making the circuit act much like a mixer. The lowpass filter will reject the sum (F_R+F_F) allowing only the difference (F_R-F_F) to pass through with increasing amplitude as the reference frequency is approached by the feedback frequency. When the two are exactly the same, a positive dc voltage is generated as the error signal. The level of the signal is directly proportional to the phase difference (ϕ) between the two signals. This value is usually expressed in electrical degrees as shown in Equation 15-1.

$$\Delta\phi = \frac{t_2 - t_1}{T} \times 360° \qquad \text{Eq. 15-1}$$

Where: $\Delta\phi$ is the phase difference in electrical degrees, t_1 is the event time in seconds from the start of one cycle on the reference frequency, t_2 is the corresponding event time expressed in seconds in the feedback signal, and T is the total time required for one cycle (360°) of the reference frequency.

Looking at Fig. 15-32, the reference frequency (F_R) has a cycle time (T) equal to 1.8 seconds. The event point (t_1) was chosen to be on the rising edge of F_R and is equal to 0 seconds. The corresponding edge of the feedback frequency t_2 is delayed by 0.3 seconds, as shown.

$$\text{Then: } \Delta\phi = \frac{0.3 - 0 \times (360^\circ)}{1.8} = \frac{0.3}{1.8} \times (360^\circ) = 60^\circ$$

It can then be said that the reference frequency in this example leads the feedback frequency by 60°, or that the feedback frequency lags the reference frequency by the same amount. In a practical phase-locked loop, the error voltage generated by the phase detector is dependent on this value, as shown by Equation 15-2.

$$V_o = K\phi \times \Delta\phi \qquad \textbf{Eq. 15-2}$$

Where: V_o is the average output voltage of the phase detector in volts, $K\phi$ is the phase detector conversion gain in volts/radian, and $\Delta\phi$ is the phase difference expressed in radians. (1 radian=180°/π= 57.295)

Practical phase detectors may be implemented using digital technology. Shown in Fig. 15-33 is an Exclusive-OR approach using the 5486 (7486) integrated circuit. This type of phase detector is usually used for input wave forms with a 50% symmetrical duty cycle. Flip-flop functions are also common to phase detectors as shown in Figs. 15-34 and 15-35. The 5474 (7474) dual flip-flop is used in these examples.

These methods are sensitive to harmonic multiples of the feedback frequency, and also if the duty cycle changes on either input, erroneous locking can result. To avoid these inconveniences, special phase detectors in IC form have been designed by various manufacturers. Figure 15-36 shows the diagram of the Motorola MC4344 (4044), which consists of two

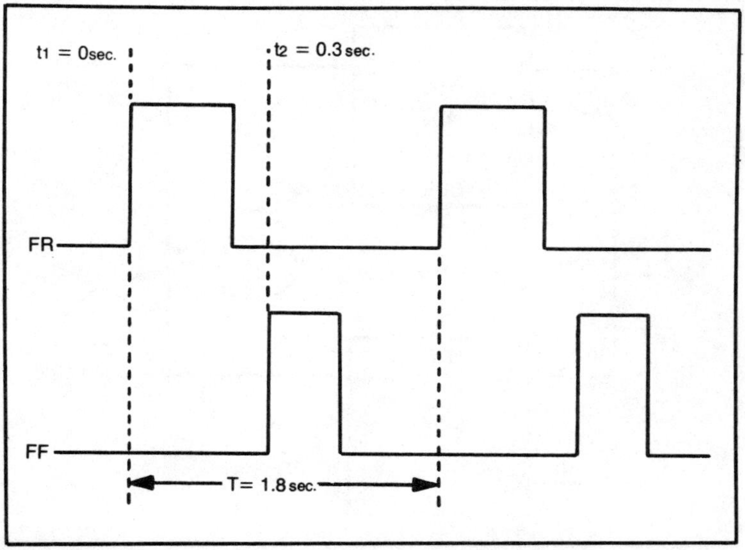

Fig. 15-32. Calculating the phase difference (ϕ) between two signals F_R and F_F.

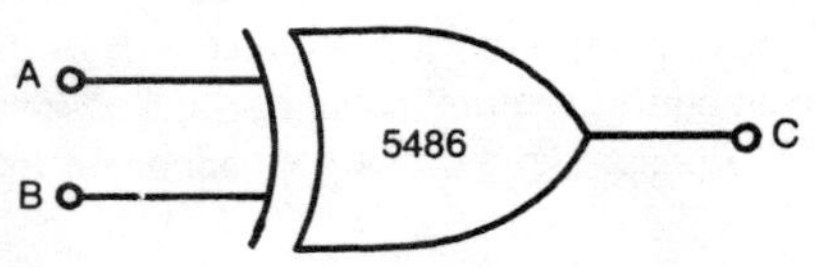

TRUTH TABLE

A	B	C
0	0	0
1	0	1
0	1	1
1	1	0

Fig. 15-33. Exclusive OR phase detector using the 5486.

digital phase detectors, a charge pump, and an amplifier. Phase detector 1 is intended for systems using negative edge triggered circuitry, where zero frequency and phase difference are required at lock. The total transfer characteristic curve of phase detector 1 (with the charge pump connected) is a sawtooth with a wide linear range of 4 radians as shown in Fig. 15-37. Typical conversion gain ($K\phi$) for this detector is 0.12 volts/radian.

Phase detector 2 is connected in parallel with phase detector 1, and can be used if quadrature lock is desired. Its outputs can also be used to drive a lock indicator for phase detector 1. Figure 15-38 shows the circuit diagram for both detectors with corresponding timing diagrams.

AMPLIFIER/LOOP FILTER TRADEOFFS

If phase detector 1 was used, as is usually the case, loop lockup occurs when both outputs U_1 and D_1 are high. Otherwise, a pulsed waveform will

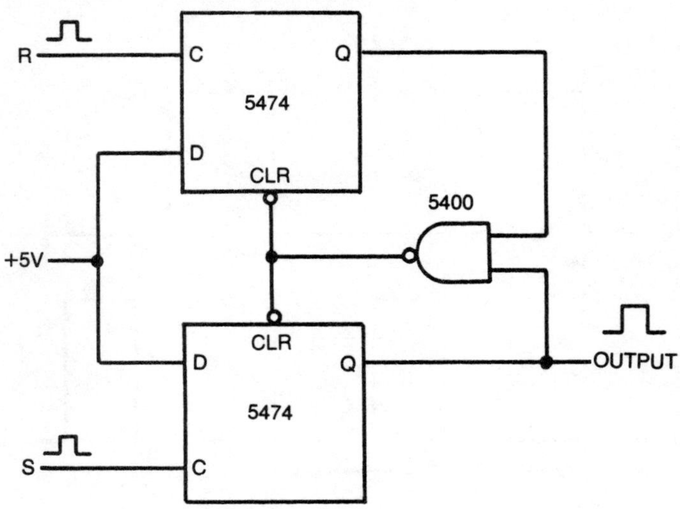

Fig. 15-34. Using flip-flops in a phase detector.

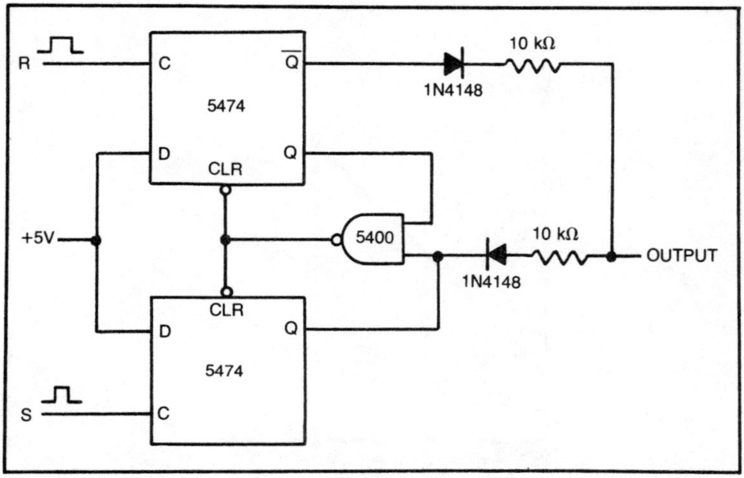

Fig. 15-35. Another approach to flip-flop phase detectors.

appear on either one of the outputs depending on the phase-frequency relationship of R and V.* These outputs are usually fed to the self-contained charge pump which is an integral part of the MC4344, as discussed earlier.

The pulsed waveform present on either PU or PD (pump up or pump down) will make the charge pump generate the right voltage in the proper direction for phase differences between R and V inputs of $\pm 2\pi$ radians as previously shown in Fig. 15-37.

This figure shows that a "no pump" condition is achieved with a phase difference ($\Delta\phi$) of zero π radians corresponding to an output voltage at UF

*(Because of the sequential logic characteristic of phase detector 1, many logic states can be observed at U_1 and D_1. For a better understanding of this phenomenon, the reader is directed to the Motorola Phase-Locked Loop System Data Book which gives a complete flow table.)

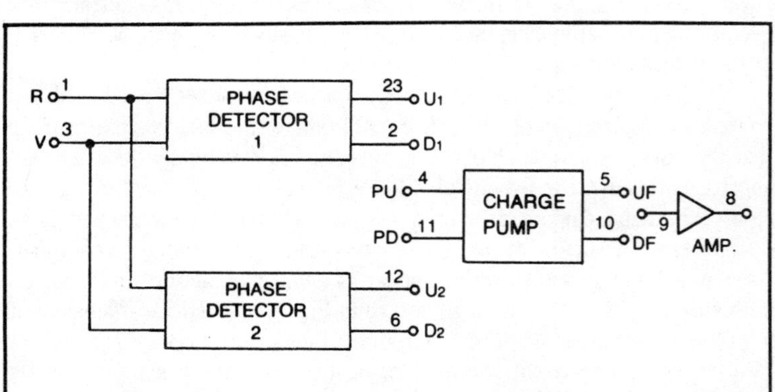

Fig. 15-36. Block diagram of the MC 4344 phase detector.

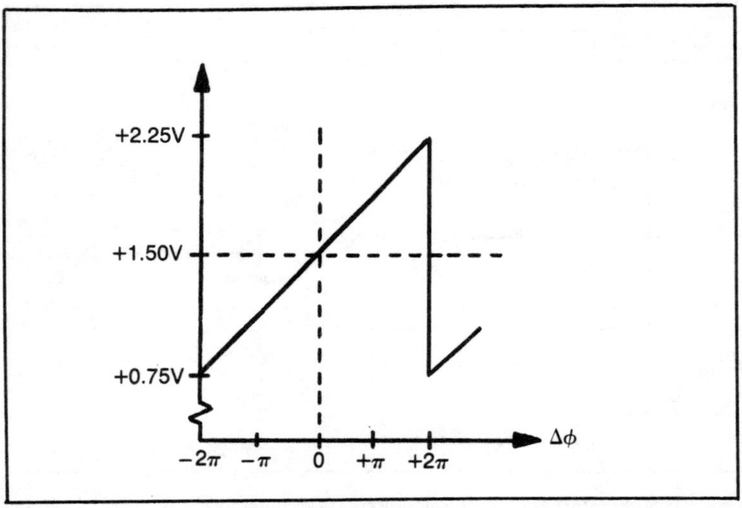

Fig. 15-37. Transfer characteristic for phase detector # 1 of the Motorola MC-4344.

and DF of 1.5 volts, which is the lock-up condition. It is this output that is used to further steer the vco in the right direction. Ideally, this signal should be a fast "sliding", pure dc voltage, in order for an inherently good vco to remain clean from a phase-noise point of view.

In reality, this signal is made of a multitude of ac signals and a dc component. The high frequency components are generated in the digital part of the phase detector. The charge pump can also add noise components to the output. In order for the phase-locked loop to perform properly, a loop filter is introduced between the output of the phase detector and the vco's input. This low-pass arrangement, has the role of suppressing ac components from the output of the phase detector while still maintaining the response requirement for the entire phase-locked loop. This contradictory requirement, usually, calls for a design having a loop response at about 1% of the reference frequency.

For example, if a 10 kHz reference was considered because of fast switching requirements, the 3 dB cutoff point would be 100 Hz, meaning that the noise performance of the vco can be improved only between 0 to approximately 100 Hz, leaving all other ac components to pass through and appear as sideband noise in the output of the vco. One of the worst problems encountered in such a system is the feed-through of the reference frequency and its harmonics which appear as angle modulation sidebands on both sides of the vco's output as shown in Fig. 15-39. These sidebands, if not filtered, are about 40 dB down from the fundamental vco output and are located on either side by the amount of the reference frequency and its harmonics. They can contribute to the total intermodulation distortion and greatly hamper the performance of a communication receiver.

The twin-T notch network, shown in Fig. 15-40 will provide an additional 40 dB attenuation for the reference frequency or its harmonics, if introduced in the output of the phase detector. This filter is usually customary in order to reject both the reference frequency and its second harmonic. Figure 15-41 shows two twin-T networks tuned to the 10 kKz reference frequency and its second harmonic at 20 kHz.

The loop filter, is also responsible for the phase-locked loop's characteristics, such as capture range, bandwidth, lock-up time and transient response. This filter is usually of the low-pass active design as shown in Fig. 15-42. Its cutoff frequency is expressed by Equation 15-2.

$$\omega LPF_{(RAD/s)} = \frac{1}{R_1 C} \qquad \textbf{Eq. 15-2}$$

Where: ωLPF is the cutoff frequency for a lowpass filter.

The loop natural frequency and the damping factor can be found from Equation 15-3 and 15-4.

$$\omega\, n_{(RAD/s} = (K\phi\ Ko\ \omega_{LPF})^{1/2} \qquad \textbf{Eq. 15-3}$$

$$\zeta^* = \left(\frac{R_2 C}{2}\right) \omega n \qquad \textbf{Eq. 15-4}$$

Where: ωn is the loop natural frequency,
$K\phi$ is the phase detector conversion gain in volt/rad.,
K_o is the vco conversion gain in rad/sec/volt
ζ^* is the damping factor.

The total transient response of the phase-locked loop is controlled by this filter. Upon a change in command (÷ N) the loop will search until the vco travels from one frequency F_1 to the other frequency F_2. This shifting is not encountered suddenly, but when frequency F_2 is reached the output of the vco will oscillate around its value until finally settling down as shown in Fig. 15-43.

The time that has passed between the given command and the output of the vco to settle within some certain limits (e.g. within 10% of F_2) determines the switching speed of the phase-locked loop and is usually referred to as the *lock-up time*.

The amount of damping ζ determines how fast this process is completed. In designing a fast-switching synthesizer for a communication receiver, Fig.'15-44 should be used for determining the value ωn for a given damping factor ζ in order to apply it in the low-pass equations presented earlier. A rule of thumb in using this graph would be to design for a certain amount of overshoot within a given settling time.

---

*Greek letter zeta.

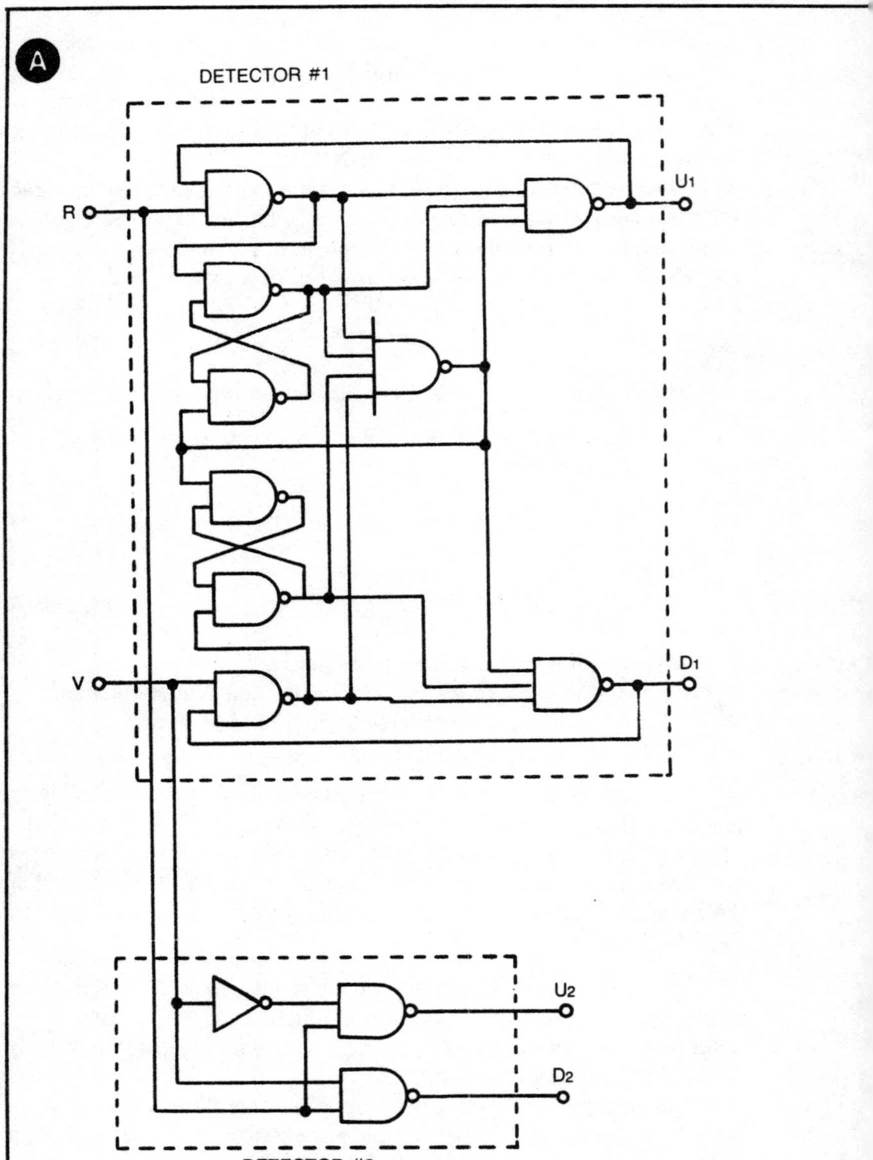

Fig. 15-38. Circuit and timing diagrams for detector #1 and #2 of the Motorola MC-4344 (4044). Detector 1 and 2 (A). R signal leads V signal (B). R signal lags V signal (C). Output at detector 2 under lock condition (R signal in phase with V signal), and out of lock condition (R signal leads V signal) (D).

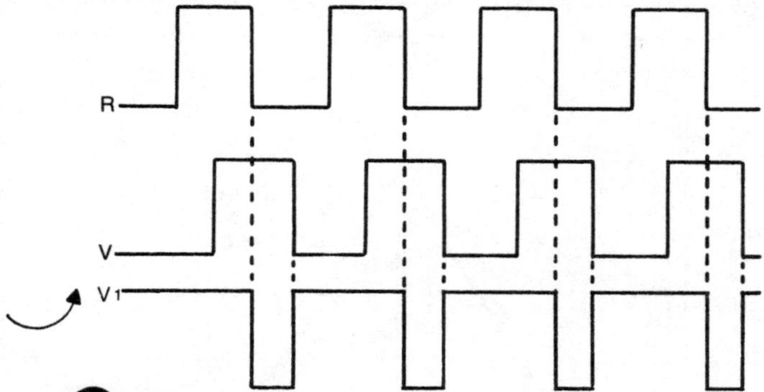

B R SIGNAL LEADS V SIGNAL

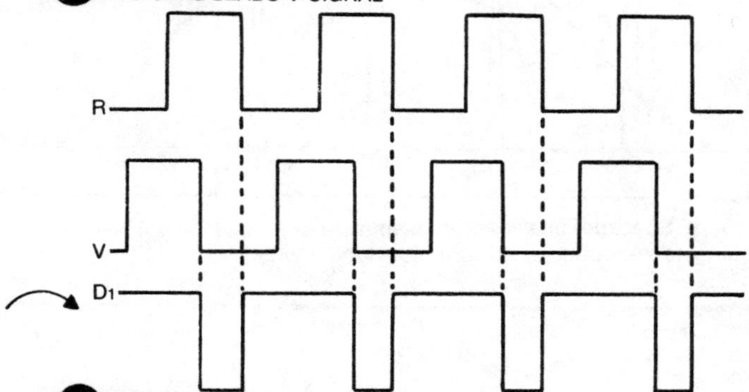

C R SIGNAL LAGS V SIGNAL

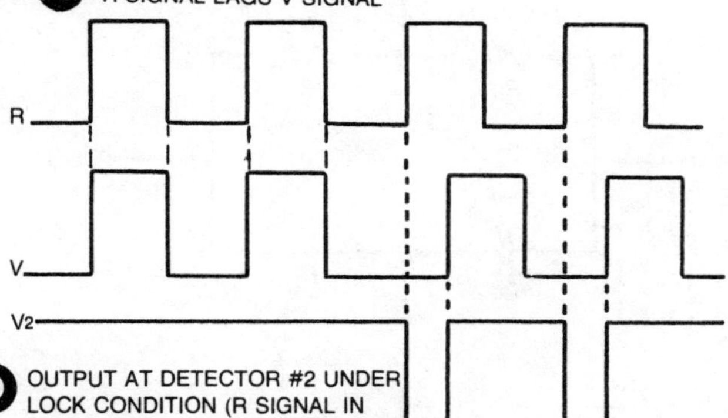

D OUTPUT AT DETECTOR #2 UNDER LOCK CONDITION (R SIGNAL IN PHASE WITH V SIGNAL), AND OUT OF LOCK CONDITION (R SIGNAL LEADS V SIGNAL).

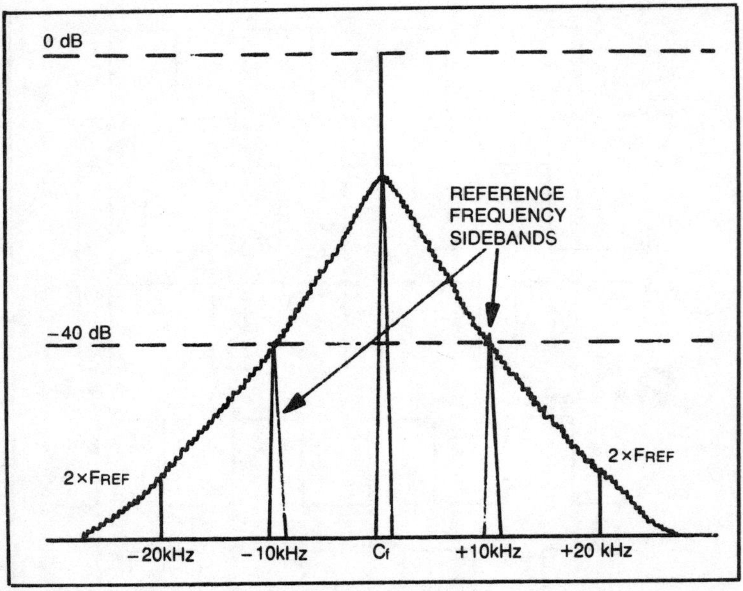

Fig. 15-39. Spectrum analyzer representation of a PLL's output with reference frequency feed-through at 40 dB below fundamental.

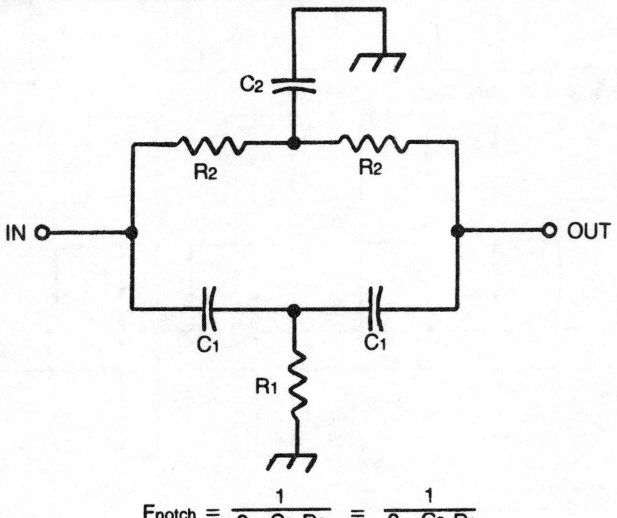

$$F_{notch} = \frac{1}{2\pi\, C_1\, R_2} = \frac{1}{2\pi\, C_2\, R_1}$$

$R_2 = 2R_1$; $C_2 = 2C_1$

Fig. 15-40. The "T" notch network can be introduced in the loop of a PLL to reduce reference feed-through at the output of the vco by 40 dB.

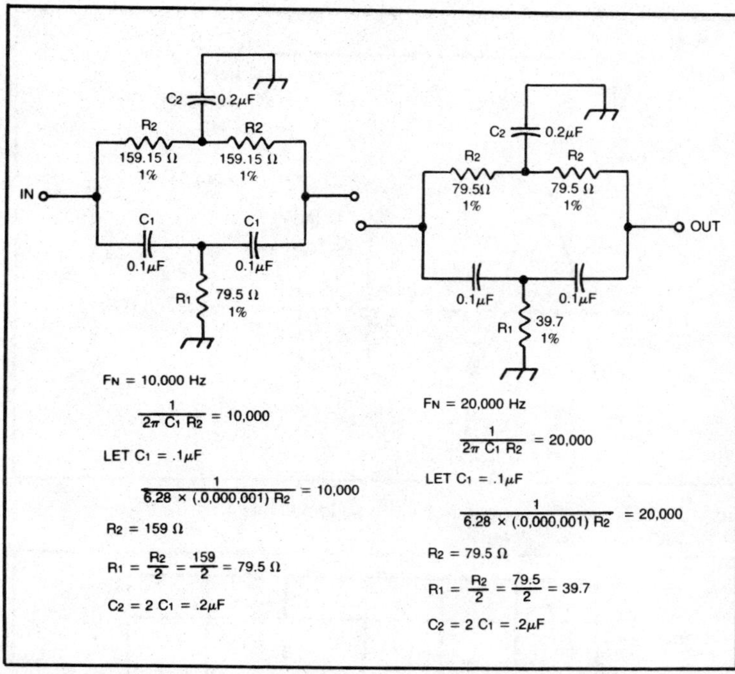

Fig. 15-41. Twin "T" networks used with a 10 kHz and 20 kHz reference frequency. 40 dB of attenuation is obtained with this arrangement.

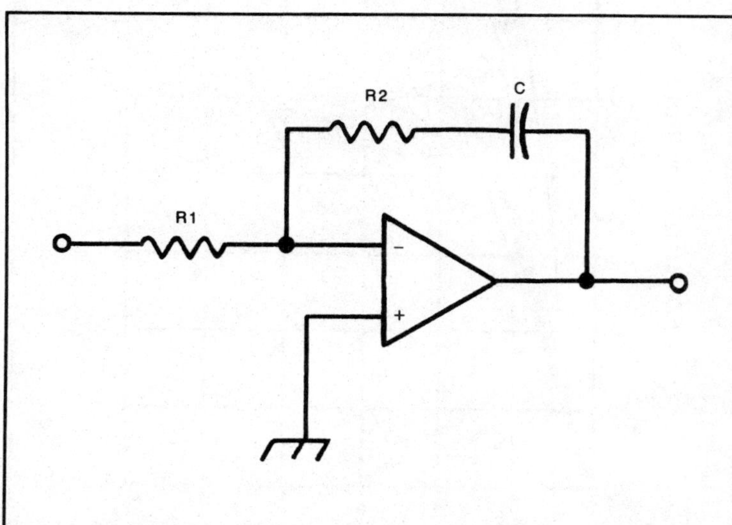

Fig. 15-42. A typical inverting low-pass filter using an operational amplifier is used for the loop filter in a PLL.

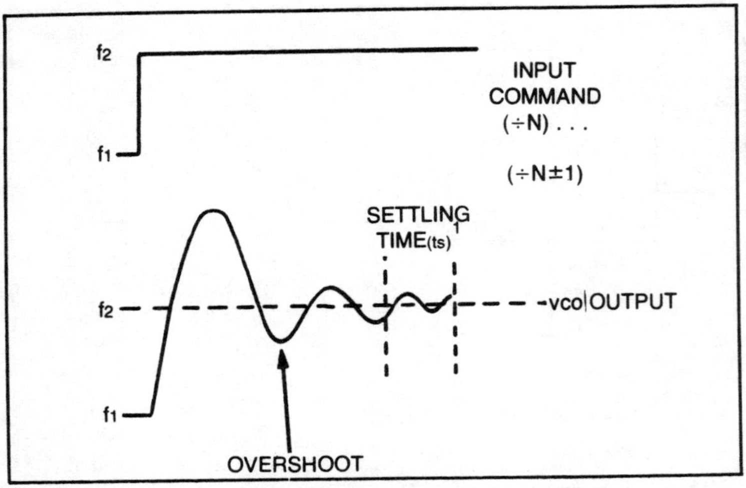

Fig. 15-43. Transient response of a phase-locked loop.

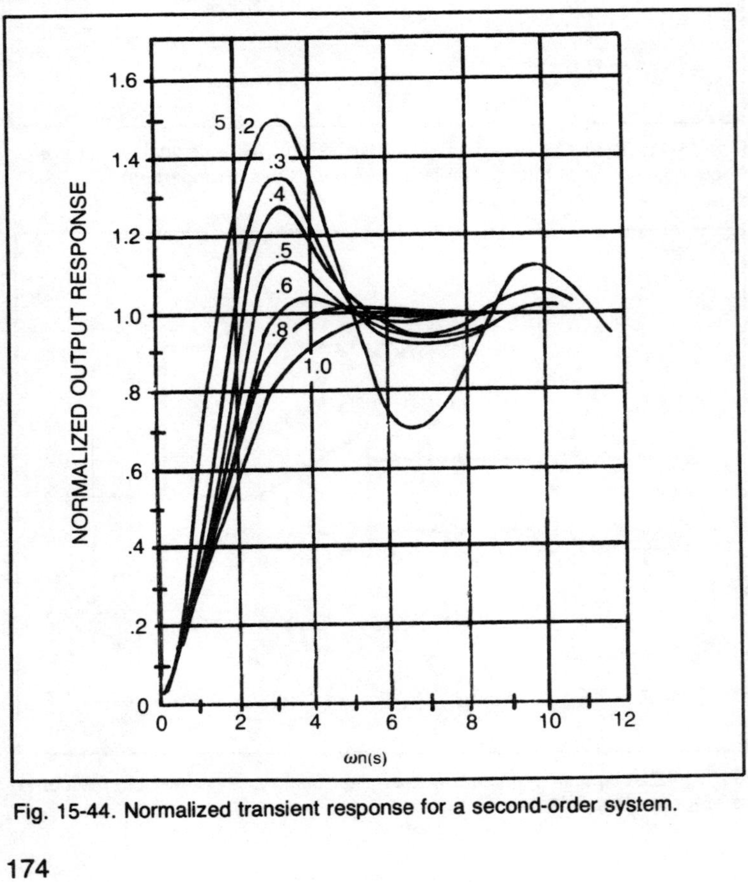

Fig. 15-44. Normalized transient response for a second-order system.

Another factor controlled by the loop filter is lock-in range (ωc). With the filter shown in Fig. 15-42, Equation 15-5 can be used to find this value:

$$\omega c = \sim \omega 1 \left(\frac{R_2}{R_1} \right) \qquad \textbf{Eq. 15-5}$$

Where: ωc is the lock-in range in radians/sec, and

ω_1 is the hold-in range in radians/sec*

From a practical standpoint, many designers prefer not to use the charge pump contained within the MC4344 because of its noise contribution to the vco's output. While still using phase detector 1, they build their own. Figure 15-45 shows how this can be implemented.

Other designers ** prefer the RCA CD4046. This CMOS device operates as a positive-edge-triggered digital memory network with a field-effect-transistor (FET) output stage. The phase noise contribution from this phase detector is claimed to be somewhat superior to that of the MC4344 equipped with the charge pump. Because of its low-current characteristics the CD4046 is an attractive choice for certain commercial

---

*Defined as to how far the input frequency to the phase-detector can deviate from a given vco frequency, ωo. This number is numerically one half the lock range.

**Modern design of Frequency Synthesizers, Ulrich L. Rohde, Ham Radio, July 1976.

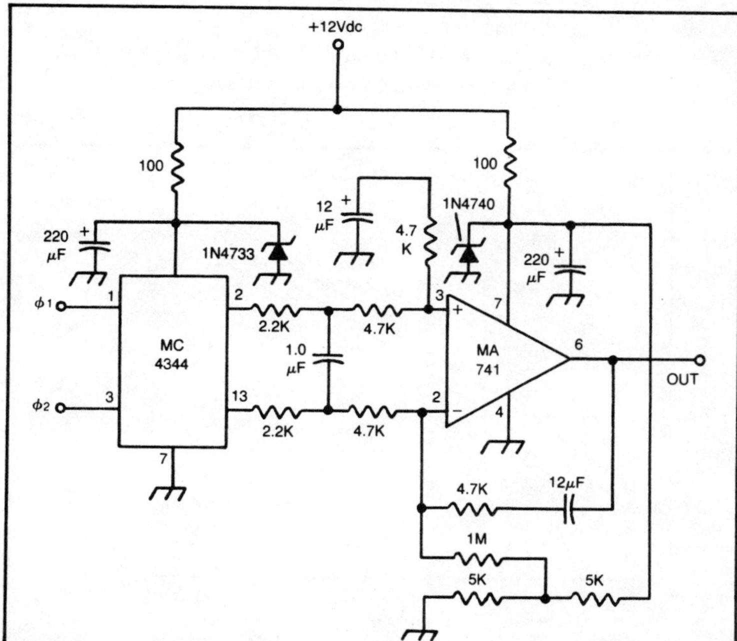

Fig. 15-45. The Motorola MC-4344 is used in this example without the charge pump in order to keep output free of transients.

applications. In most military applications, the MC4344 remains the workhorse of phase comparators.

THE VOLTAGE CONTROLLED OSCILLATOR (VCO)

The next circuit required in a phase-locked loop is the *voltage controlled oscillator* (vco). This circuit takes the voltage commands from the loop filter and outputs the required sinewave frequency necessary for the receiver's conversion. This signal is also used to feed the programmable counters which in turn, close the loop at the second input to the phase detector as we previously discussed.

Ideally, the voltage controlled oscillator supplies an output frequency directly proportional to its input dc voltage control over the required temperature range. Figure 15-46 shows test results of the linearity over the temperature range of – 50° C to +50° C for a well-designed vco. This performance was measured with a free-running vco whose input voltage was supplied by a programmable power supply that can generate precise dc voltages. It can be seen from this discussion that the vco can be referred to as a *voltage-to-frequency converter* (Fig. 15-47).

Its transfer function characteristics can then be expressed by Equation 15-6.

$$\omega_o = K_o V_f \qquad \textbf{Eq. 15-6}$$

Where: ω_o is the vco output in rad/sec.
V_f is the vco control voltage provided by the loop filter
K_o is the vco conversion gain in rad/sec/volt.

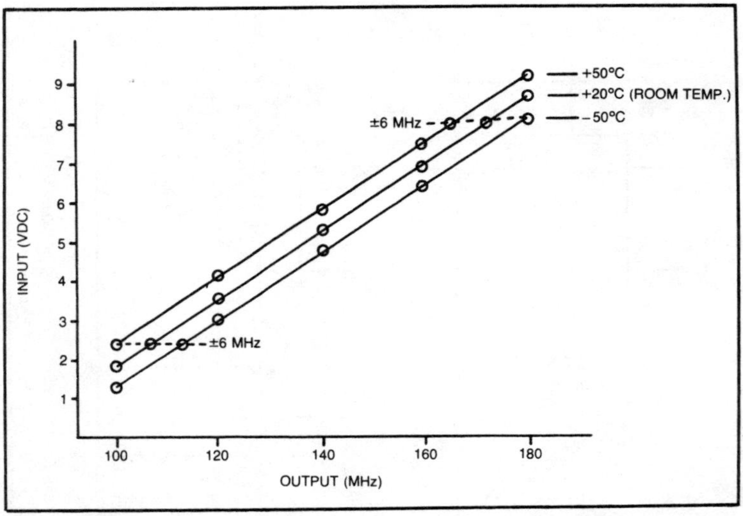

Fig. 15-46. Linearity of a well-designed vco over the temperature range of – 50°C to +50°C.

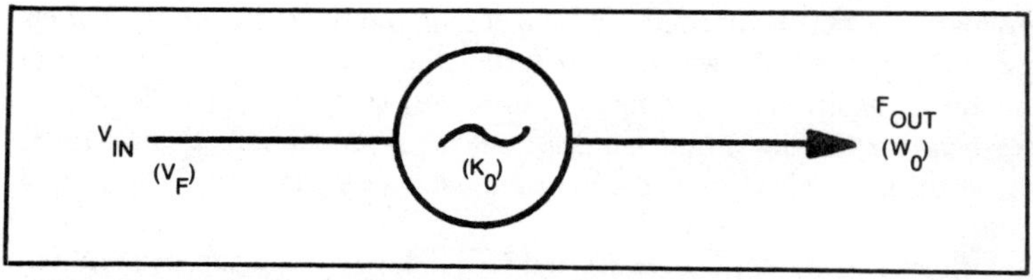

Fig. 15-47. Block diagram of a basic voltage controlled oscillator.

The output frequency expressed in Hz is given by Equation 15-7.

$$f_o = \frac{\omega_o}{2\pi} \qquad \textbf{Eq. 15-7}$$

Where: f_o is vco frequency in Hz
ω_o is vco frequency in rad/sec
$\pi = 3.14$

From a practical standpoint, the vco much resembles Vfo technologies, with the tuning being achieved through the use of voltage-variable capacitors (varactors) in the oscillator tank circuit. Low-noise vco's can be designed today by using field-effect transistors, if frequencies do not exceed 500 MHz. For higher frequencies, bipolar transistors and GaAs FETs are being used.

Often, the frequency range requirement over which a vco has to perform is extremely wide, with the result that it would take a long time for the loop to lock. This search would also be accompanied by extensive phase noise even after the loop has settled. This noise would be transferred into the receiver's circuits, accounting for serious intermodulation distortion problems and creating the so-called blocking phenomenon.

Some designers use several limited-range vco's which are automatically switched in by complex digital circuitry. A block diagram example of this approach is shown in Fig. 15-48. A better method of achieving low-noise operation over a wide frequency range is shown in Fig. 15-49.

In this example, the correct vco is selected in rough steps of 1 MHz by using a digital-to-analog converter, which brings the loop into the required frequency excursion, speeding the final lock-up process, and improving the noise performance of the system. This method has been used extensively by manufacturers and offers good results from a noise standpoint.

A less expensive method of achieving coarse steering and low-phase noise performance in a vco is shown in Fig. 15-50. In this example, limited frequency ranges are switched into the tank circuit of the vco by placing the selected capacitance across it. The switching is accomplished under micro-computer control with the use of PIN diodes used as current-controlled switches. The Hewlett/Packard HP 3081 diode is used in this example.

Similar results can be obtained by switching inductors instead of capacitors. Either technique can be implemented by building a printed-circuit board which will contain a bank of tuned circuits together with their switching mechanisms which could include the digital decoder. Because of the low Q requirements for the inductors, they could be etched directly on the board providing an economical means of production. Figure 15-51 shows the principles behind creating a film inductor bank in a phase-locked loop.

Equations 15-8, 15-9, 15-10, and 15-11 show how to approximate the inductance in such an approach.*

$$S = D^2 \qquad \textbf{Eq. 15-8}$$

$$N = \left(\frac{D}{2}\right)(q + p)^{-1} = \left(\frac{D}{2p}\right)(1 + r^{-1})^{-1} \qquad \textbf{Eq. 15-9}$$

$$1 = 4\left(\frac{D}{2}\right) N = \left(\frac{D^2}{p}\right)(1 + r^{-1})^{-1} \qquad \textbf{Eq. 15-10}$$

$$L = 85 \times 10^{-10}\, D N^{\frac{5}{3}} = 27 \times 10^{-10} \left(\frac{D^{\frac{8}{3}}}{p^{\frac{5}{3}}}\right)(1 + r^{-1})^{\frac{5}{3}} \qquad \textbf{Eq. 15-11}$$

*Designing Inductors for Thin-Film Application, H.G. Dill, Electronic Design, Vol. 12, No. 4, pp., 52-59; 1964.

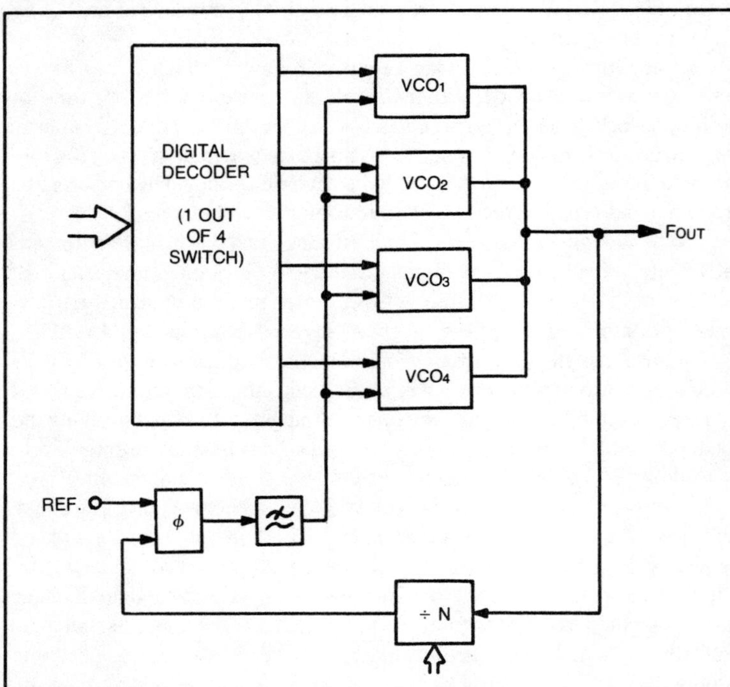

Fig. 15-48. This method of switching vco's in a PLL allows superior noise performance, but is expensive.

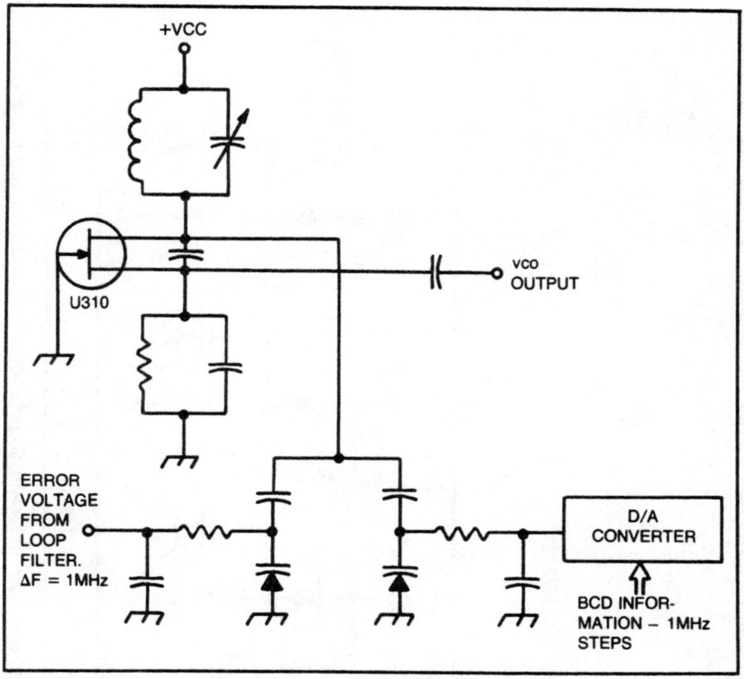

Fig. 15-49. Method of steering a vco by using a D/A converter.

Where: D is dimension in cm, S is surface area of coil (D_1=0), N is number of turns, L is inductance in henries, 1 is length of spiral in cm., p is spiral width in cm., and q is distance between the outer edges of two adjacent spirals.

$$\frac{p}{q} = r$$

Shown in Fig. 15-52 is the schematic diagram for a vco using film inductors in order to accomplish coarse steering. Such a design exhibits a typical power spectral density of 120 dBm/Hz at 1 kHz from the main carrier. This noise can vary however, and is highly dependent on the layout of the circuit. This is one of the reasons why the actual implementation of such a vco is usually considered a propietary matter by the manufacturers. Creating film capacitors on a printed circuit board has been also implemented with similar techniques, although not as often as the film inductor.

Practical vco's take many forms. Figure 15-53 shows a basic FET approach using two 2N4416 transistors followed by a 2N5179 buffer amplifier. This vco will tune the frequency range of 110 to 165 MHz. The layout of the tank circuit can greatly influence this range. The harmonic

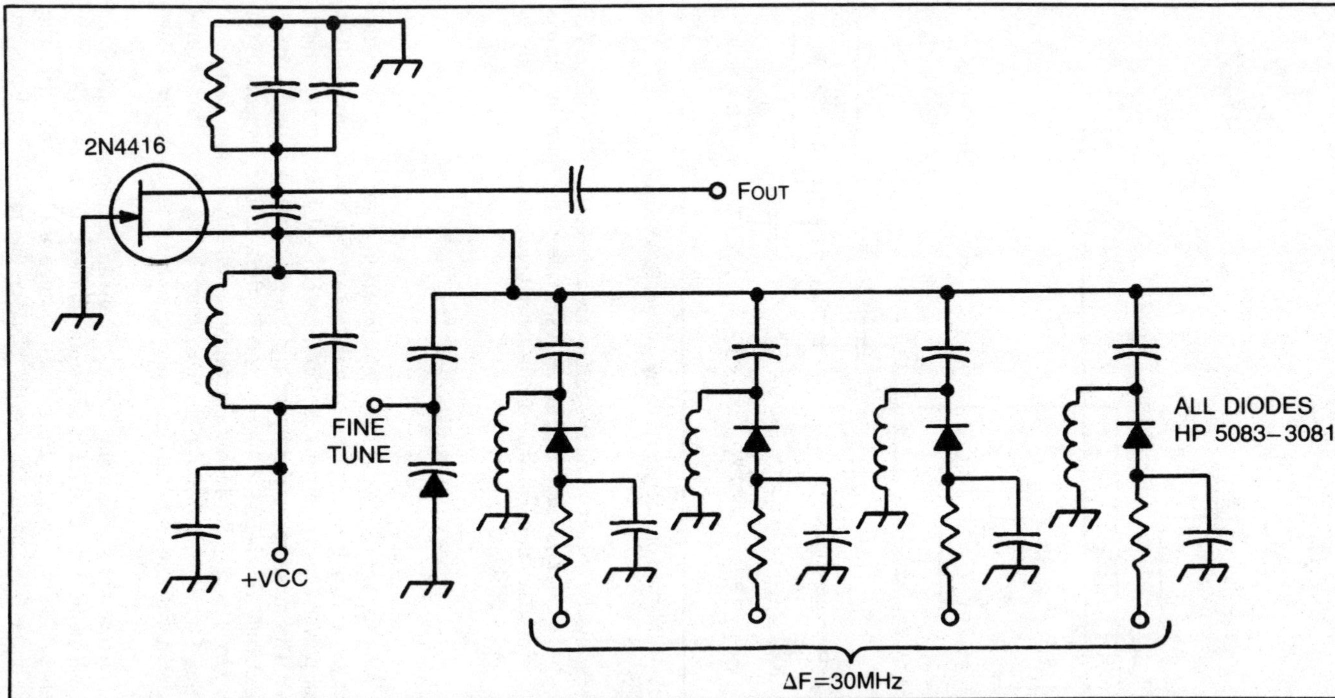

Fig. 15-50. A very low-noise oscillator, over a wide range can be obtained with automatic switching of capacitors across the tank circuit of a vco.

content of this vco is shown in Fig. 15-54. No output filtering was used to obtain this information. A simple lowpass filter can reduce these harmonics by at least 20 dB making the vco's output extremely clean from a harmonic standpoint. See Figs. 15-55 and 15-56.

Integrated circuits intended for use as vco's are available from manufacturers. Figure 15-57 shows a circuit implementation for the Motorola MC1648. In this example, a very wide-range (3:1) varactor is used to achieve a total coverage of 30 MHz. A buffer amplifier is also shown. This amplifier is required to condition the ECL level of the MC1648 to the TTL level necessary to feed the programmable dividers which close the loop in a PLL. Although the simplicity of such a vco is quite obvious, IC vco's are seldom used in communication receivers.

THE PROGRAMMABLE DIVIDER AND THE DUAL MODULUS APPROACH

The remaining circuit function requirement for a phase-locked loop synthesizer is the programmable divider. Its function is to provide the right count to the phase detector so that the output of the vco will be the correct frequency according to the PLL equation (15-1) presented earlier. It can be seen from this equation that the step resolution of the PLL depends on the division number N. In a phase-locked loop synthesizer, this arithmetic function is performed with the help of digitally programmable dividers. These devices accept the vco's output (once it is transformed to a square wave) and divide it by a predetermined value (N) which can be controlled by the operator at the programming inputs. Figure 15-58 shows the schematic diagram of the 54192 (74192) used in this function.

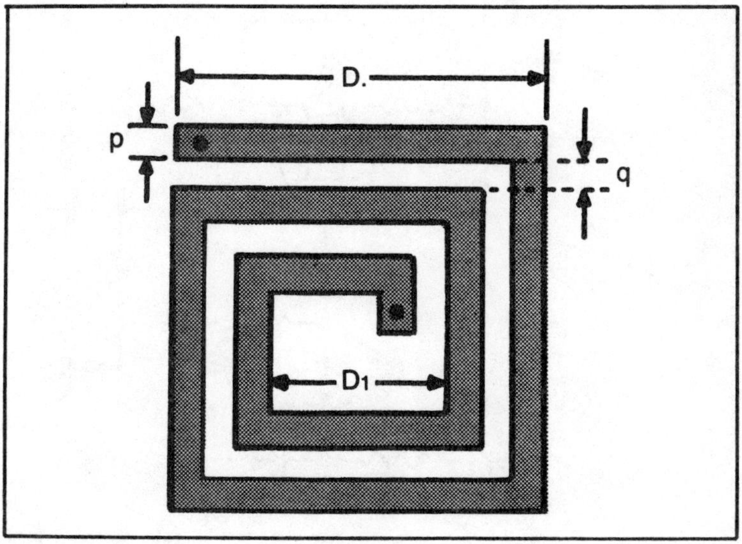

Fig. 15-51. Creating a film inductor for the steering mechanism.

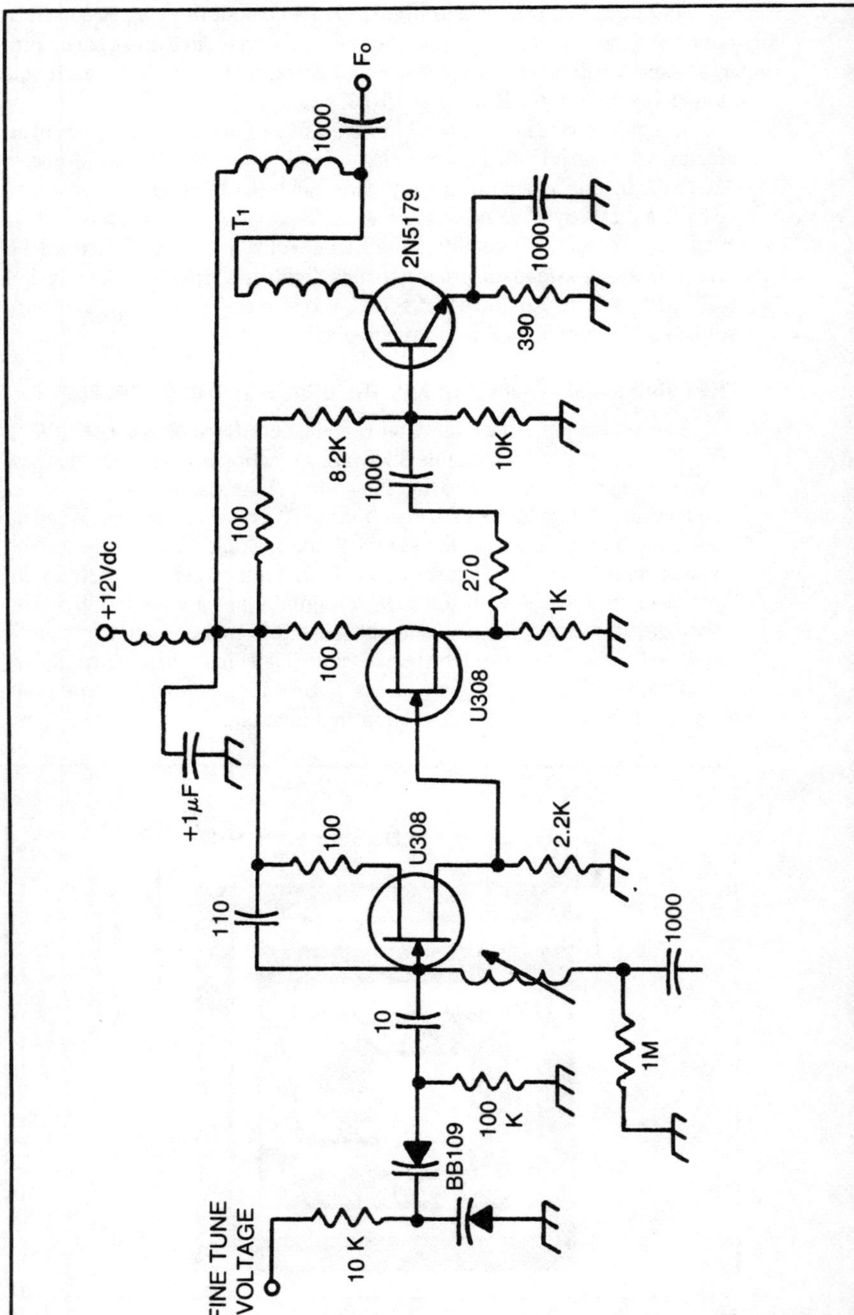
Fo
1000
T1
2N5179
1000
390
+12Vdc
100
8.2K
1000
10K
270
100
1K
U308
+1μF
100
U308
2.2K
110
1000
10
1M
100
K
BB109
FINE TUNE
VOLTAGE
10 K

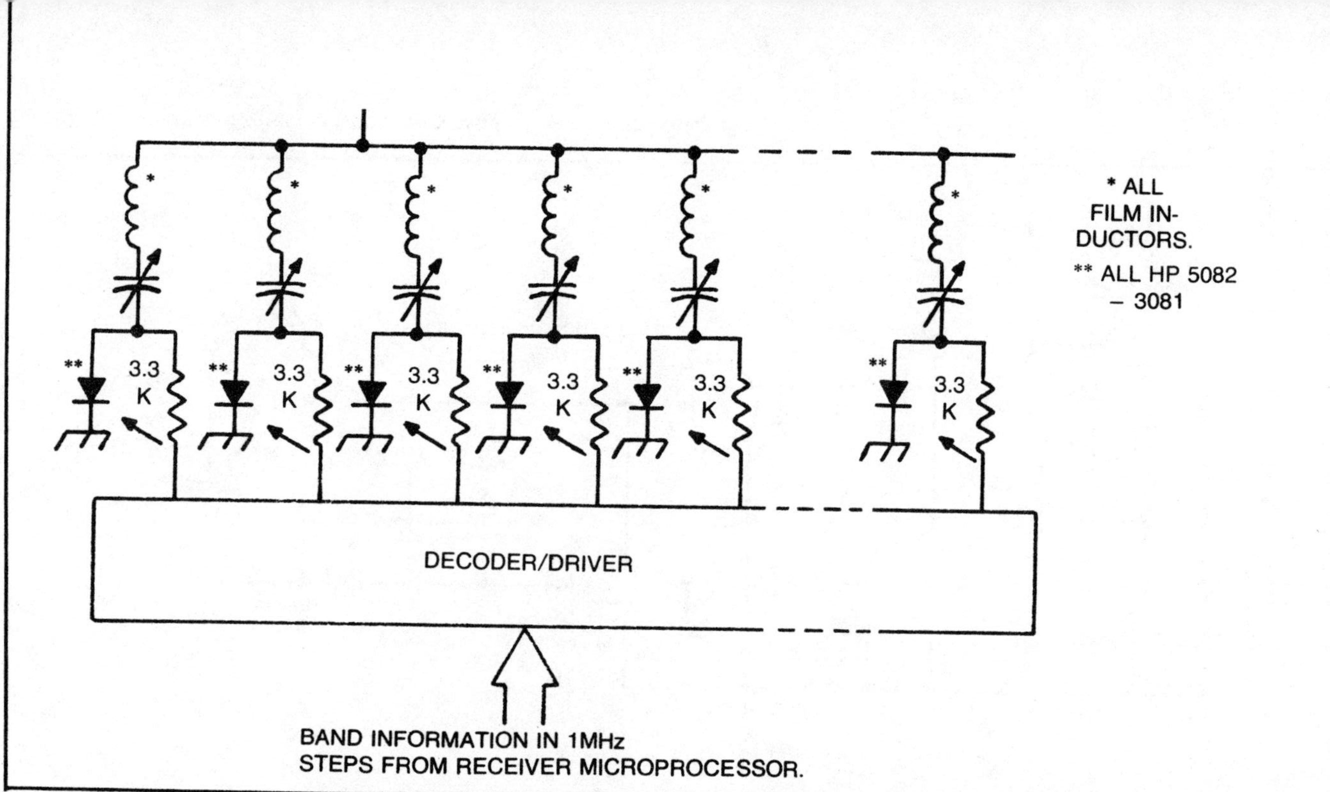

Fig. 15-52. Steering mechanism and vco for a high performance communications receiver. Switched film inductors are used with this method.

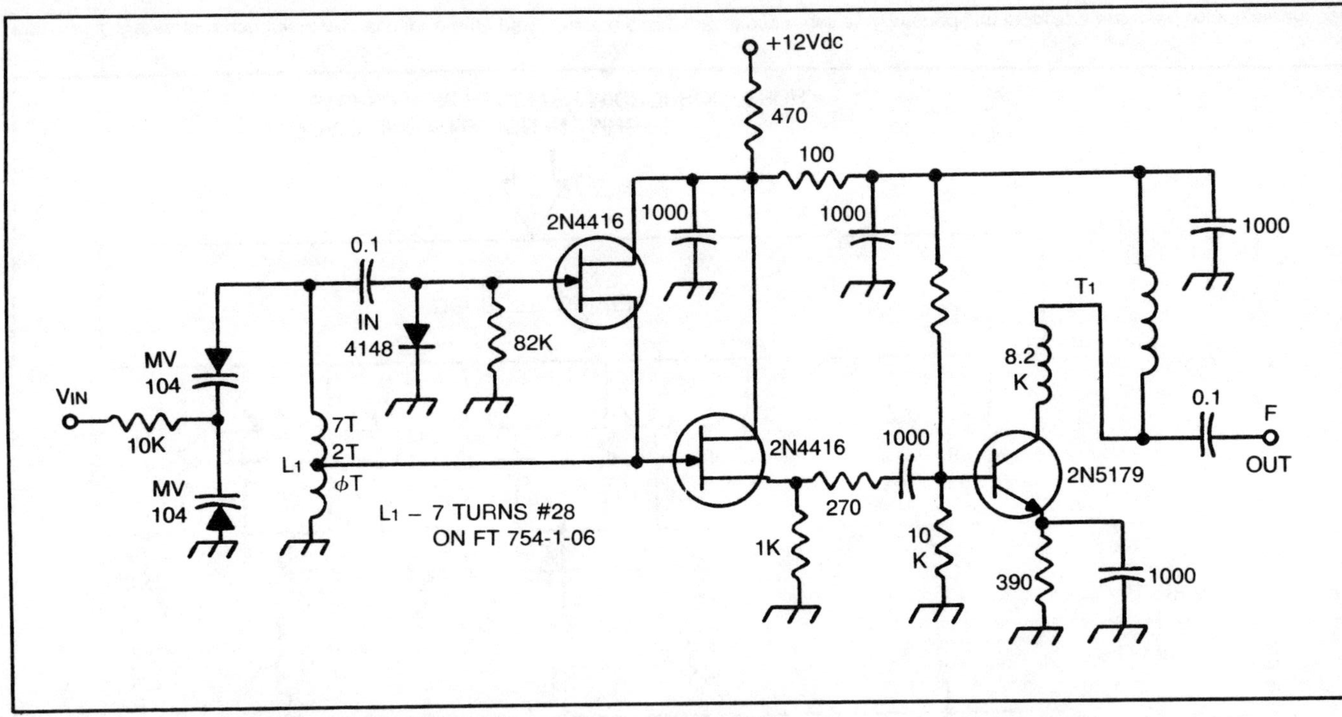

Fig. 15-53. Schematic diagram of a vco using two FET's and a transistor. Frequency range is 110 to 165 MHz.

The modulus N is implemented by making true (positive logic level equals 1) the binarily weighted inputs A, B, C, and D as shown by the truth table. For numbers larger than 10, more devices can be cascaded as shown in Fig. 15-59. This arrangement shows three 54192's cascaded and programmed to divide by N=239. The programming is accomplished by a thumb-wheel switch which is a mechanical memory device that can directly output a binary number on four lines for each decimal number selected by the operator (N_{max} = 999). Figure 15-60 shows such a switch as used in a communications receiver.

Other dividers can also be used to create divide-by-N functions. The Motorola MC4016 is such an example. This IC was expressly created for phase-locked loop circuit applications. Figure 15-61 shows the implementation of a programmable divider with this IC. Although somewhat simpler to use than the 54192, the MC4016 will accept a maximum frequency of only 8 MHz at its input. If the vco's frequency exceeds the input frequency of the divider chain, a prescaler has to be used. The prescaler would also have to be a programmable counter which will accept the vco's output, sometimes in uhf frequencies. Because of technological drawbacks, programmable devices have not yet been fabricated for uhf operation. The solution to this problem is to combine the properties of a high-speed fixed counter with the functional properties of the slower programmable counter. This device is referred to as the *dual-modulus prescaler* and the technique used in implementing it was named the *pulse swallowing technique**. A PLL Synthesizer using this approach will therefore be referred to as a *dual-modulus*

*Pulse Swallowing, J. Nichols and C. Shinn, EDN, October 1, 1970.

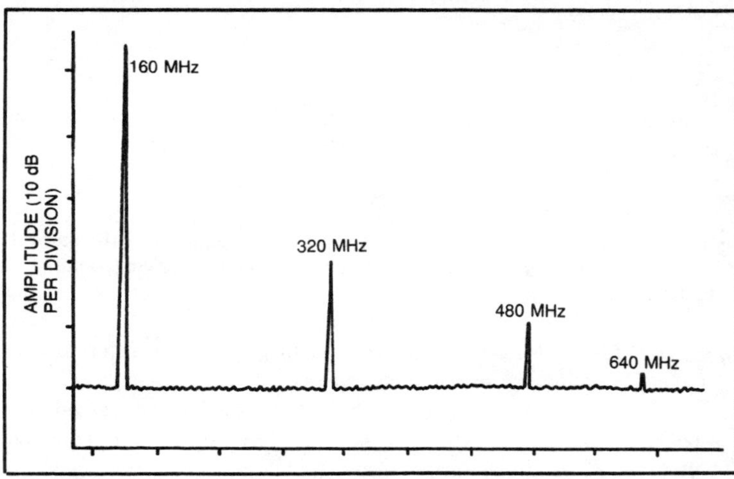

Fig. 15-54. Spectral purity of the vco shown at A, indicate that the second harmonic is 38 dB below the fundamental, the third harmonic is 48 dB below, and the fourth harmonic is more than 50 dB below the fundamental. No filtering was used at the output of the vco in making the test measurements.

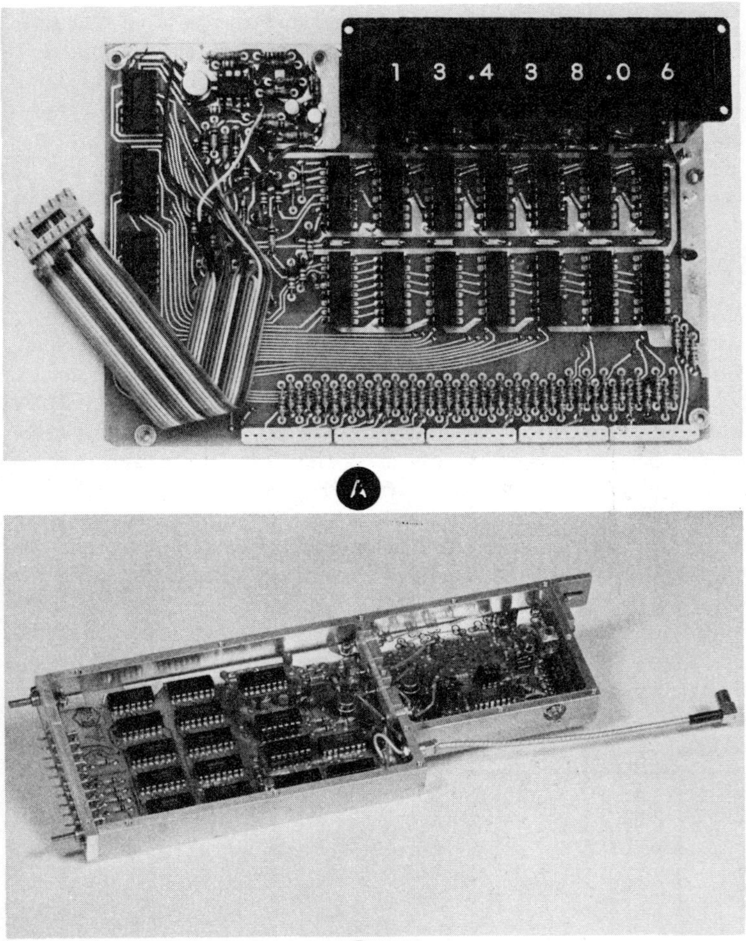

Fig. 15-55. Frequency control module (A) and typical synthesizer (B) (shown with cover removed) of the Cubic HF-1030 communications receiver (courtesy of Cubic Communications).

prescaler–PLL synthesizer. To understand this technique, refer to Fig. 15-62 and the following explanation.

Assume that an output frequency (F_o) of 150.020 MHz is expected with a reference frequency of 5 kHz. From the phase-locked loop equation (15-1): $F_o = F_R \times N$,

$$N = \frac{F_o}{F_R} = \frac{150.020 \times 10^6 \text{ Hz}}{5 \times 10^3 \text{ Hz}} = 30{,}004$$

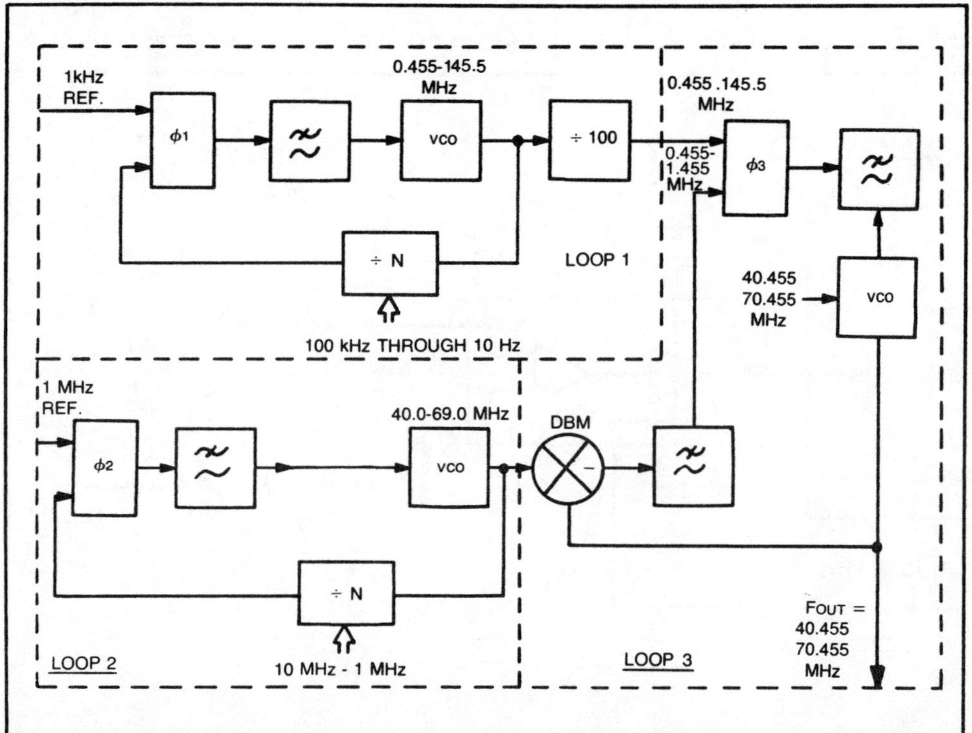

Fig. 15-56. Block diagram for the frequency synthesizer of the Cubic Communications HF 1030 Receiver. Variable references are provided by loops 1 and 2 to loop 3 in this approach. This is another method of producing high resolution in a synthesizer.

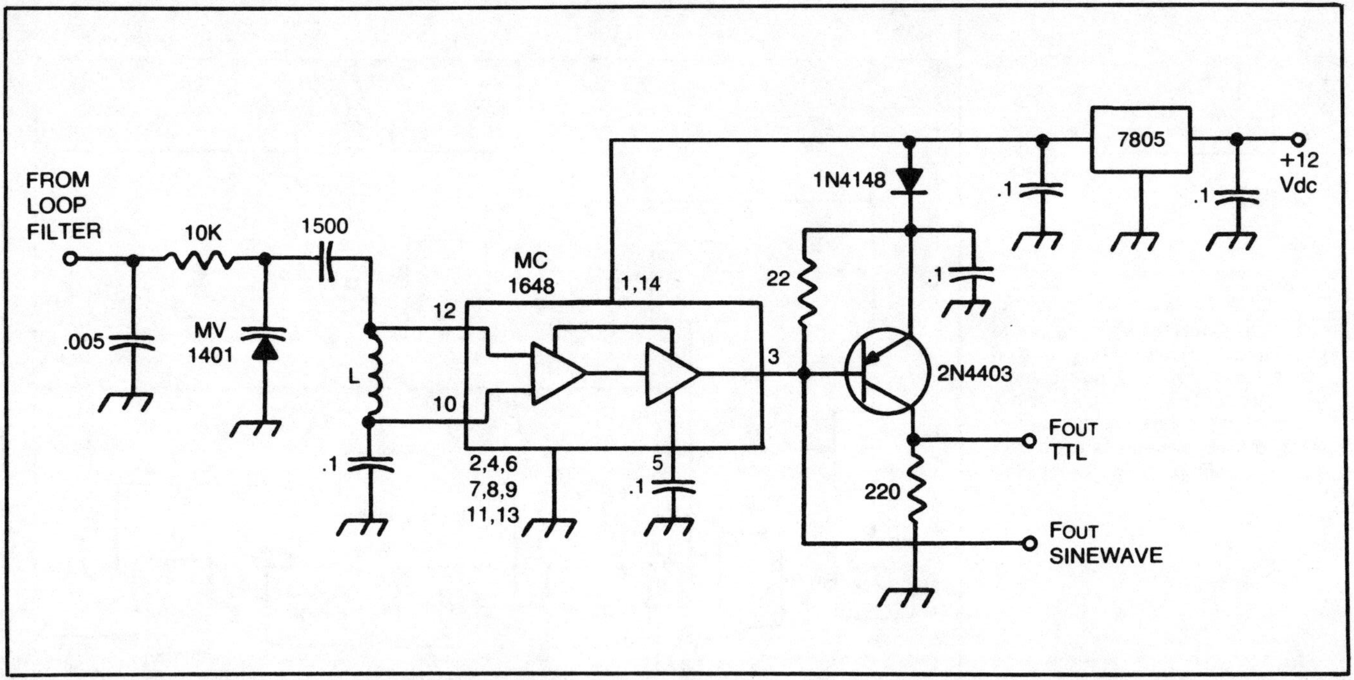

Fig. 15-57. Implementing a vco with the Motorola MC-1648 IC. The phase-noise performance for this type of vco, is usually inferior to that of discrete FET types.

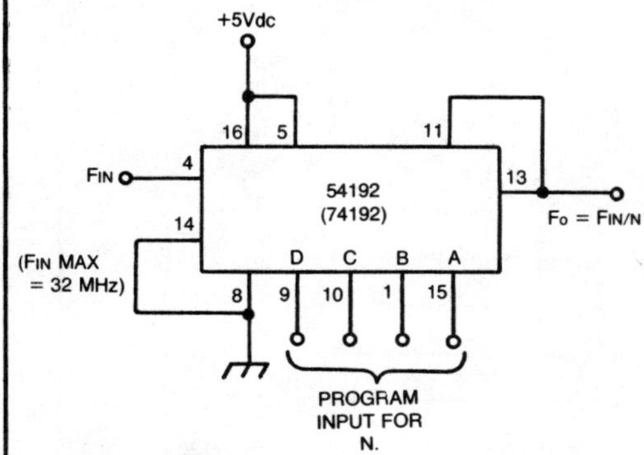

TRUTH TABLE

N	A	B	C	D
	1	2	4	8
1	1	0	0	0
2	0	1	0	0
3	1	1	0	0
4	0	0	1	0
5	1	0	1	0
6	0	1	1	0
7	1	1	1	0
8	0	0	0	1
9	1	0	0	1
10	0	1	0	1

Fig. 15-58. Any division between 1 and 10 can be achieved with the programmable divider.

Assume the dual modulus values P and P+1 are the numbers 20 and 20 + 1 = 21* if N is divided by the P modulus, the resulting quotient is the n program value and the A program value is the remainder.

$$\frac{N}{P} = \frac{30{,}004}{20} = 1500 \text{ and a remainder of } 4$$

$$n = 1500 \qquad \text{and } A = 4$$

The key to the operation of this synthesizer is the P + 1 function of the dual-modulus prescaler. If the above values for n and A are programmed into the synthesizer as shown in Fig. 15-62 the loop will perform in the following manner.

The P/P + 1 divider is initially set up to divide by 21 and will output a pulse to the divide-by-n counter as well as to the divide-by-A counter for every 21 pulses it receives. Since divider A is programmed to divide by 4, this cycle will continue for another three times (each cycle equals 21 vco pulses at the P/P + 1 counter), at which time divider A outputs a command (count of four) to the control logic which instructs divider P/P + 1 to change the divisor to P = 20. Upon execution of this change, the control logic outputs an inhibit signal back to the divider A, preventing it from any further count until the total cycle is repeated. Meanwhile, divider n which has already counted to four, continues to receive pulses from divider P/P + 1 at a ratio of 20 : 1. After the remaining 1496 pulses have been received (1500–4), divider n finally outputs a pulse to the phase comparator as well as to the control logic which resets all counters and the process repeats.

*This choice is arbitrary and depends on the type of device available.

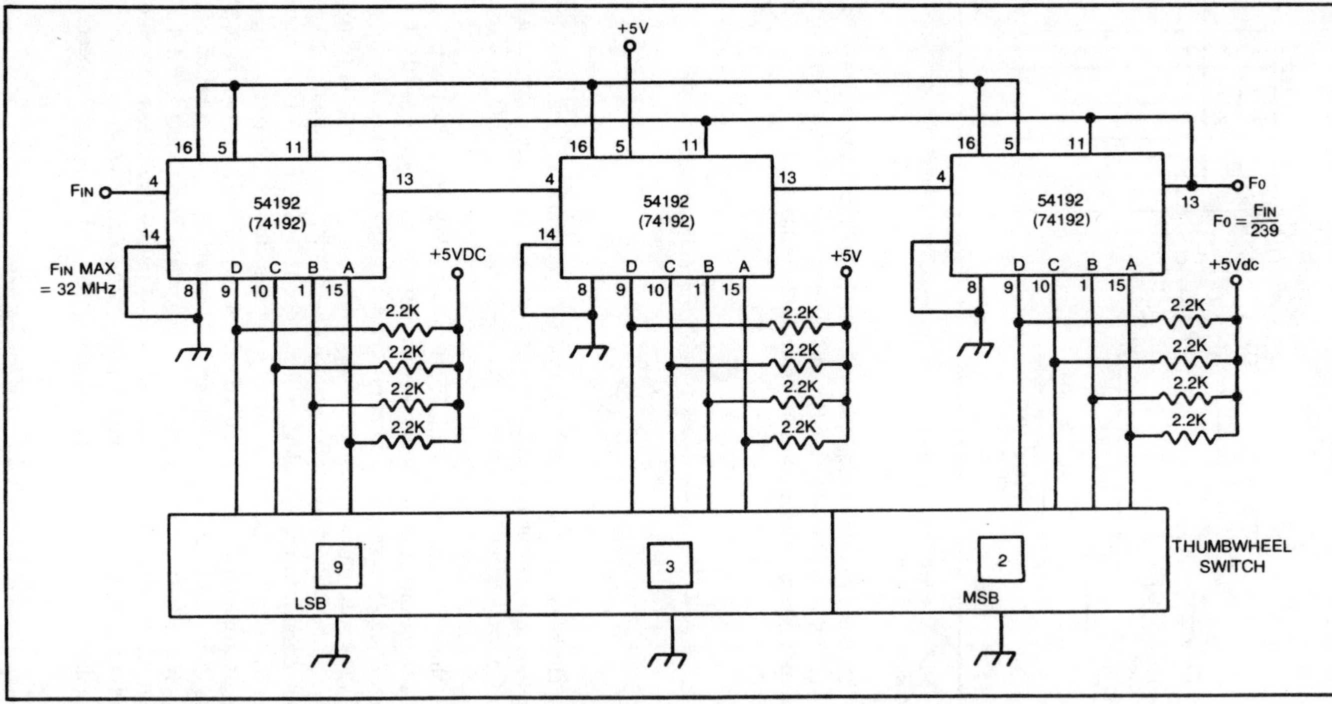

Fig. 15-59. Implementing a divide-by-N = 239 with the 54192 IC.

Fig. 15-60. A thumbwheel switch can be used to program the programmable divider in a phase-locked loop.

It can be seen from this explanation how the "swallow" terminology has been chosen to describe this technique. To verify the mathematics involved in this example, we'll use the following equation:

$$N = (P + 1)A + (n - A)P \qquad \textbf{Eq. 15-9}$$

$$N = 30{,}004 \text{ as found previously}$$

$$\therefore\ 30{,}004 = (21 \times 4) + (20 \times 1496)$$

This method allows direct programmability over the entire range of a vco operating at frequencies beyond the capabilities of conventional non-prescaled counters. Devices are available today which allow other dual-modulus (P/P + 1) divisions as shown in Table 15-2.

Figure 15-63 shows the circuit implementation for a 650 MHz, fully-programmable dual-modulus prescaler that could be used between a vco with such frequency characteristics and a phase detector. This circuit uses the Fairchild 11C90 integrated circuit which is a ÷ 10/11 prescaler. The up-counter should be of the ECL type in order to allow for ultra-high frequency operation.

Figure 15-64 shows a similar prescaler which uses the Fairchild 95H90, a similar ÷ 10/11 dual-modulus prescaler which allows operation to 250 MHz. Schottky TTL Logic (93S10) was used following the device in order to reduce the delay at the $\overline{PE}$ input. Other similar prescalers are also available from manufacturers such as Motorola and Plessey.

OTHER CONCEPTS IN SYNTHESIZERS

There are several other concepts of frequency synthesis used today. Some are derived from the phase-lock approach, while others are frequency-lock techniques, as well as digital techniques.

THE MIXER PLL SYNTHESIZER

When the dual-modulus prescaler approach is not economically practical, or the frequency of the vco is so high that there is no digital component

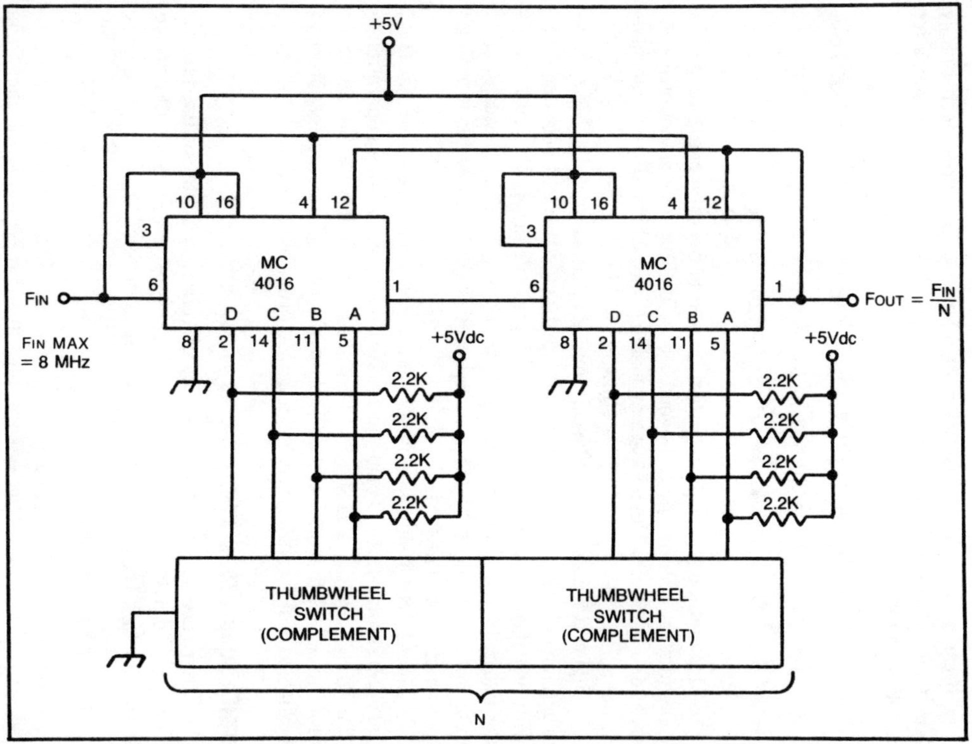

Fig. 15-61. Using the Motorola MC-4016 to obtain divide-by-N functions in a PLL.

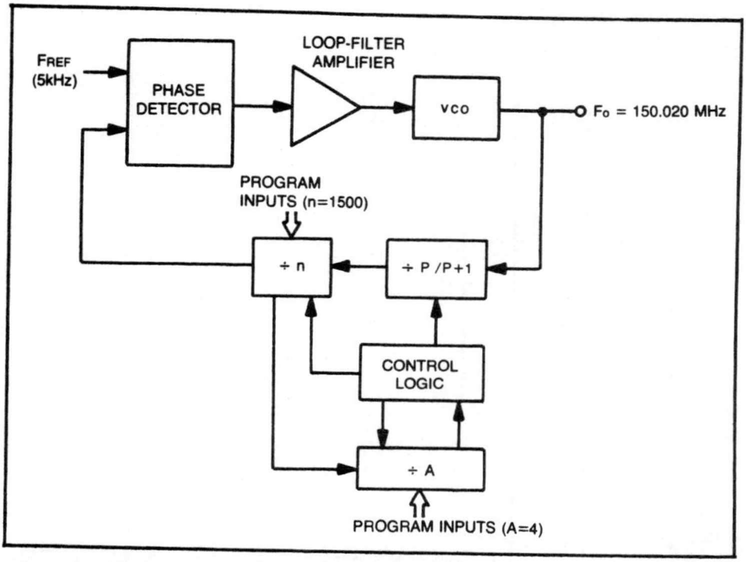

Fig. 15-62. The dual-modulus prescaler approach.

available to perform this function, as in the case of microwave synthesizers, the mixer phase-lock technique is usually used, as shown in Fig. 15-65.

This approach allows the operation of the vco at frequencies, several order of magnitude higher than the speed capability of the programmable divider, providing for inexpensive parts to be used in this part of the system. In a communications receiver, this advantage is two-fold due to the inherently quiet nature of the CMOS counters that can be used in this type of synthesizer.

Referring to Fig. 15-65, the vco's output operating at vhf or uhf frequencies is mixed in M_1 with the fixed frequency F_i which is usually greater than the maximum frequency of the vco. This injection frequency can be derived by means of frequency multiplication of the crystal-controlled reference oscillator. Another way to obtaining this frequency would be to use a fixed PLL which is locked to the same reference. The

Table 15-2. Other Dual-Modulus (P/P+1) Values Can Be Obtained in IC Form from Semiconductor Manufacturers.

P/P + 1	P/P + 1
5/6	40/41
8/9	64/65
10/11	80/81
15/16	100/101
32/33	128/129

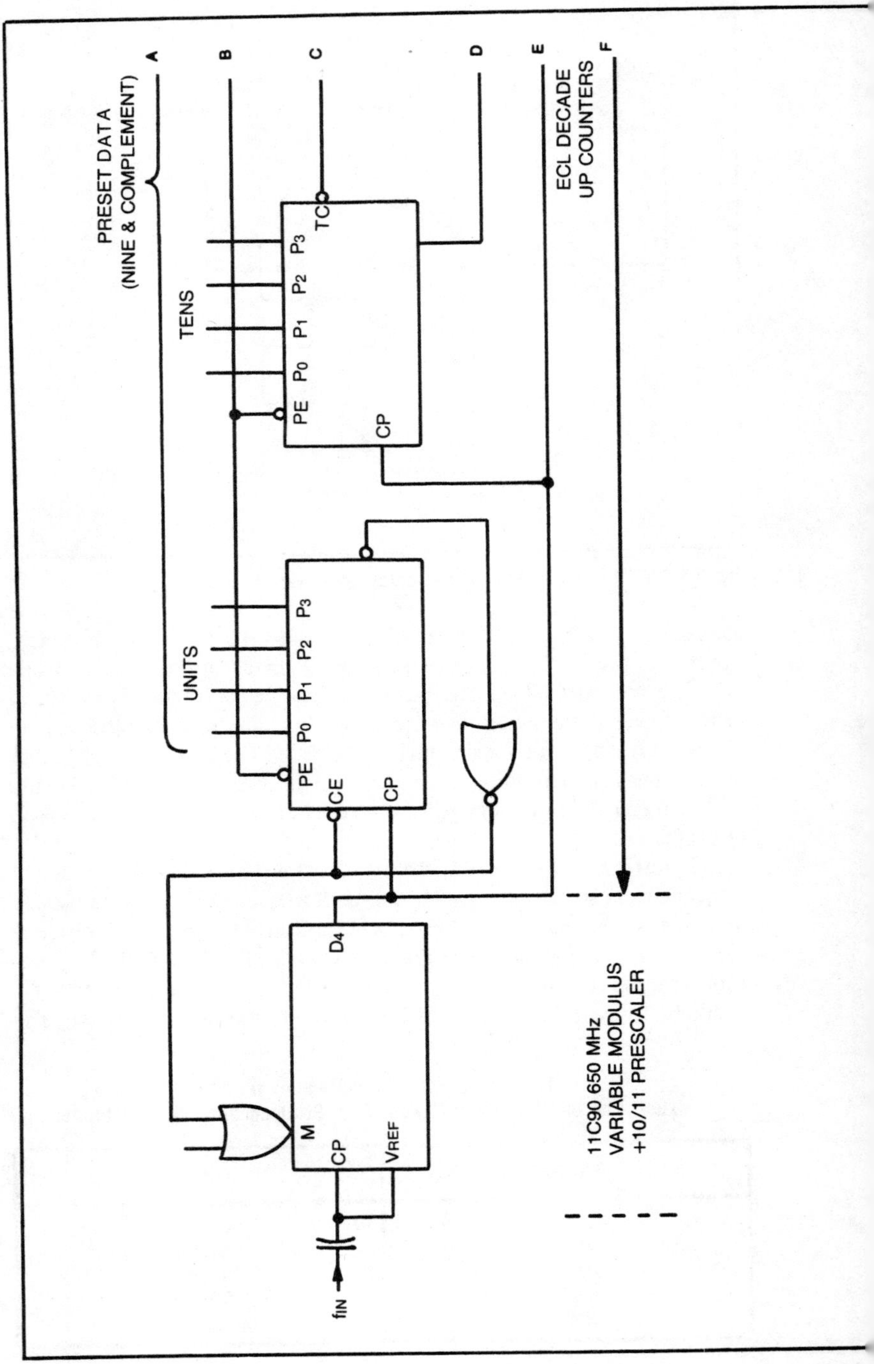
PRESET DATA
(NINE & COMPLEMENT)
A
B
C
D
E
F
TENS
UNITS
PE
P0
P1
P2
P3
TC
CP
CE
D4
M
VREF
fIN
ECL DECADE
UP COUNTERS
11C90 650 MHz
VARIABLE MODULUS
÷10/11 PRESCALER

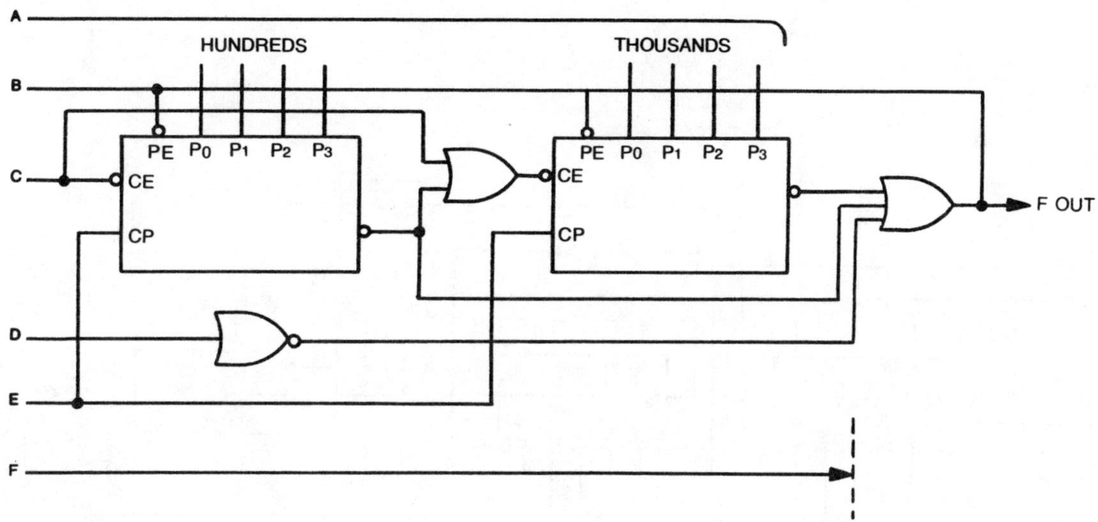

Fig. 15-63. Circuit diagram for a 650 MHz dual-modulus prescaler using the 11 C 90 dual-modulus prescaler.

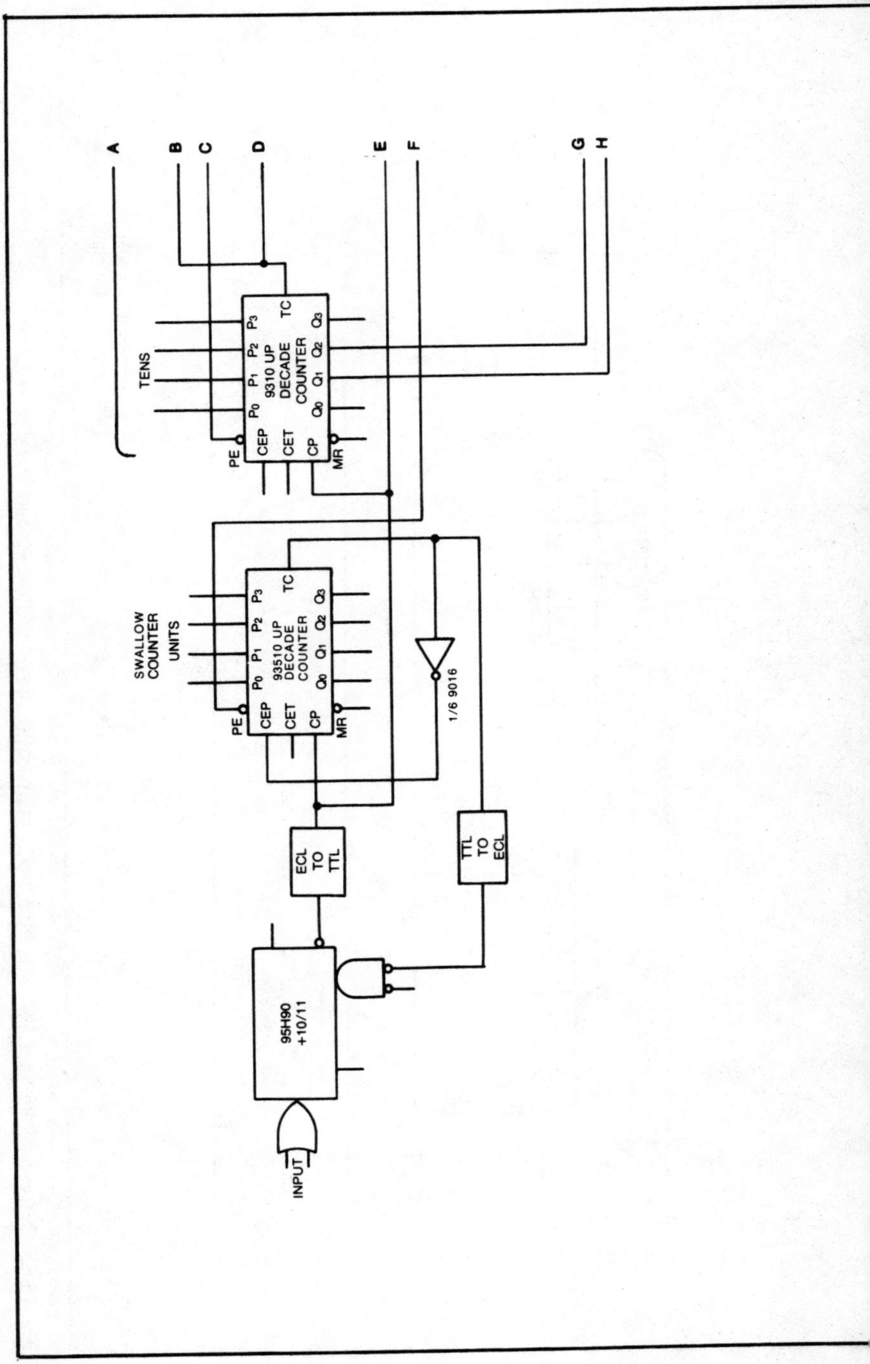

A
B
C
D
E
F
G
H
TENS
9310 UP DECADE COUNTER
SWALLOW COUNTER
UNITS
93510 UP DECADE COUNTER
PE
CEP
CET
CP
MR
TC
1/6 9016
ECL TO TTL
TTL TO ECL
95H90 ÷10/11
INPUT

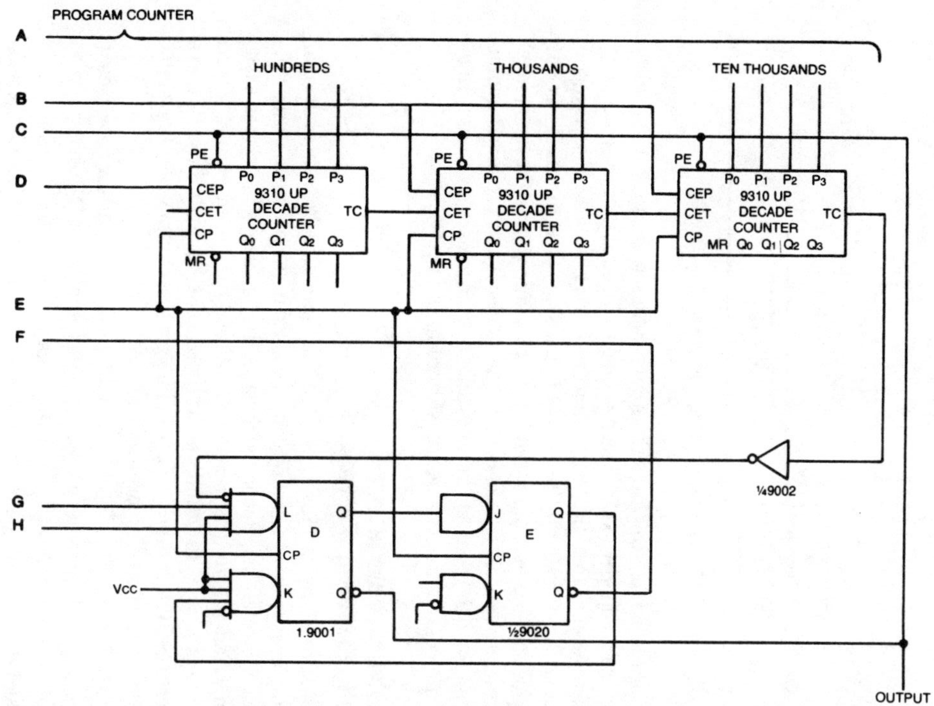

Fig. 15-64. A 250 MHz implementation using the 95H90, 10/11 prescaler.

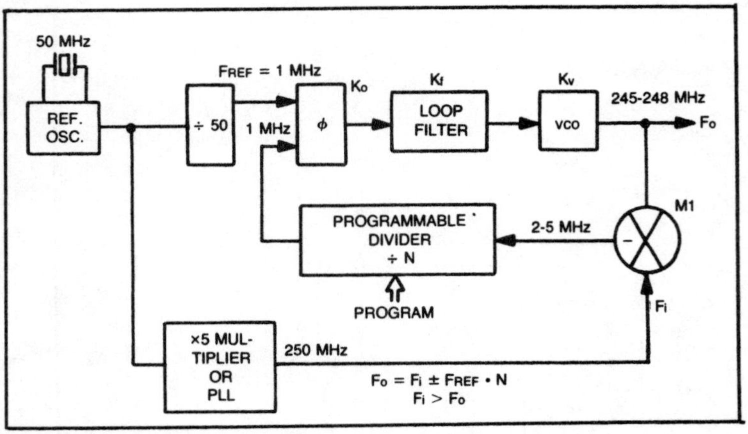

Fig. 15-65. Mixer PLL synthesizer block diagram.

output of the mixer which operates in a subtracting mode is fed to the programmable divider, and the loop is finally closed at the phase detector.

It can be seen from this discussion that the N is reversed by the mixing process so that it changes (increases) in the same direction as K_v, greatly improving the loop dynamics (reducing ωn variations over the loop range). The disadvantage of this concept is the necessity of providing another signal, and also the common intermodulation problems associated with the mixing process, which can be transferred to the vco's output.

THE DIGIPHASE® SYNTHESIZER AND THE FRACTIONAL N APPROACH

The problem of achieving high resolution with relatively high reference frequencies in a PLL has been discussed in the previous paragraphs. Because the reference frequency dictates the frequency interval in such a system, relatively low frequencies would have to be used. However, such an approach proved to be impractical, and complex multi-loop PLL synthesizers resulted. It is possible to build a synthesizer capable of resolution much finer than its reference frequency.

If we consider the system in Fig. 15-66, an accumulator increases its number (N_A) by the number presented at its control lines (N_c) every cycle of the reference frequency (F_{REF}). The output number is the desired number of phase cycles for the synthesizer as seen at some point in the reference cycle. It is this digital number that is compared with the number produced by a digital frequency counter which is continuously monitoring the vco's output (N_{dc}). The comparison is performed by a digital comparator function which switches a current I_p into the loop filter whenever the accumulator number (N_A) is greater than the counter number (N_{dc}).

This relatively simple method is the basis for the Digiphase® technique which allows high resolution which is independent of the reference frequency. Figure 15-67 shows the block diagram for a Digiphase® synthesizer. In this approach, the accumulator register is intentionally built longer than the digital counter or the digital comparator registers. The control number N_C is comprised of a main number n_1 and a fractional number n_2 which represents the resolution. Although the total number (N_c) is not compared with the content of the digital comparator because of the lack of resolution in it, the overflowing of n_2 in the accumulator results in a current I_p, which is summed together with I_{p_1} allowing for small current increments to be presented to the loop filter, and providing fine resolution for the vco. (The accumulator and the counter have to recycle.)

A resolution of 1 Hz can be achieved with a reference frequency of 100 kHz by using this method. Such a synthesizer has been reported* and exhibits relatively clean phase noise performance with the near-in spurious sidebands of less than − 70 dBc (decibels from carrier).

A further improvement of the Digiphase® method has been reported and is called the fractional-N synthesis. Although, this method resembles the Digiphase® approach, it looks like a phase-locked synthesizer as shown in Fig. 15-68. Further improvements to the Digiphase® method of synthesis has been introduced by Racal (patents pending). This consists of a further reduction of the low-frequency components at the output of the phase detector by using digital techniques. The output of the accumulator in Fig. 15-68 is further integrated in a similar size accumulator. When an overflow occurs from the second accumulator, a pulse is removed from N_1. By adding this pulse to the next count, a greater phase is obtained for one reference

*Frequency synthesis by Phase-Lock, Egan. Willey.

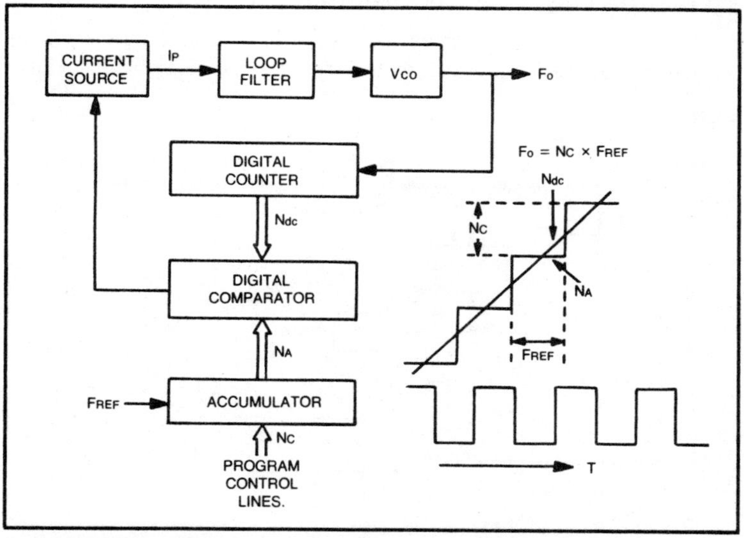

Fig. 15-66. Digital synthesizer using comparator and counter.

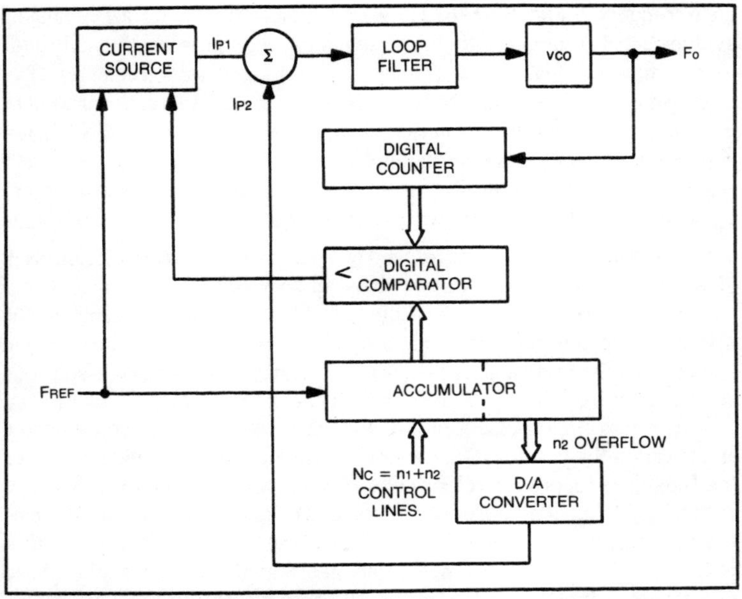

Fig. 15-67. The Digiphase® approach.

period, balancing out the continuously decreasing phase due to $(F_o / N_1) > F_{REF}$. The Digiphase and the fractional N synthesizer are finding increased use in communication receivers today.

SYNTHESIZER PITFALLS IN COMMUNICATIONS RECEIVERS

Because of the vast meaning of the word synthesizer, there are many misunderstandings about what a synthesized radio receiver should be. It is

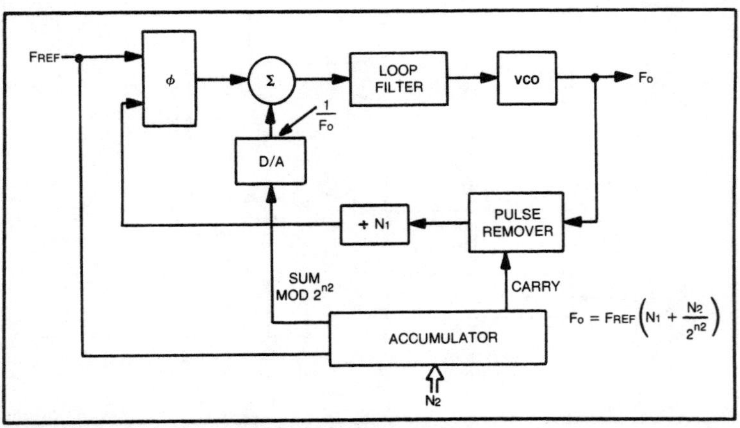

Fig. 15-68. Fractional N synthesizer block diagram.

clear from the previous paragraphs that a synthesizer derives its stability from the reference oscillator whose stability has further impact on the receiver's performance circuitry. Unless all other frequency sources used in the receiver are also slaved to the same reference frequency, the receiver is considered as not fully synthesized. There are many receivers claiming frequency synthesis on the market today. While a phase-locked loop is usually used in the first conversion to provide large steps in resolution (usually 1 MHz), a variable-frequency-oscillator (Vfo) or some other form of free-running oscillator (permability tuned oscillator-PTO) is used to fine tune each range. The Vfo has a relatively unstable characteristic over the temperature range and tends to spoil the total frequency stability of the receiver.

Some receivers use vco's in the second or third conversions. Fine increments of dc voltage are provided through a D/A converter and a network of precision resistors to the vco's varactor. Although this method accepts digital commands which can be fed from the same thumb-wheel control switch that feeds the rough tuning synthesizer, the stability of the entire receiver depends on the temperature characteristic of the resistors used in the D/A converter. This method is not superior in frequency stability to a Vfo approach, because of its open loop nature. The design and development cost of a fully synthesized receiver, far exceeds that of a quasi-synthesized approach. One may ask why does anyone need a few Hz stability after all?

In specialized radio communications, a radio receiver can be asked to do a variety of tasks, such as tune several ranges of frequencies in a particular manner selecting different bandwidths. It can also be programmed to be active at certain frequencies and/or time. Upon receiving a wanted signal, a second receiver can be activated to analyze that frequency, while the first receiver continues its predetermined task. The importance of being on the right frequency is vital, especially if narrow frequency-shift-keying (FSK) is the mode to be received. If the receiver wasn't fully synthesized and inherently very stable, errors would result at the output of the FSK demodulator decreasing the probability of intercept in a system which is already affected by fading and noise phenomena.

THE SR-2090 SYNTHESIZED COMMUNICATIONS RECEIVER

Figure 15-69 shows the basic SR-2090, synthesized communications receiver, which is one of several configurations achieved with the SR-2000 series, manufactured by Norlin Communications. These receivers can provide full coverage from 0.5 MHz to 1200 MHz by using a synthesizer with a resolution of 10 Hz (standard resolution is 1 kHz).

The basic system yields a double-conversion approach with the first i-f at 678.6 MHz*, and the second i-f at 21.4 MHz. Rf inputs from the antenna

*This frequency was chosen 35% higher than the highest frequency to be received of 500 MHz (a down converter is used for higher frequency coverage).

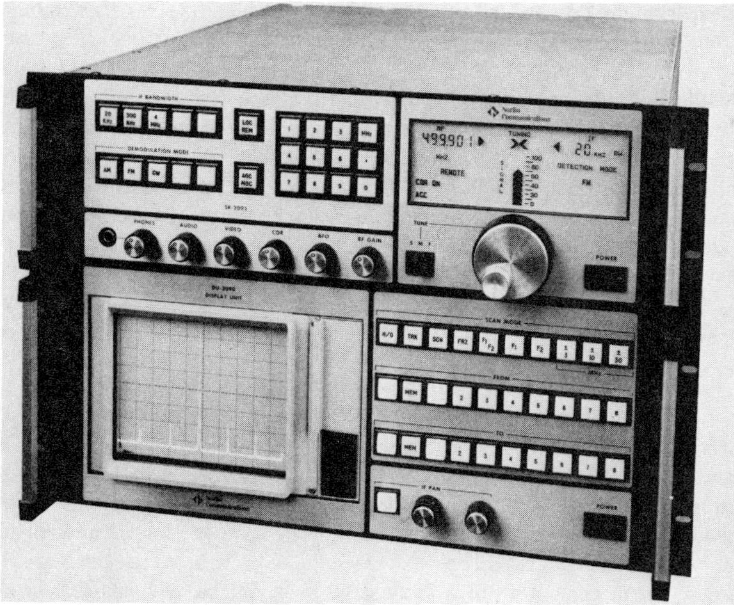

Fig. 15-69. The SR-2093 synthesized communications receiver shown with the DM-2090 display unit (courtesy of Norlin Communications).

are filtered by one of several microprocessor-controlled, sub-octave preselectors, which are designed in order to reject any second-order intermodulation products, followed by a broadband, low-noise amplifier with a very high IMD intercept point which in turn is inputted to the first mixer. The mixer is of the high-LO-level doubly-balanced type in order to keep intermodulation distortion to a minimum.

The synthesizer operating between 698.6 MHz to 1178.6 MHz is fed to the LO port of the mixer, and the resultant i-f at 678.6 MHz is further filtered by a 12 MHz wide filter, and amplified in the first i-f amplifier which has just enough gain to overcome mixer loss while keeping the intermodulation distortion low.

This signal is further mixed with the 700 MHz phase-locked loop frequency which provides the fine resolution necessary for tuning. A converted 21.4 MHz signal at the second i-f is produced as a result, and filtered in the microprocessor-controlled filter bank, which provides receiver final selectivity. The microprocessor also controls the audio/video amplifiers, as well as the agc signal for the previous stages.

Since the receiver is entirely synthesized, it can be remotely controlled by an external computer, or other digital commands, with the addition of an interface module. Several receivers can be remote controlled, providing full master/slave capabilities, not only for the frequency control part, but for all functions involved including several memories for recalling preprogrammed, or microprocessor-calculated stations.

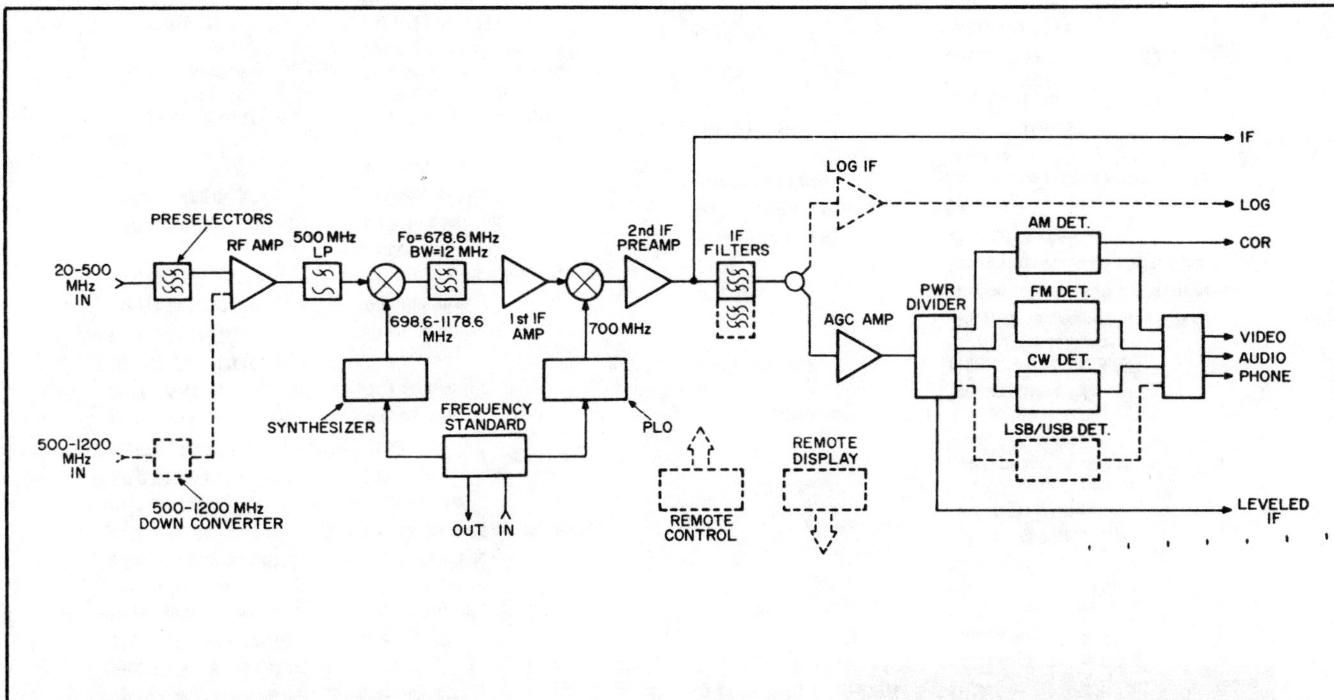

Fig. 15-70. Block diagram of the SR-2090 receiver.

Table 15-3. Specifications of the SR-2090 Receiver.

Specification	Value
FREQUENCY RANGE:	20 - 500 MHz
FREQUENCY RESOLUTION:	1 kHz
TYPE OF RECEPTION:	AM, FM, CW
IF BANDWIDTHS: (Any 3 Standard)	10, 20, 60, 100, 300, 500, kHz 1, 2, 4, 10, MHz
FREQUENCY STABILITY:	1 X 10^{-6} per day
AGC RANGE:	100 dB for 6 dB output variation
AGC ATTACK:	10 msec, max
AGC DECAY:	200 msec, typical
NOISE FIGURE:	13 dB max.
INPUT INTERCEPT POINT:	-5 dBm, min.
INTERNAL SPURIOUS:	-107 dBm, max.
IMAGE REJECTION:	95 dB, min.
IF REJECTION:	90 dB, min.
LO RADIATION:	-90 dBm, max.
BLOCKING:	-15 dBm for 3 dB reduction in output
BFO RANGE:	± 2 kHz minimum
INPUT IMPEDANCE:	50 ohms, nominal
VSWR:	3:1 max.
OUTPUTS:	
Preselective IF	21.4 MHz, rear panel BNC 20 dB gain above input
Postselective IF	21.4 MHz, rear panel BNC leveled to -13 dB min.
Video (2)	1 V RMS into 75 ohms 50 Hz - 5 MHz
OUTPUTS CONTINUED:	
Audio 1	100 Hz - 15 kHz terminal block, 1 watt, 8 ohms
Audio 2	100 Hz - 15 kHz terminal block; balanced, 10 mw, 600 ohms
Phone	100 Hz - 15 kHz, Front Panel Jack, 20 mw, 600 ohms
AGC Monitor	0 - 10 V, proportional to signal strength, terminal block, 10K ohms
COR Contacts	.5 amp, 100 V Terminal Block Pair
INPUT POWER:	115/230V/240 AC, ±10%; 47 - 400 Hz; 140 watt (Alternate fan required for 400 Hz)
FINISH:	Brushed Aluminum/Black front panel
EMI:	Designed to MIL-STD - 461, Class 1A
TEMPERATURE:	0° to 50° C, Storage
RELATIVE HUMIDITY:	0 to 95%
PRESELECTOR:	Six suboctave filters automatically switched

DIMENSIONS:

	Inches	cm
Height	5.25	13.3
Width	19.0	48.3
Depth	21.0	53.3

WEIGHT: 65 lbs. (29.5 kg)

The SR-2090 synthesizer switches in less than 500 u Sec so that the receiver can be rapidly step-scanned when used in conjunction with the DU-2090 display unit. It's single sideband noise spectrum, related to reciprocal mixing, is shown below.

TYPICAL RECEIVER SSB PHASE NOISE PERFORMANCE

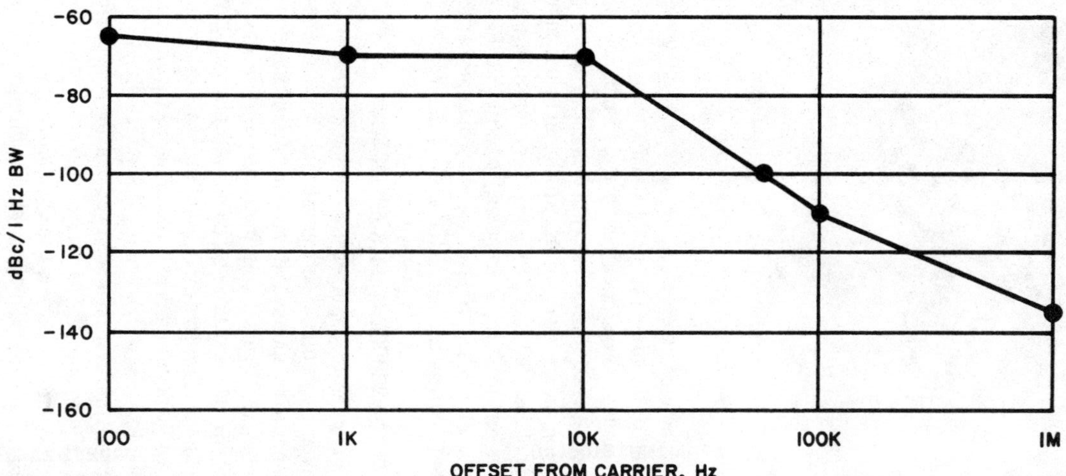

Fig. 15-71. An opto-encoder (courtesy of Disc Instruments).

for all functions involved including several memories for recalling preprogrammed, or microprocessor-calculated stations.

One human operator can control "hands-off" several slave receivers preprogrammed to do a variety of tasks at a variety of times, with computer precision, in an environment which might not allow for individual hand tuning of each receiver. This receiver is modular in packaging, allowing for full system flexibility in frequency coverage, with a minimum of electrical changes. Figure 15-70 shows the block diagram of the SR-2090 receiver and Table 15-3 lists the characteristics of this receiver.

THE OPTO-ENCODER AND ITS APPLICATION

In order to duplicate the tuning feel of the Vfo, the synthesized receiver uses a digital optical encoder. It has been seen that a synthesizer responds to computer-like encoded information to step from one frequency to another. To input this information, it is easiest to use a detented thumb-wheel encoder as we previously discussed. While such a method is fully adequate for some fixed applications, or for rough tuning, fine tuning a receiver with this method is rather impractical from an operator standpoint. A knob-like dial which will provide a feel similar to the familiar old Vfo is desired. This function is achieved with the help of the opto-encoder. Figure 15-71 shows an opto-encoder manufactured by Disc Instruments.

The opto-encoder is usually a film or metal disc with holes in certain locations. An infrared beam shines through the holes. Two photocells detect when the light is interrupted, as well as the direction of rotation. Information in the form of pulses is presented to a chain of up/down counters which are commanded to count in the direction of movement of the disc. This information is finally presented in a parallel format to the synthesizer's dividers, and an imitation Vfo feel is realized by the operator. This process is illustrated in Fig. 15-72.

DESIGNING GOOD VFO'S

Let's return now briefly to the Vfo. Whether buying or designing a receiver, Vfo stability can be as important as dynamic range. Several years ago, many receivers had to be turned on at least one-half hour before operating to obtain adequate stability. Component warm-up especially in vacuum tubes, made excessive drift a fact of life even though these Vfo's were temperature-compensated with the use of NPO and NTC capacitors.

However, finding the right values and temperature coefficients for compensation was a long and tedious task. Today we need not worry about drift when using synthesizers. Modern Vfo's use FETs instead of tubes for better frequency stability. Polystyrene capacitors have proven to have extremely good stability characteristics, replacing the NPO and NTC capacitors.

Whenever designing a Vfo, such as the one shown in Fig. 15-1, remember one rule of thumb: several small polystyrene capacitors in parallel will provide much better stability than one large capacitor. Another rule is to buffer the Vfo and to use a good filter in the output in order to eliminate harmonics.

Bipolars should be avoided in Vfo's. Because of their typical low impedance levels, they demand a high L to C ratio, resulting in variable capacitors as high as 500 pF for inductances as little as 1 μH in a 5 MHz Vfo, for example. A change of a fraction of a μH can cause severe drift in such a case, and this is entirely possible with inductors which rely on slug-tuned powdered iron or ferrite as core material. Perhaps one of the worst cores that can be used in this application is a toroid, because of the large area being exposed to temperature changes. Experience has shown that a ceramic form without any core is the best for winding Vfo coils on. Increased stability is obtained by covering the finalized coil with paraffin or wax.

There is no substitute for a quality variable capacitor (Fig. 15-73). The use of capacitors with aluminum plates should be avoided. They can be noisy and microphonic. It has been found that the smoothness of the plates in a variable capacitor had a significant effect on the Q and the noise floor of a Vfo. A mirror-like surface would be preferred. Finally, the mechanics should not be ignored. Springy couplers if they are used can introduce backlash. Noise in Vfo's and phase noise in synthesizers can account for a lot of intermodulation products which will participate in the receiver's dynamic range, making it sound "mushy" and lowering the effective intercept point. Much has been written on Vfo's but the one shown in Fig. 15-1 is a good example of a clean and stable modern design.

It is possible to build a digital Vfo by using the phase-locked loop techniques described earlier. Figure 15-74 shows the block diagram for a 5 to 5.5 MHz system. The vco operates at 200 MHz in order to allow for a relatively high reference frequency of 1 kHz, which in turn provides fast switching for a final resolution of 10 Hz at the output of the synthesizer. This approach could be used with any of the presently manufactured

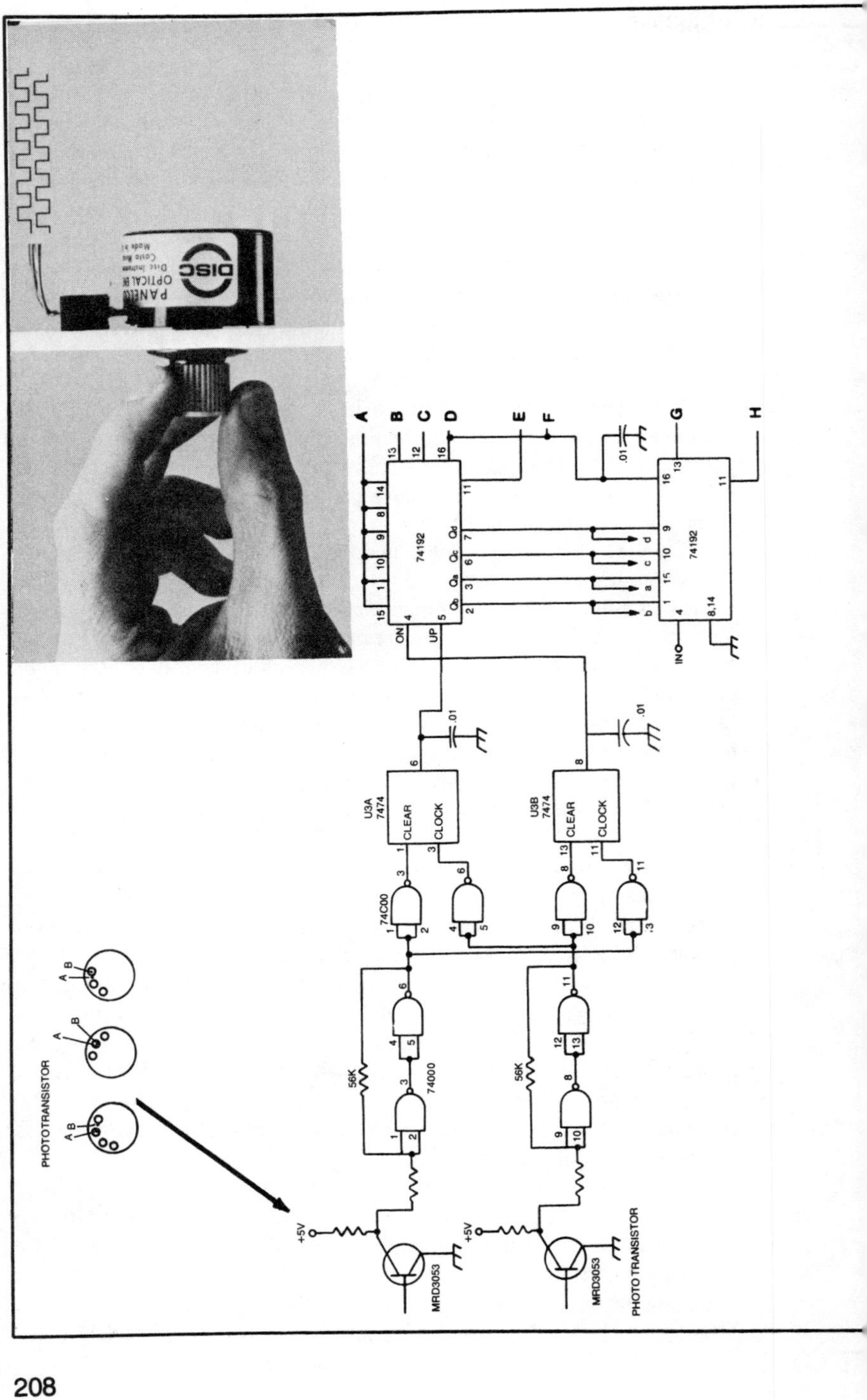
PHOTOTRANSISTOR
A B
+5V
MRD3053
PHOTO TRANSISTOR
56K
74000
74C00
U3A
7474
U3B
7474
CLEAR
CLOCK
.01
74192
ON
UP
IN
A
B
C
D
E
F
G
H
DISC

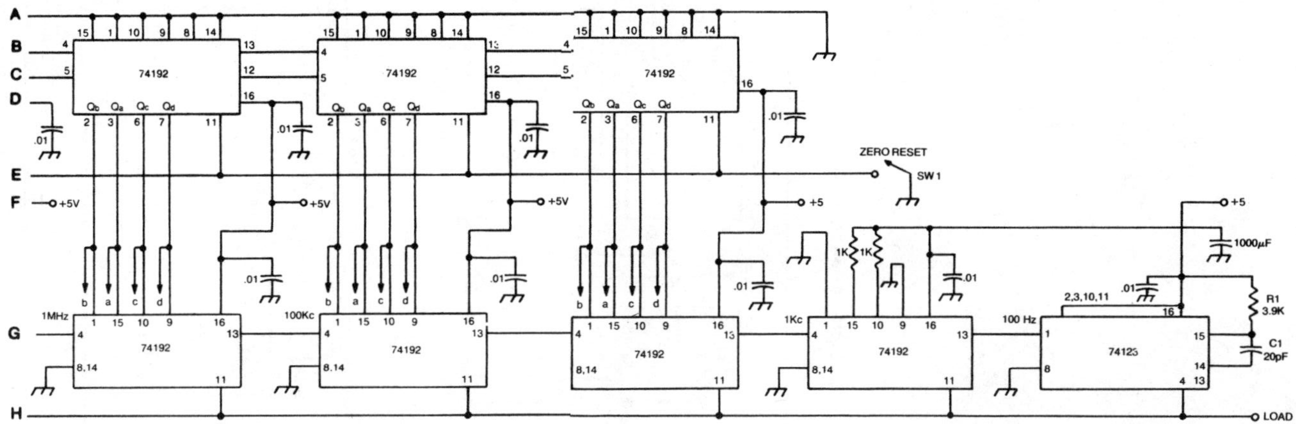

Fig. 15-72. An optical encoder is used in this example to provide fine tuning for a communications receiver (courtesy of Disc Instruments).

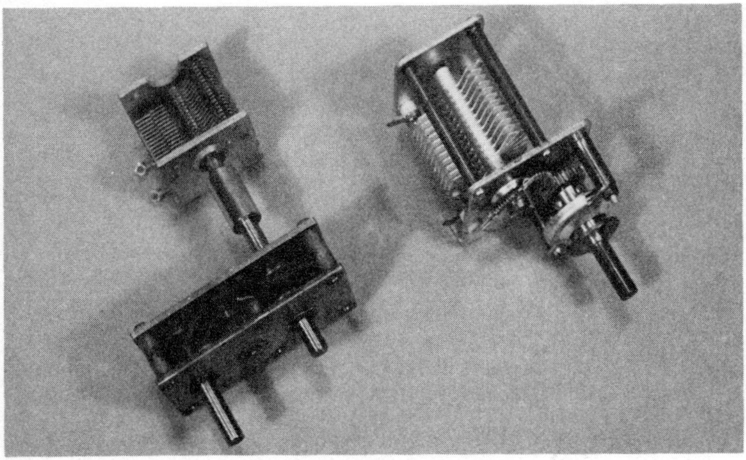

Fig. 15-73. Typical Vfo capacitors showing mechanical gears.

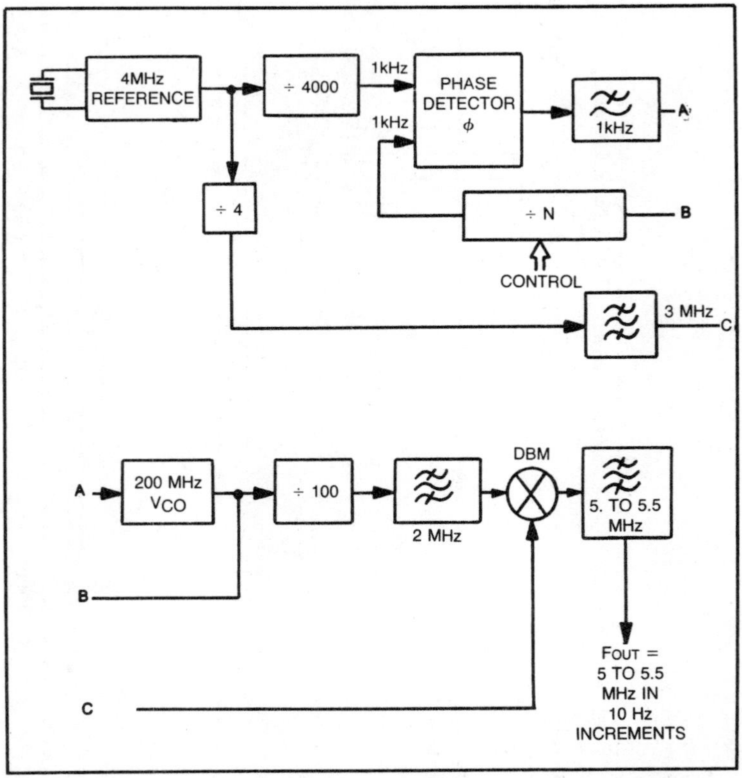

Fig. 15-74. Block diagram of a synthesized Vfo. Output is 5 to 5.5 MHz in 10 Hz steps.

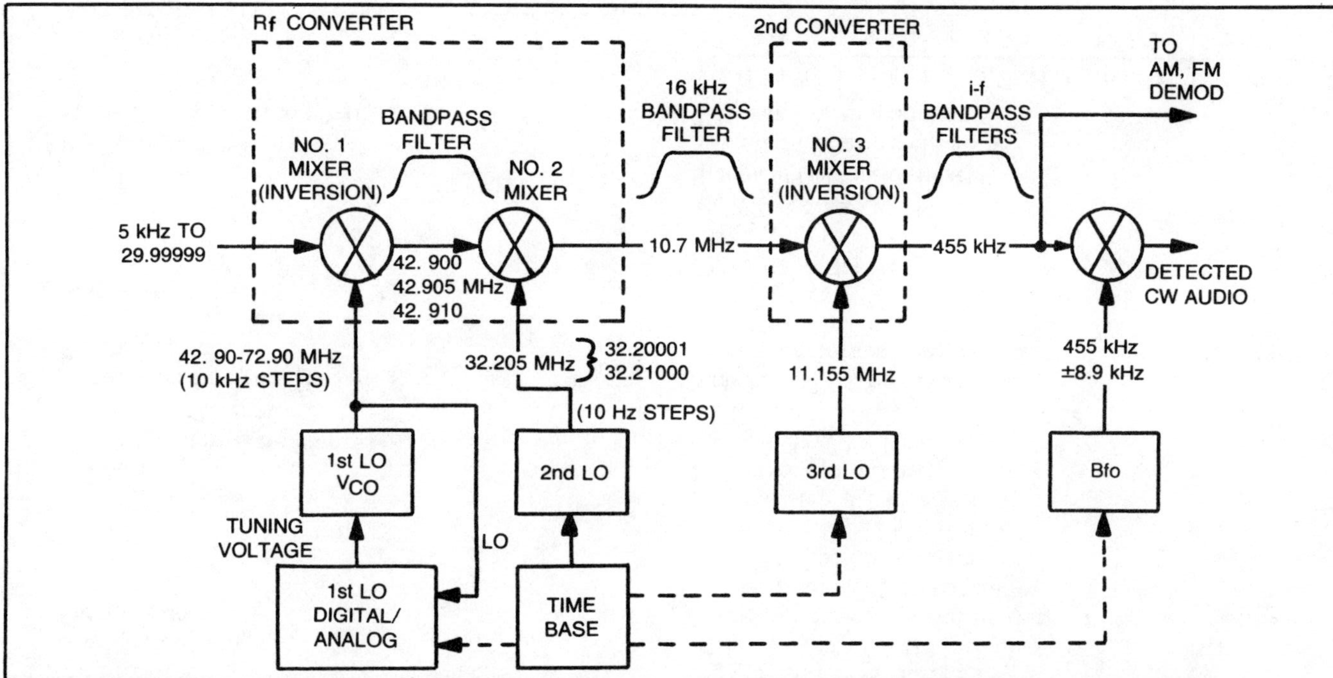

Fig. 15-75. Block diagram of the WJ-8718 HF communications receiver manufactured by Watkins-Johnson Company. All LO's are slaved to the same time base.

Table 15-4. Specifications of the WJ-8718 Receiver.

Specification	Value
Tuned Frequency	5.0 kHz to 29.99999 MHz
Display	7 Digit yellow LED 1/2 inch high (see options list)
Tuning Speeds	Four front panel, pushbutton selected 10 kHz step, 1.2 MHz/turn 1 kHz step, 120 kHz/turn 100 Hz step, 12 kHz/turn 10 Hz step, 1.2 kHz/turn
Resolution	10 Hz (see options list)
Stability (Internal Reference)	6 x 10^{-8}/day, 2 x 10^{-6}/year
External Reference	1 MHz, 50 to 500 MV rms into 50 Ω
Synthesizer Lock-up Time	3 ms typical, 10 ms maximum
Optional	Calculator format, keypad entry of frequency (optional with MFP)
Detection Modes	AM – A3-A4A FM – F1-F2-F3-F4 CW – A0-A1 MCW – A2-A4A Sideband (see options list)
I-f Bandwidths	5 standard front panel selected
Shape Factor (3 dB to 60 dB)	3 dB bandwidth minimum

0.3 kHz	1 kHz	3.2 kHz	6 kHz	16 kHz
7.0:1	4.5:1	2.5:1	2.3:1	2.0:1

I-f Output	455 kHz 20 mV minimum into 50Ω for an input signal > 3 μV (see options list)		
Gain Control Mode	Manual, AGC fast and slow		
Range	100 dB minimum		
AGC threshold	3.0 μV typical		
AGC attack time	15 ms typical		
AGC release time	Fast – 25 ms maximum Slow – 4 seconds maximum		
BFO	±8 kHz range		
Display	2 digit		
Resolution	100 Hz (see options list)		
Optional	Keypad entry of frequency offset, 10 Hz resolution		
Sensitivity, 200 kHz - 30 MHz	IF BW	Input Signal	(S+N)/N at Audio Out
CW	0.3 kHz	0.40 μV	16 dB
AM	6 kHz	1.7 μV/50% mod. 400 Hz	10 dB
FM	16 kHz	2.5 μV/400 Hz mod. 4.8 kHz peak dev.	17 dB
CW Sensitivity, 5 kHz – 200 kHz (0.3 kHz IF Bandwidth)			
50 kHz – 200 kHz	A 0.63 microvolt signal will produce at least a 16 dB (S+N)/N ratio at the Audio output.		

Table 15-4. Specifications of the WJ-8718 Receiver. (Continued from page 213.)

15 kHz – 50 kHz	A 1.4 microvolt signal will produce at least a 16 dB (S+N)/N ratio at the Audio output.
5 kHz – 15 kHz	A 63 microvolt signal will produce at least a 16 dB (S+N)/N ratio at the Audio output.
Sideband	See options list
Signal Handling Capabilities	
3rd Order Input Intercept Point	+20 dBm, minimum for signals separated by > 30 kHz. (Performance may degrade below 3MHz)
Reciprocal Mix	I-f BW 3.2 kHz, desired sig. of 25 μV (–79 dBm) with undesired signal of 79 mV (–9 dBm) Δf from desired >30 kHz, noise ratio (S+N)/N⩾20 dB.
Cross Modulation	I-f BW 1.0 kHz, desired signal of 10 μV (–87 dBm) with undesired signal 31.6 mV (–17 dBm) Δf from desired > 50 kHz and 30% AM, <10% cross modulation.
Internal Spurious Responses	All internal spurious responses are less than –114 dBm referred to the input above 50 kHz tuned frequency.
I-f Rejection	Greater than 90 dB
Image Rejection	Greater than 90 dB
Audio Outputs	
Power Output	For a 3μV 30% AM signal 600 Ω unbalanced Line Audio 1W minimum.
Amplifier response	±1.5 dB from 100 Hz to 8 kHz, 1 kHz reference frequency
Distortion	< 5% at rated output.

Headphone Output	30 mW minimum into 600 ohm phones.
FM/CW Output	DC coupled low level output from FM/CW detector is provided on the rear panel.
Power Interrupt	Programmed and operating parameters are retained during power interrupts up to 48 hours and restored upon resumption of power.
Analog Metering	Front panel selectable line audio or signal strength monitoring.
Antenna Input	
Impedance	50Ω unbalanced, nominal
Protection	Will withstand the effects of RF power up to +30 dBm and static build up. The circuit automatically resets.
Conducted Oscillator Radiation	–87 dBm, maximum
Operating Temperature Range	0°C to +50°C.
Power Requirements	115/220 V AC ±15%, 48-410 Hz, 70 watts nominal 90 watts with maximum options installed.
Weight	Approximately 35 pounds (15.75 kg)
Size	5.25 inches high (13.34 cm) 19 inches wide (48.26 cm) 19.4 inches deep (49.28 cm)

Note: All μV measurements referenced to 50Ω impedance.

Table 15-5. Specifications of the R 4245 Receiver.

Frequency Coverage: 10 kHz to 30 MHz continuous plus 8 possible crystal controlled channels.

Sensitivity (SSB, CW): (2.3 kHz BW) 1.8 to 30 MHz less than .20 μV for 10 dB (S + N)/N with preamp on (typically .15 μV) (Noise floor typically −134 dBm) Less than .50 μV for 10 dB (S + N)/N without preamp (typically .30 μV) (Noise floor typically −128 dBm). 0.01 to 1.5 MHz less than 1.0 μV for 10 dB (S + N)/N.

Sensitivity (AM): (30% Mod; 4.0 kHz BW) 1.8 to 30 MHz less than 1.2 μV for 10 dB (S + N)/N @ 30% modulation, preamp on. Less than 2.0 μV for 10 dB (S + N)/N @ 30% modulation, preamp off. .01 to 1.5 MHz less than 4.0 μV for 10 dB (S + N)/N @ 30% modulation.

Selectivity:

	−6 dB BW	−60 dB BW	Ultimate Rejection
a.	4 kHz	8 kHz	−100 dB
b.	2.3 kHz	4.4 kHz	−100 dB
c.	1.8 kHz	3.6 kHz	−100 dB
d.	500 Hz	1.1 kHz	−100 dB
e.	300 Hz	700 Hz	−100 dB

IF and Image Rejection: Greater than 80 dB (48.05 MHz 1st IF, 5.645 MHz 2nd IF, 50 kHz 3rd IF).

Power Requirements: 100/120/200/240 VAC ± 10%, 50/60 Hz, 60 watts or 11.0 to 16.0 VDC (13.8 VDC nominal), 3 amps.

AGC Performance: Less than 4 dB audio output variation for 100 dB input signal change above AGC threshold. AGC threshold is typically .8 μV with preamp off and .25 μV with preamp on.

Blocking: >145 dB above noise floor.

Intermodulation: (100 kHz tone spacing)

Preamp Off:

Two-tone Dynamic Range: 99 dB 1.8 to 30 MHz

Third Order Intercept Point: + 20 dBm

Preamp On:

Two-tone Dynamic Range: 95 dB 1.8 to 30 MHz

Third Order Intercept Point: +10 dBm

AGC Attack Time: 1 millisecond. Three selectable release times: **Slow:** 2 sec—**Medium:** 400 m sec — **Fast:** 75 m sec. Also, "Off" position is provided.

Antenna Input Impedance: Nominal 50 ohms. (200 ohms on .01 to .5 MHz Range.)

Audio Output: 2.5 watts with less than 10% T.H.D. into nominal 4 ohm load.

Dimensions: 12.5″D (31.75 cm) excluding knobs and connectors, 19″W (48.26 cm), 5.25″H (13.33 cm).

Weight: 19 lbs (8.6 kg)

Available with cabinet, size approx.:
14.5″D × 20.75″W × 7.5″H (36.8 × 51.4 × 19 cm).

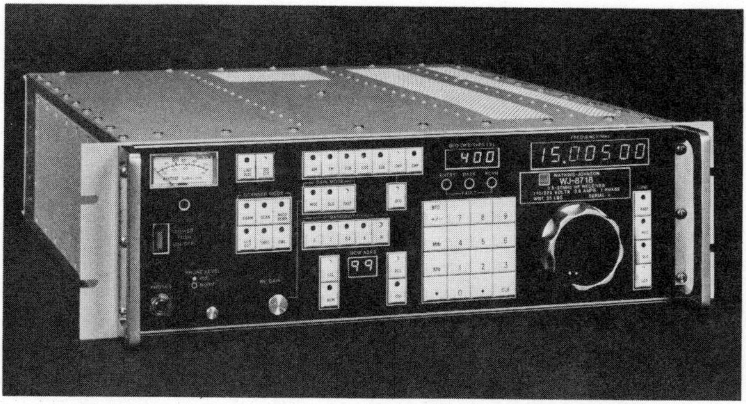

Fig. 15-76. The Watkins-Johnson WJ-8718 is a highly versatile, fully synthesized (including the bfo) hf receiver which covers the frequency range of 5 kHz to 30 MHz in one band, with a resolution of 10 Hz and a typical lock-up time of 3 ms (courtesy of Watkins-Johnson Company).

transceivers requiring a 5 to 5.5 MHz Vfo, however, this modification will not transform the rig into a truly synthesized apparatus because of the requirement for additional mixing with crystal oscillators which are internal to the transceivers. Unless these sources are also slaved to the same reference frequency as the Vfo synthesizer, the system is said to be a quasi-synthesizer approach.

In conclusion, we discussed several ways of implementing local oscillators in communications receivers, with the accent on several forms of synthesizers which can allow for all local oscillators in a radio receiver to be coherent with a highly stable reference frequency. Figure 15-75 shows the block diagram of the WJ-8718 high frequency communications receiver, manufactured by Watkins-Johnson. It can be seen from this example that all

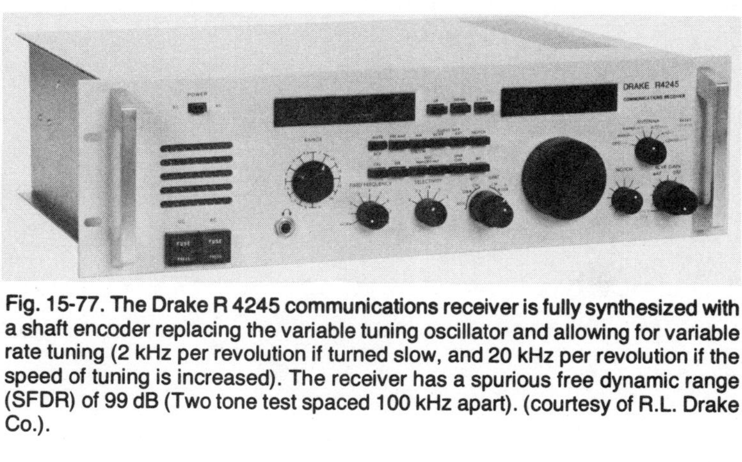

Fig. 15-77. The Drake R 4245 communications receiver is fully synthesized with a shaft encoder replacing the variable tuning oscillator and allowing for variable rate tuning (2 kHz per revolution if turned slow, and 20 kHz per revolution if the speed of tuning is increased). The receiver has a spurious free dynamic range (SFDR) of 99 dB (Two tone test spaced 100 kHz apart). (courtesy of R.L. Drake Co.).

local oscillators in this triple-conversion approach are slaved to the time base generator. Also, see Figs. 15-76, 15-77 and Tables 15-4 and 15-5.

Although several books alone can be written on the subject of synthesized local oscillators, this chapter was intended as a triggering mechanism for those interested in furthering the cause of technical excellence.

—•••—

Chapter 16

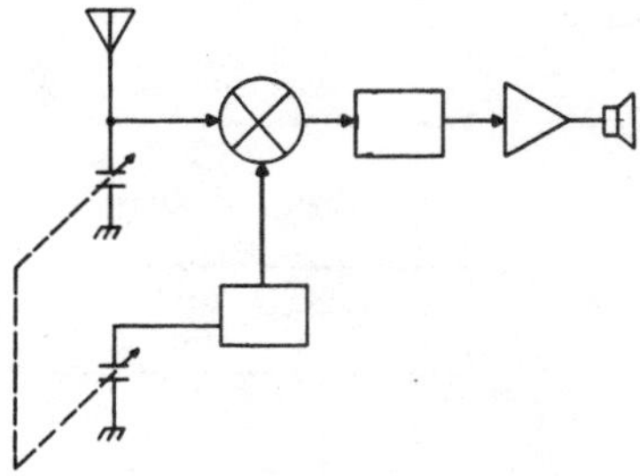

Digital Frequency Counters and Radio Receivers

Modern communications receivers use a frequency readout to indicate tuning. In digitally synthesized approaches, this information is collected and displayed directly from the binarily-coded inputs. In an open loop system such as the one involving a Vfo, the frequency is displayed as a result of a frequency counter which measures the number of cycles outputted by the Vfo, and/or all other local oscillators, against a highly stable reference frequency. The result indicates with high accuracy, the tuning frequency of the receiver directly in MHz. Figure 16-1 shows a basic block diagram for such a frequency counter. It can be seen from this example that the programmable counter (as modified by the band and i-f correction inputs) is allowed to count the conditioned Vfo frequency, only as gated by the reference divider, resulting in a display of the actual frequency being received. Although this rather simplistic approach to a frequency counter has been adopted in many designs, the net result of such a system is the "flickering" phenomenon resulting in the ambiguous digit problem.

The timing diagram in Fig. 16-1, shows that the counter samples the Vfo frequency half the time and it displays it the other half. This is known as the 50/50 sampling procedure, and can be found widely in the literature and in some commercial equipment. Because the frequency of the Vfo is fully unsynchronized with the reference frequency, the counted result will always hesitate between two values, displaying a flickering least significant digit. The 50/50 sampling technique will only aggravate this flickering.

A well-designed frequency counter will sample for a very short time and display for a relatively long time. Shown in Fig. 16-2 is the timing diagram for such a counter. The Vfo frequency (data count) is counted for 100 msec and 10 msec after this count is finished, a short latch enable pulse strobes the display, thus changing the number to be viewed by the operator

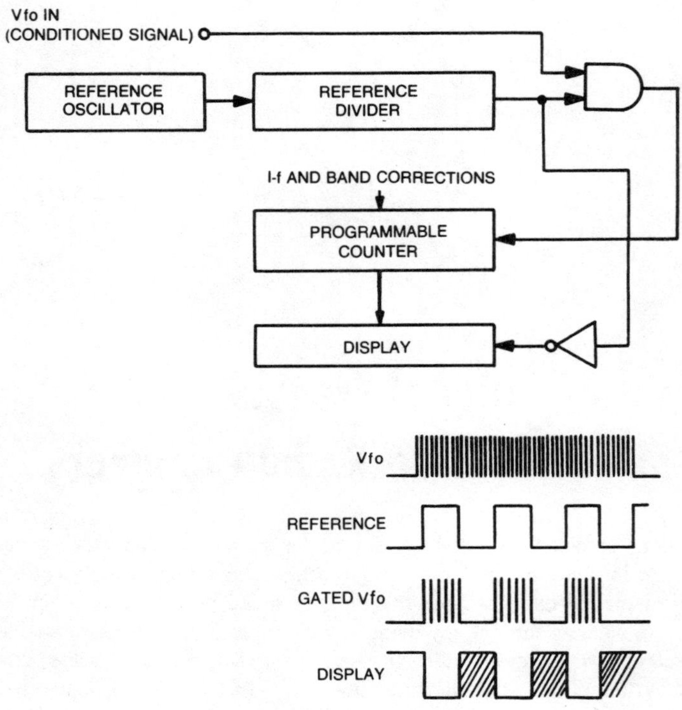

Fig. 16-1. Block diagram for a basic frequency counter.

to a new number. This number will remain displayed for the next 140 msec while a new latch enable pulse is generated. In the meantime, the counter was reset by the count load line and a new frequency can be displayed. Because of the very short sampling period and the relatively long retention of the displayed information, the flickering effect will be greatly minimized.

Figure 16-3 shows the schematic diagram for such a frequency counter as applied to the 9 MHz i-f communications receiver. Offset i-f frequencies of ± 9 MHz, as well as band corrections are automatically programmed into the counter by the band switch so that the display will read actual received frequencies. Blanking of the first digit and the decimal point are also designed in.

This is accomplished through a diode matrix which is also set up to correct for the proper sideband corresponding with the particular band used. A reverse mode switch is also included if abnormal sideband reception is required by the operator. This entire interlock system should also control the bfo frequencies. This counter performs extremely smooth with no visible flicker. Further improvement in minimizing the flicker phenome-

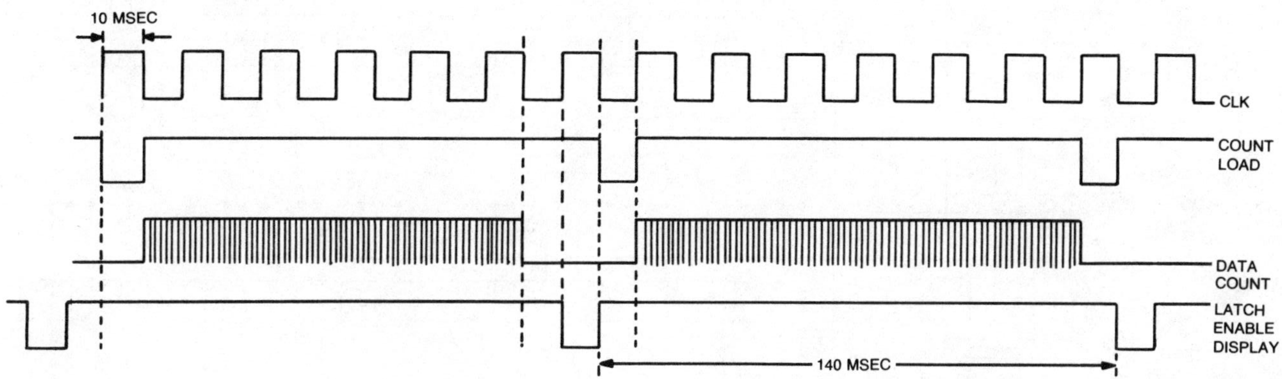

Fig. 16-2. Timing diagram for the digital frequency counter.

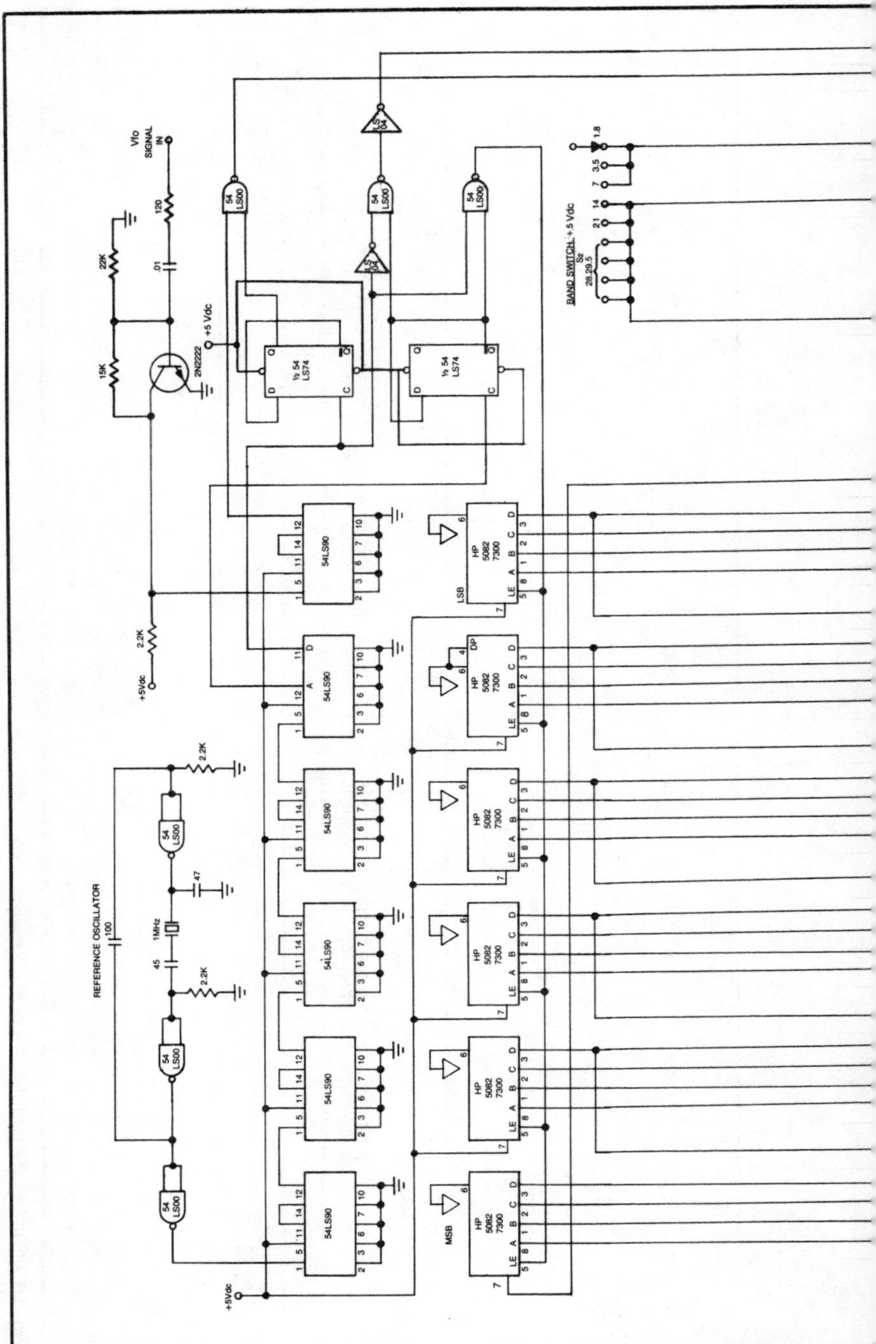
Vfo
SIGNAL
IN
120
.01
22K
15K
2N2222
+5 Vdc
½ 54
LS74
54
LS00
LS
04
BAND SWITCH +5 Vdc
S2
1.8
3.5
7
14
21
28.29.5
54LS90
2.2K
+5Vdc
HP
5082
7300
LSB
MSB
DP
REFERENCE OSCILLATOR
100
45
1MHz
47
2.2K

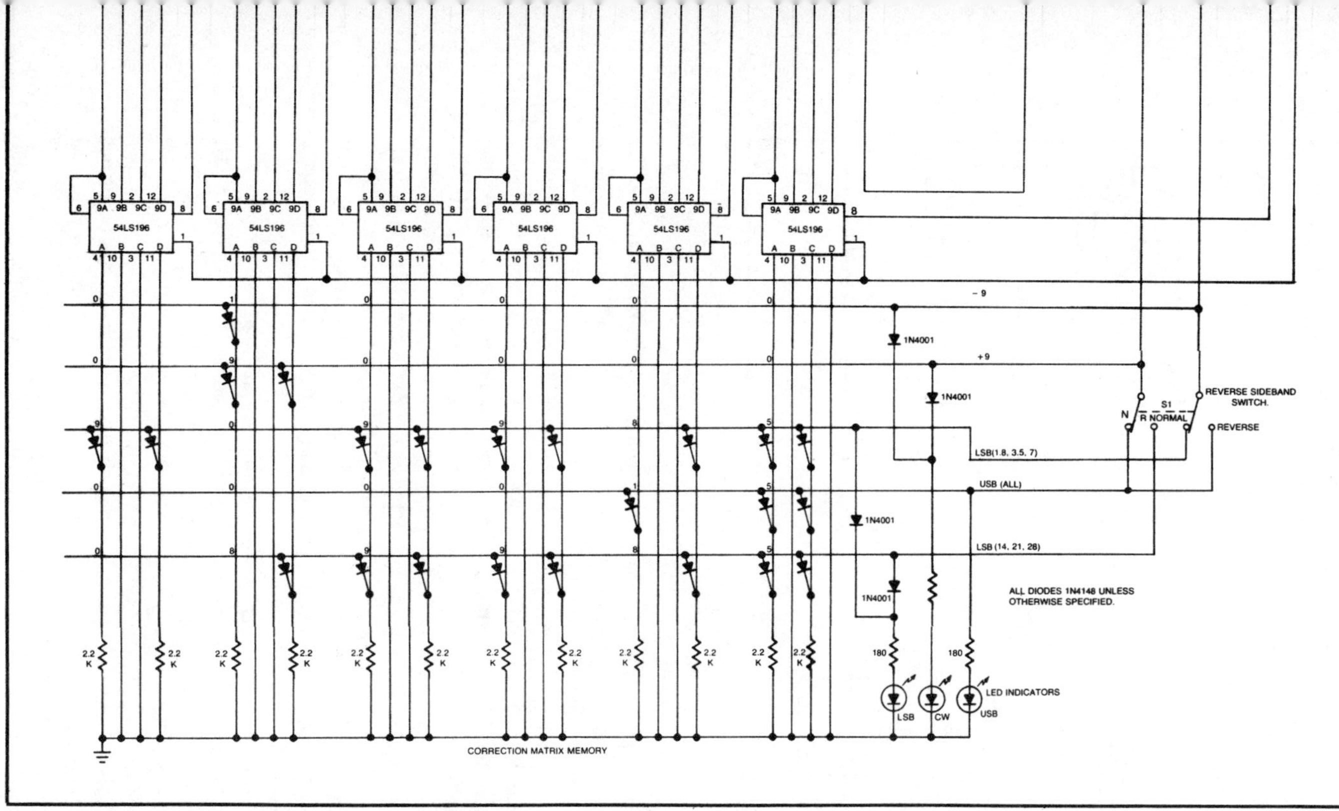

Fig. 16-3. Schematic diagram of a frequency counter used in a communications receiver. Automatic band and sideband corrections are made through the diode matrix memory.

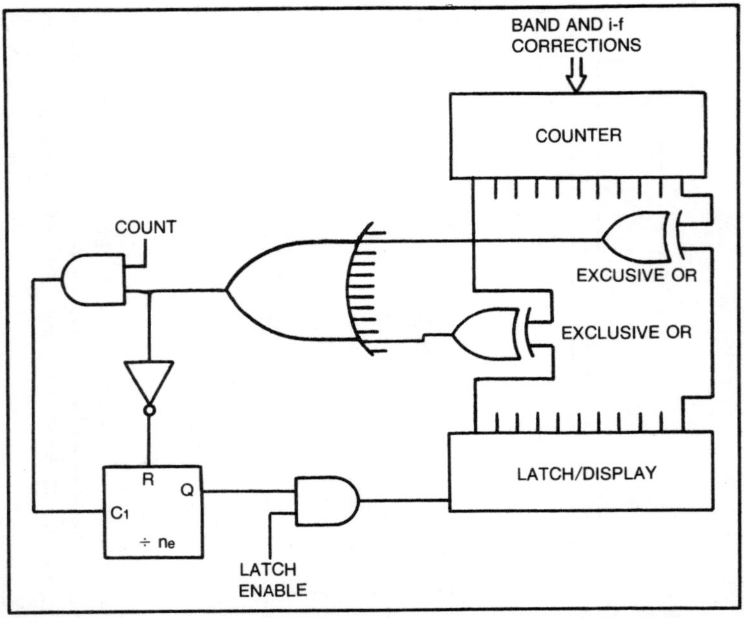

Fig. 16-4. An ambiguous digit detector can greatly improve the problem of flickering in a counter.

non in a frequency counter could be obtained if the command to load the latch would be only given by the system when the value in the counter is really a new value.

If the operator didn't change the frequency there is obviously no reason to refresh the display, allowing for flicker-free operation. This is accomplished by comparing the new count value with the latch/display content through the use of exclusive-OR gates as shown in Fig. 16-4.

In this approach any discrepancy between the displayed information and a newly counted number is detected by the Exclusive OR comparators, making the error counter increment by one. Although the latch enable pulse will be present at the end of the count as we previously discussed, the AND gate will not allow it to go through unless the error counter further detects changes in the counter content.

In the case of the flicker phenomenon, the last digit will jump back to the previous value if there was no change in frequency. This value is exactly that contained in the display, and when compared with it, will output all zeros to the OR gate, which in turn will reset the error counter through the inverter R.

Unless there is a significant change in frequency which was confirmed several times by the error counter (the value of n_e determines how sure we want to be), the display will solidly stay on the received frequency, almost totally ignoring the flickering phenomenon. I built a counter with a resolution of 100 Hz and connected it with a Vfo which had a drift of ± 40 Hz/hr at room temperature. The Vfo was locked in one position and no change in the readout frequency of the counter was observed over a period of one and a half hours.

The trend toward digital frequency counters in communications receivers leads to the introduction of a wide variety of digital components expressly made for this application. It is possible today to buy integrated circuits which will perform this function. Figure 16-5 shows the functional diagram of the Mostek, MK 50395 integrated circuit. This device includes a six-decade synchronous up/down-counter, storage-latches, a comparator register, and display driver circuits. A relatively inexpensive frequency counter can be built around this device. However, the multiplexed output

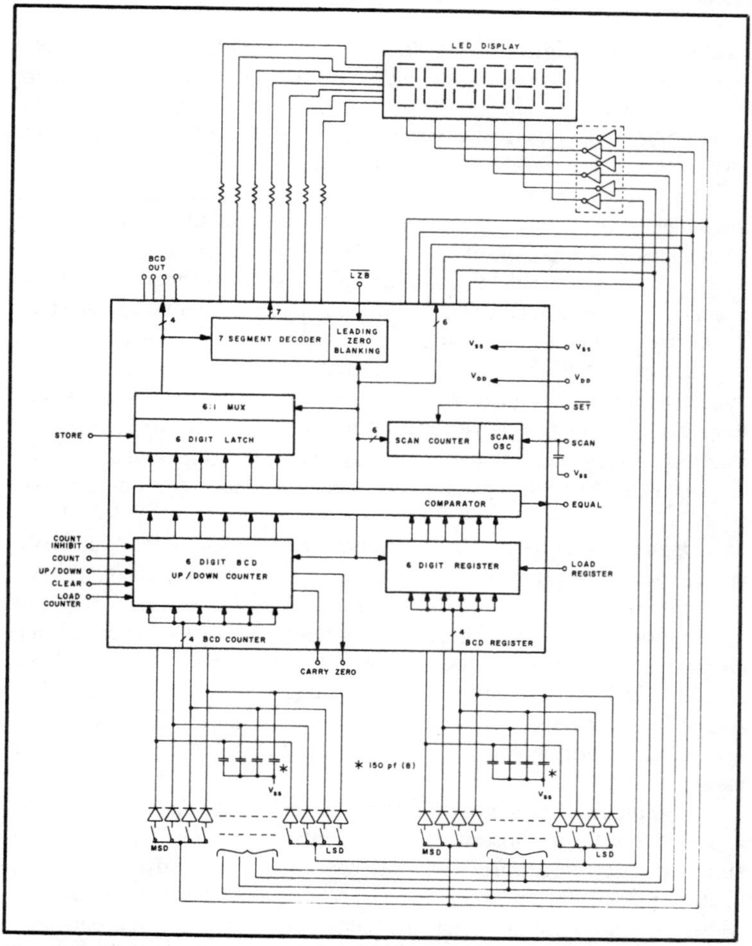

Fig. 16-5. Functional diagram of the Mostek MK 5039 digital counter.

Fig. 16-6. Several forms of numerical displays used in a communications receiver.

can be of a concern in a communications receiver since its noise can directly interfere with the receiver. The static display design presented earlier allows for minimum radiation to the receiver and is superior to the multiplexed mode which generates various unwanted spurious frequencies on each edge of the square-wave signals.

To further improve on the noise performance of frequency counters, low speed digital devices should be used wherever possible. CMOS, L and LS technologies make ideal choices in implementing frequency counters. Proper layout, ample shielding and careful bypassing will help considerably in avoiding spurious interference. When buying a receiver equipped with a digital frequency readout, listen for a multitude of noises at different frequencies with the antenna disconnected. The choice of display is also important in the design of a frequency counter for a Communications Receiver. Figure 16-6 shows several forms of displays, from neon discharge, through electroluminescent, LEDs, and liquid crystal display (LCD) types. The most popular display today is still the light emitting diode type.

Open collector output stages are usually used to drive these displays, and the only external component used is a current-limiting resistor, however this type of display can produce unwanted noise due to relatively high current transients. Fluorescent gas discharge displays are used, but driving circuits become more complex, as high voltage is required to trigger these devices. The liquid crystal display is becoming increasingly popular today. Although this type of display requires an ac drive mechanism for the segments, the very low current requirement allows for low noise performance.

New multiplexed sinewave devices are being introduced today. They allow for transfer of information to the display at the zero crossing of the sinewaves providing nearly noise-free operation. The disadvantage of this method is that the segment drivers must be capable of driving peak currents which exceed typical light emitting diode (LED) current requirements.

In conclusion, simple and efficient frequency displays can be designed and built with the techniques presented here. Almost noise-free operation can be realized with the systems using sinewaves, and the flickering phenomenon is a problem of the past, when using ambiguous digit detectors.

—•••—

Chapter 17

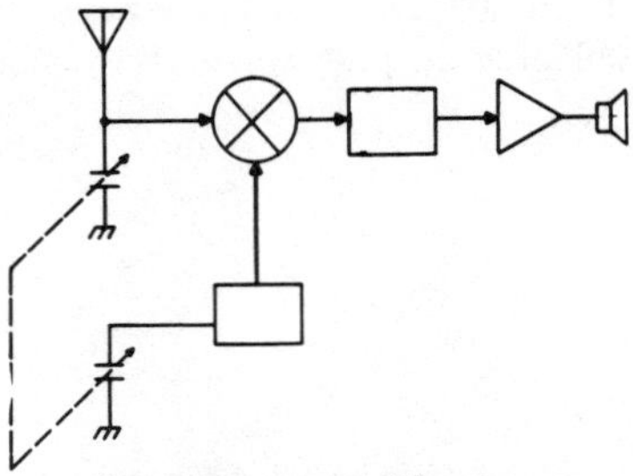

Intermediate Frequency (I-f) Amplifier Design

We previously discussed the importance of low intermodulation distortion in the front end and the first mixer of a communications receiver. The diplexer was also considered, as a means of maintaining the low intermodulation distortion, all the way to the intermediate frequency amplifier. In addition, the noise performance of the i-f amplifier is crucial to the performance of the receiver. Without this, the receiver's noise figure could be dominated by the noise figure of the i-f system and the loss of the crystal filters used in it.

Figure 17-1 shows an i-f amplifier designed for the 9 MHz i-f communications receiver. In this design, the traditional diplexer has been eliminated, as the input resistance of the Crystalonics CP-643 can be precisely matched with the 50 ohm output of the preceding mixer by adjusting the 250 ohm potentiometer. A total of 95 dB of gain is achieved with this approach, although an additional 15 dB can be obtained from the 1590's if needed, but with increased intermodulation distortion.

While the MC-1590 has a theoretical noise figure of about 6 dB, the i-f presented in Fig. 17-1 has a practical noise figure of 12 dB. This is compatible with the rest of the 9 MHz i-f receiver. It is interesting to note that some of the noise could be produced in the very high-gain stages following the crystal filter in spite of all the precautions taken in the layout.

In order to eliminate some of this noise, a second crystal filter which is matched with the first one in frequency, could be introduced at the output of the i-f as shown. This filter will almost completely eliminate the noise sidebands of the i-f amplifier.

Noise in an i-f can also be created by noisy local oscillators as we previously discussed, especially in synthesized systems which exhibit relatively poor phase noise performance. The effect can be noticed by gradually detuning a receiver with a steep-skirt crystal filter until the audible signal of a strong cw station falls outside the bandpass of the filter. If

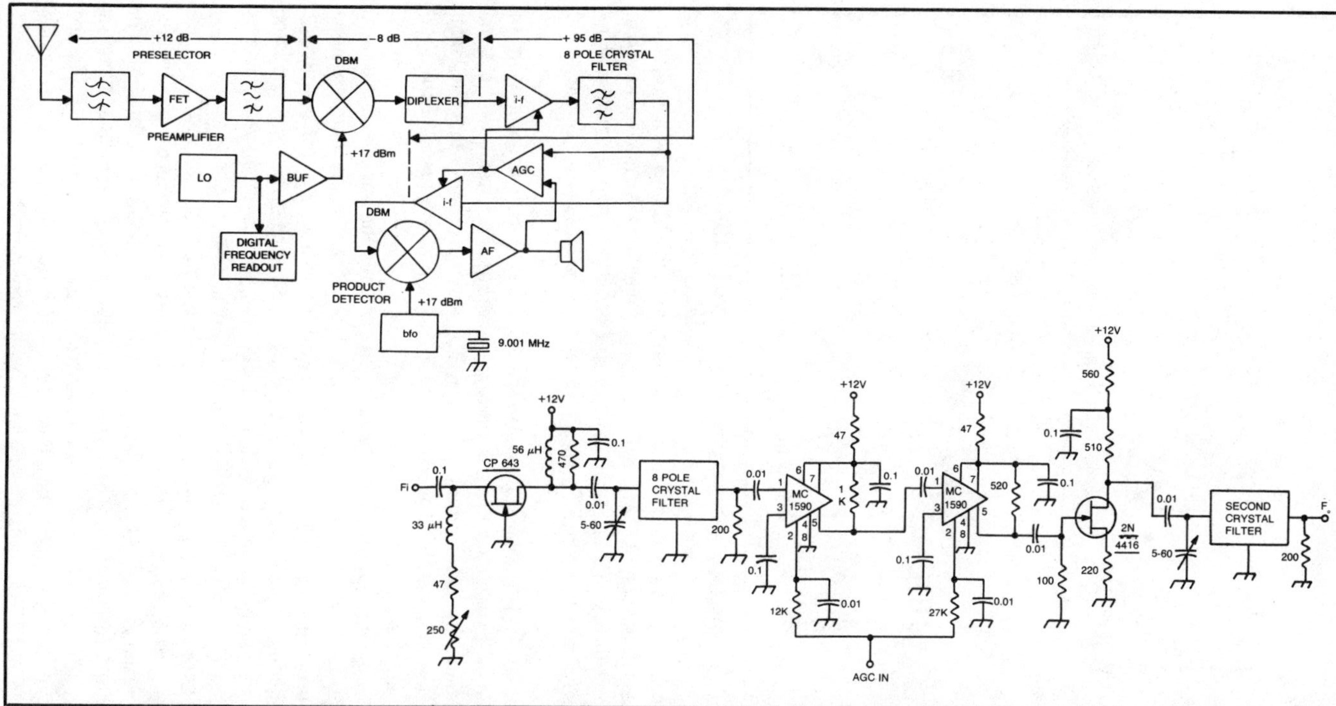

Fig. 17-1. A typical i-f amplifier as used in a 9 MHz i-f receiver.

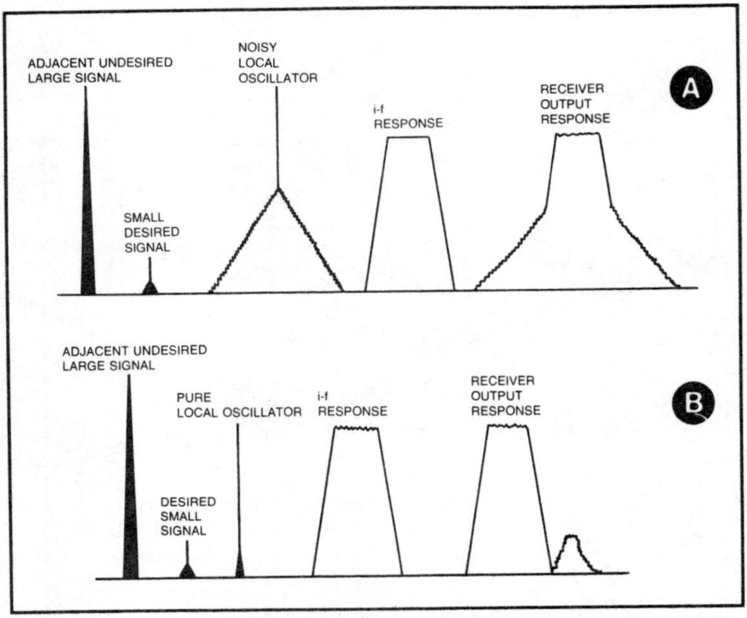

Fig. 17-2. Selectivity with noisy LO (A) and pure LO (B).

the receiver has a noisy local oscillator, an audible hiss would still be present at the output of the receiver, every time the cw signal was keyed in spite of the fact that the receiver is tuned to a different frequency. This "mushy" sound can greatly interfere with a small signal which we would like to copy.

Figure 17-2A, illustrates the impact of a noisy local oscillator on the i-f output of a communications receiver, while Fig. 17-2B, shows the same situation with a clean local oscillator. This example shows that the small signal is completely obscured in the i-f by the phase noise of the local oscillator.

It is believed that noise modulation occurs in the mixer and is caused by the mixer's balance. Using a balanced type mixer usually improves upon this problem. Although this is true to a certain extent, the real solution for eliminating noise modulation of an i-f is good filtering in the local oscillator but this can be quite a problem if a wide-range multi-loop synthesizer is used at the first local oscillator in a communications receiver.

Other popular integrated circuits such as the RCA CA3028 or the Plessey SL612 are often used in the i-f. They provide more than satisfactory results for many commercial and military receivers. FET amplifiers in balanced configurations are used for a lower noise figure and higher dynamic range. Figure 17-3 shows such an example. Two CP643 FETS are used in push-pull following a high-level mixer. Additional gain is provided by a dual FET amplifier using the 3N200 transistor.

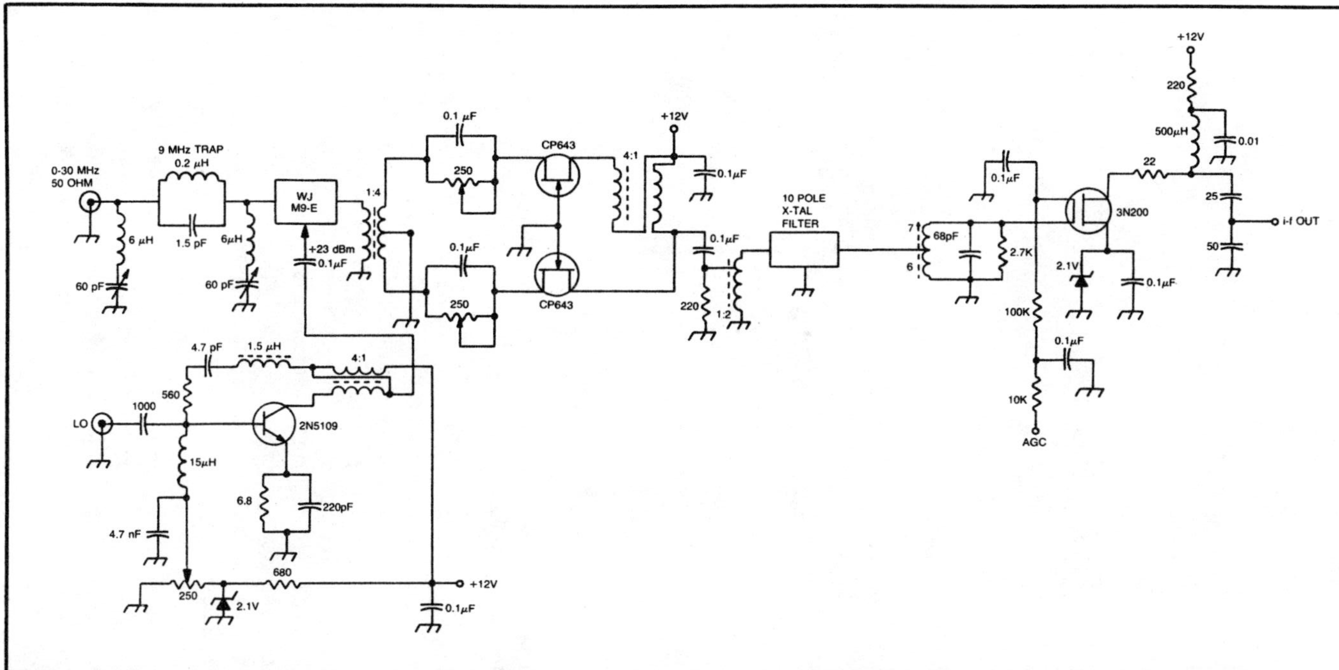

Fig. 17-3. First mixer and i-f amplifier for a communications receiver. Two CP643 transistors are used in push-pull following a class III high level mixer (WJ-M9 D). The 3N200 is used following the crystal filter.

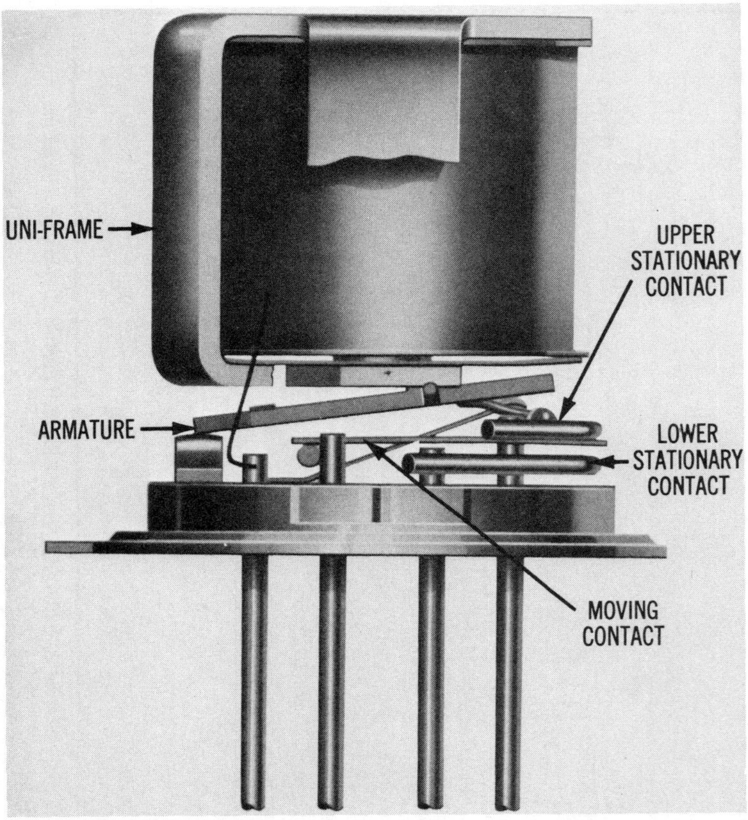

Fig. 17-4. Internal configuration of an rf relay which can be used to switch filters in the i-f of a communications receiver (courtesy of Teledyne Inc.).

Often, communication receivers require several bandwidths to be switched in and out in the i-f amplifier chain. The switching of different crystal or mechanical filters is typically accomplished with diodes in a way similar to that of preselector filters. Some receivers use FET switches for better isolation. Leakage and intermodulation distortion caused by the diodes becomes extremely important factors in the performance of a communications receiver, and some of the most modern radios today use relays to perform this function. Low capacity, high reliability rf relays are used to perform this switching.

Figure 17-4 shows the internal configuration of a Teledyne rf relay which was specifically designed for such an application. The relay is housed in a TO-5 package and can be mounted immediately adjacent to the filters. Two relays are usually used for each of the filters involved.

Other relays can also be used to perform this function economically as shown in Fig. 17-5. In this example, two double-pole single-throw relays are used to select between two mechanical filters.

Variable bandwidths can also be implemented in a communications receiver as shown in Fig. 17-6. In this approach, the bandwidth of the 455 kHz filter is slid through the bandwidth of the 9 MHz filter (by varying the 455 kHz bfo) and only the information contained in the overlapped portion of the two filters is passed to the audio amplifier.

NOISE BLANKERS

The i-f amplifier is usually the place where noise blankers are implemented. The ideal function of a noise blanker is to discriminate against all forms of electrical noise which could interfere with the wanted information. The degree of descrimination is limited by the nature of the noise and because of it, noise blankers take many forms. Shown in Fig. 17-7 is the oldest form of diode limiter. This method of reducing pulsating noise is usually implemented in the audio portion of a receiver. Although an i-f type could be implemented as shown in Fig. 17-7. This low-cost method provides relatively poor results compared to the more sophisticated approaches.

Shown in Fig. 17-8 is another method of producing a noise blanker circuit. The i-f amplifier chain is sampled for pulsating noise prior to the crystal filter, allowing for amplification and detection of fast rise-time pulses, which are amplified by the dc amplifier, which in turn will make transistor Q3 provide the ground path for the i-f signal during the pulse.

The net result of this effect is that the receiver will be blanked during the pulse, obscuring the noise, and providing much improved copy of the wanted signal. Figure 17-9 shows a chart recording of two receivers attempting to copy the same code information "CQ" through very heavy pulsating noise conditions, directly off the air. The audio output of the receivers were rectified through a simple half-wave detector and fed directly into the inputs of the chart recorder. The receiver at the top was

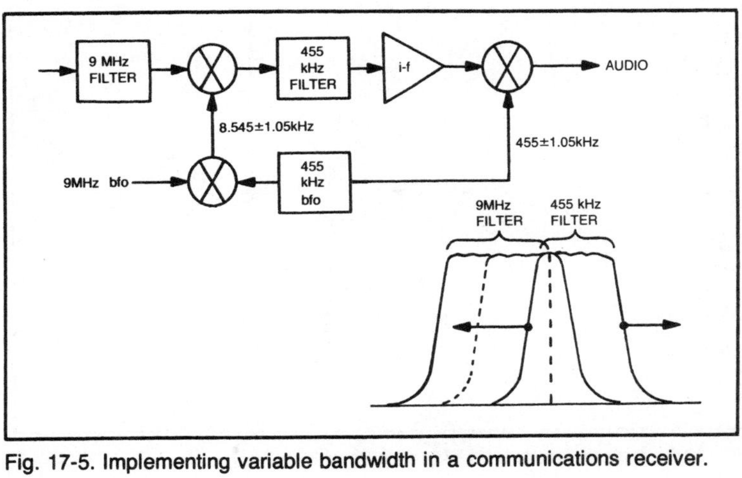

Fig. 17-5. Implementing variable bandwidth in a communications receiver.

Fig. 17-6. Two double-pole double-throw relays are used to select mechanical filters in this i-f amplifier.

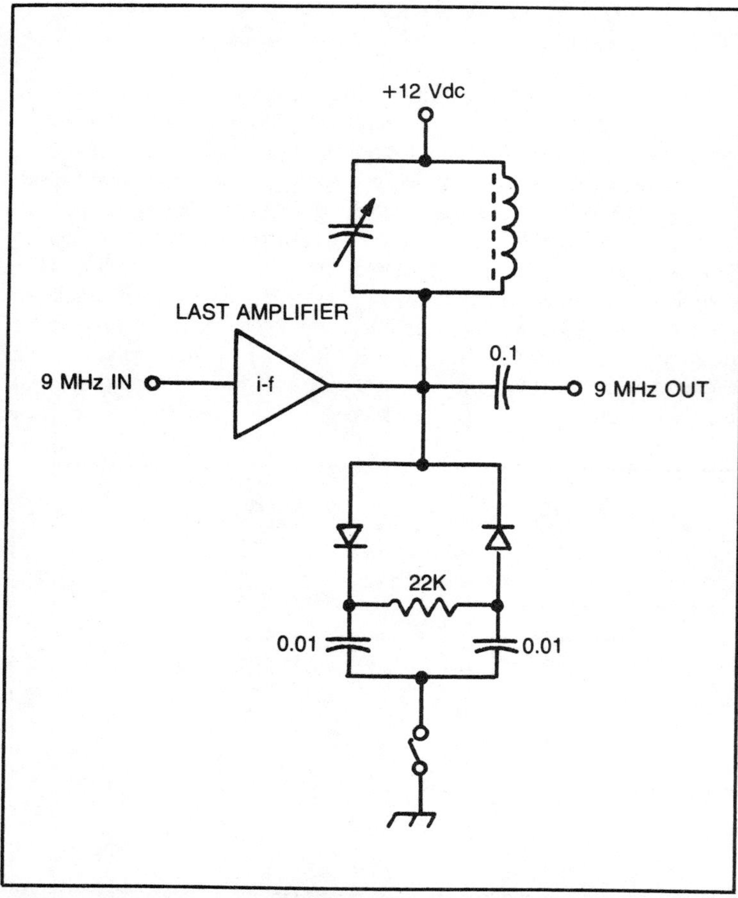

Fig. 17-7. A diode limiter noise blanker used in a 9 MHz i-f.

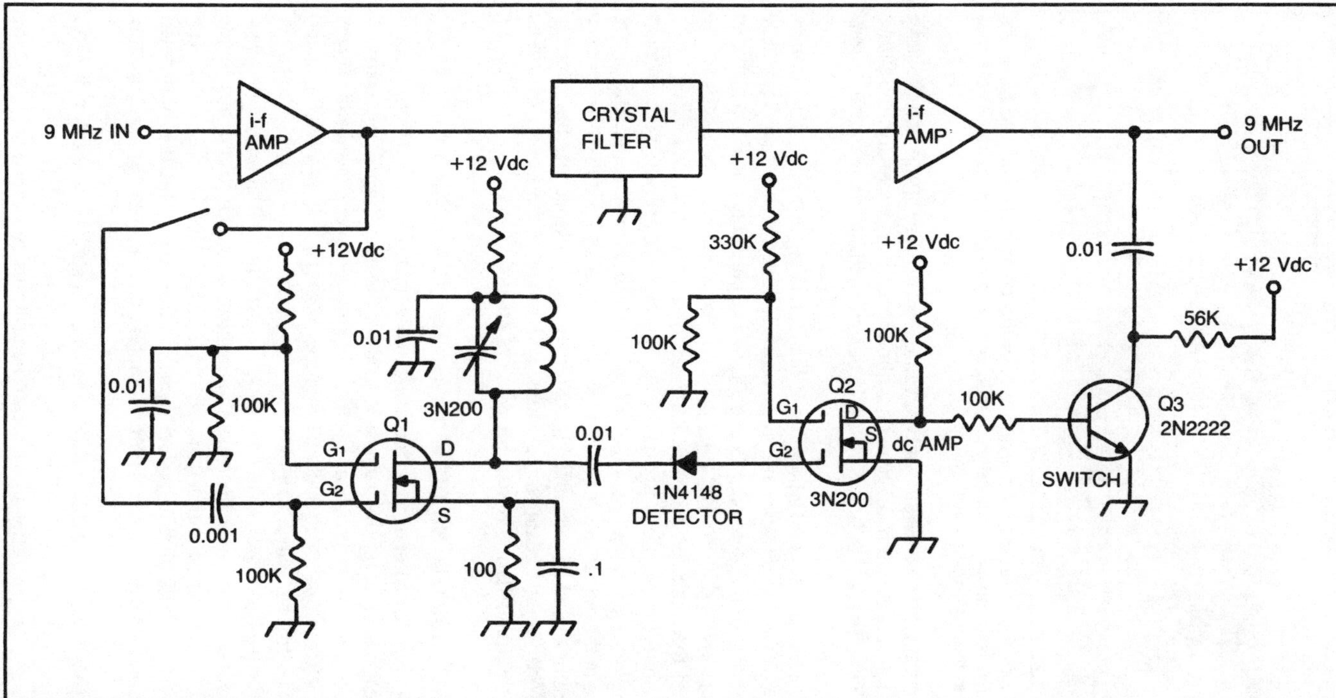

Fig. 17-8. A noise blanker circuit.

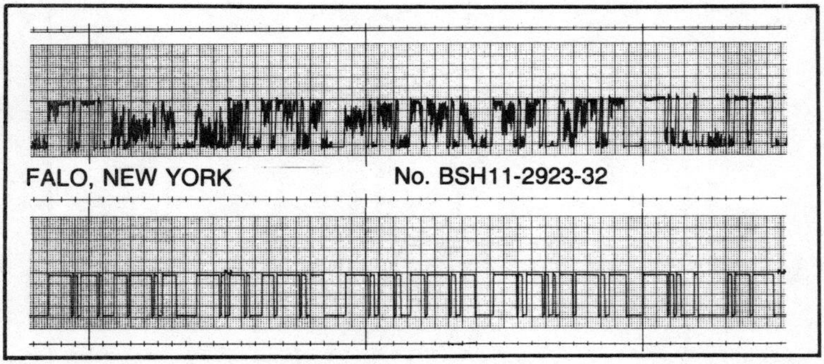

Fig. 17-9. Chart recorder information showing performance of two receivers copying the same information through pulsating noise. Receiver at the bottom is equipped with a noise blanker.

not equipped with a noise blanker, while the receiver at the bottom had a noise blanker circuit of the type described here. This demonstration shows that the noise blanker is a worthwhile design addition to a communications receiver.

There are several other noise blanker designs today. The Collins approach uses a separate receiver which is tuned to a frequency where there are no intelligent signals expected. The pulsating noise is processed through a superheterodyne approach and the detected signals are used to shut off the main receiver much the same as previously described.

While the methods presented in this chapter are effective in most noise situations, sometimes they do not perform as well depending on the nature of the pulsed signals received. New and complex noise blankers are sometimes used in communications receivers. They process the information received and adapt to the nature and length of the signals providing selective cancellation of certain types of pulses.

THE NOTCH FILTER AND THE BANDPASS TUNING MECHANISM

The notch filter is an interesting and very useful function in a communications receiver. Its purpose is to allow the operator to cancel out one or more unwanted signals which fall within the receiver's passband. This is usually accomplished at audio frequencies with the help of a tunable audio rejection filter of very high Q.

Figure 17-10 shows the schematic diagram for a variable Q, wide range notch filter which can be used in a communications receiver and can provide 40 dB of attenuation within its passband (300 Hz to 3 kHz). The design is a twin "T" network filter using feedback. The frequency control is performed by R2 and R3 which are two potentiometers coupled together on the same shaft.

Shown in Fig. 17-11 is the frequency response of the notch filter showing an unwanted cw interfering signal located by 100 Hz away from the received carrier in the i-f of a communications receiver. This example

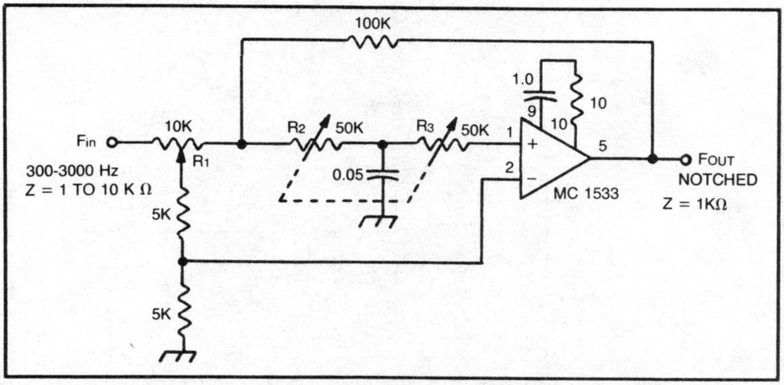

Fig. 17-10. Circuit diagram of a variable frequency audio notch filter.

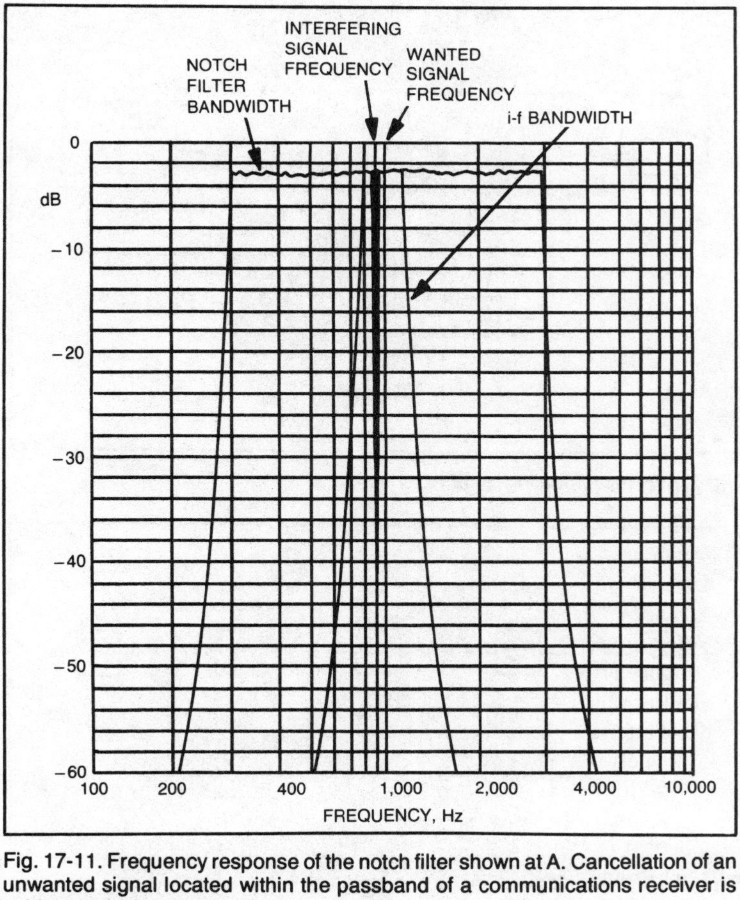

Fig. 17-11. Frequency response of the notch filter shown at A. Cancellation of an unwanted signal located within the passband of a communications receiver is evident in this example.

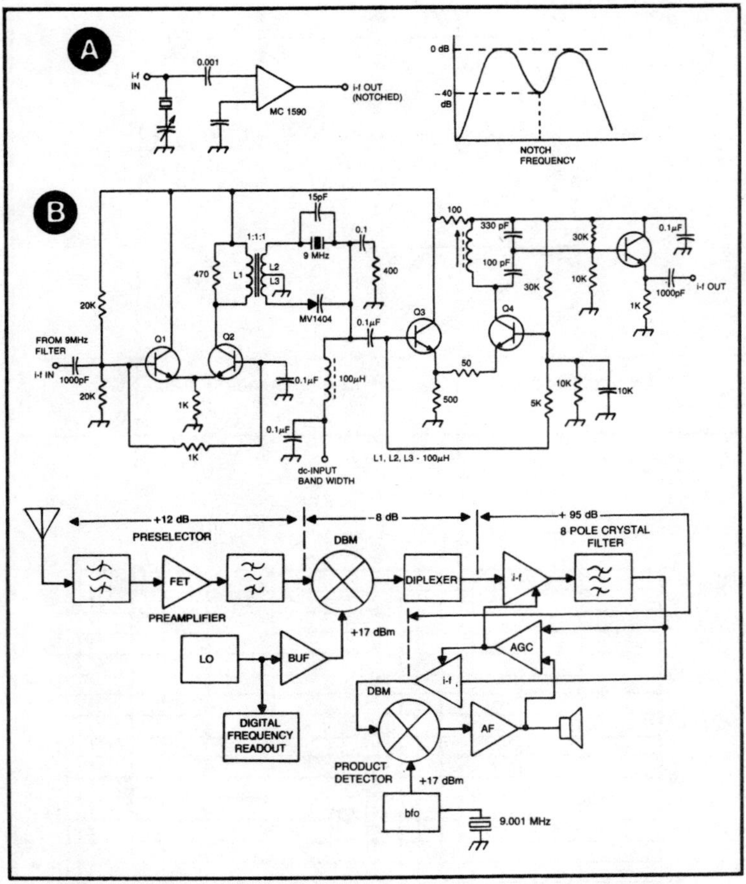

Fig. 17-12. Implementing an i-f notch filter.

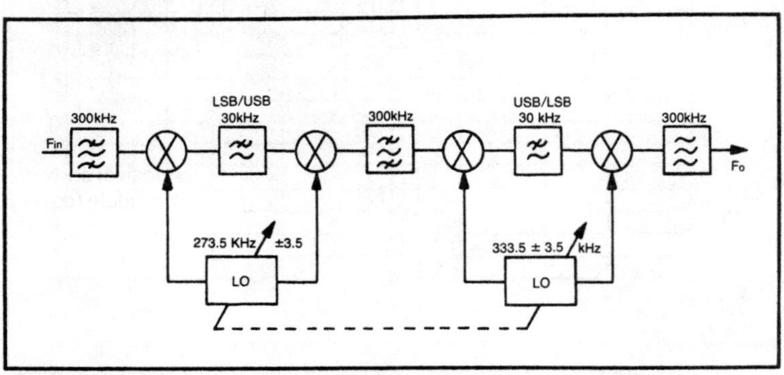

Fig. 17-13. Block diagram of the band-pass tuning process.

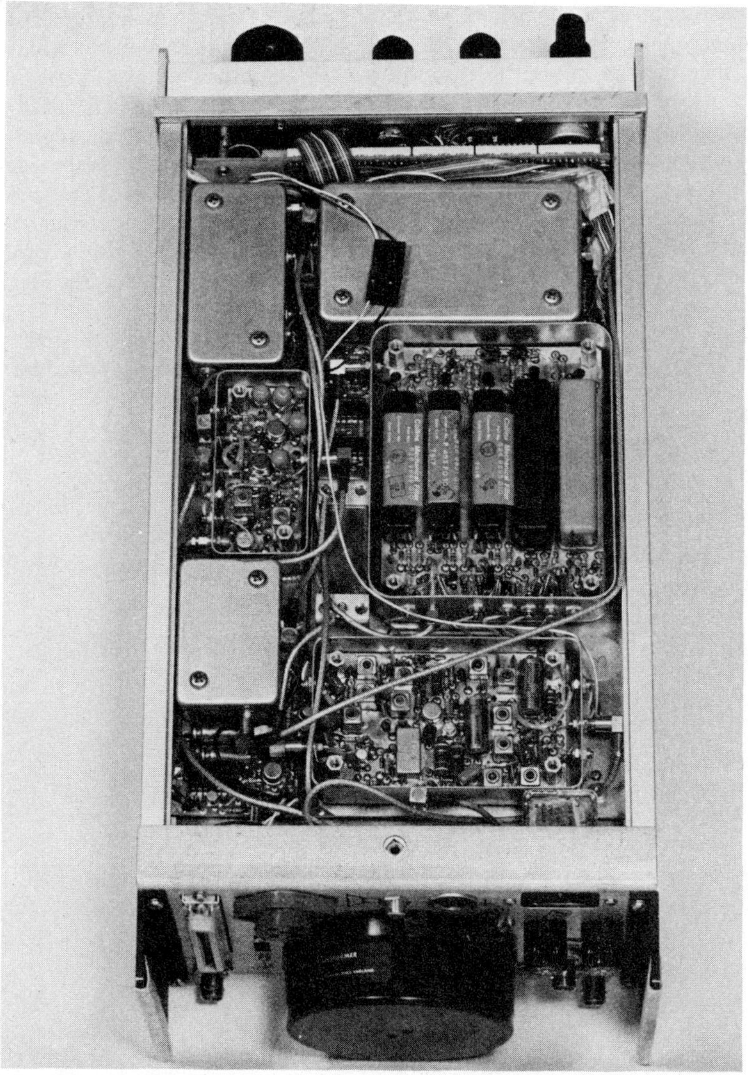

Fig. 17-14. Bottom view of the Cubic HF-1030 communications receiver showing the rf input module, the AM/agc module and the 455 kHz i-f filter module (covers removed) (courtesy of Cubic Communications).

shows that the notch filter is a powerful tool in combating unwanted interference. A more efficient way of implementing notch filters is to design them directly at i-f frequencies prior to AGC-ing.

Figure 17-12A, shows a means of pulling a 9 MHz crystal in an i-f providing a variable notch over the bandpass of the i-f. About 40 dB of cancellation can be obtained with this method. A more sophisticated approach is shown in Fig. 17-12B, where a dc voltage is used with a wide-range varactor in a series-resonant crystal configuration.

Another method of eliminating unwanted interference in a communications receiver is bandpass tuning shown in Fig. 17-13. This technique was introduced by Collins Radio Co. and was used extensively by several manufacturers. It provides continuous shifting of the audio passband response through the i-f bandwidth, allowing rejection of some unwanted signals. The system contains four mixers, three band-pass filters, two lowpass filters and two local oscillators which track together. Also, see Fig. 17-14.

This example shows that implementing bandpass tuning in a Communications Receiver is a complex matter. High-level mixers should be used to minimize intermodulation distortion which could be created quite easily in a multi-convertion system.

—•••—

Chapter 18

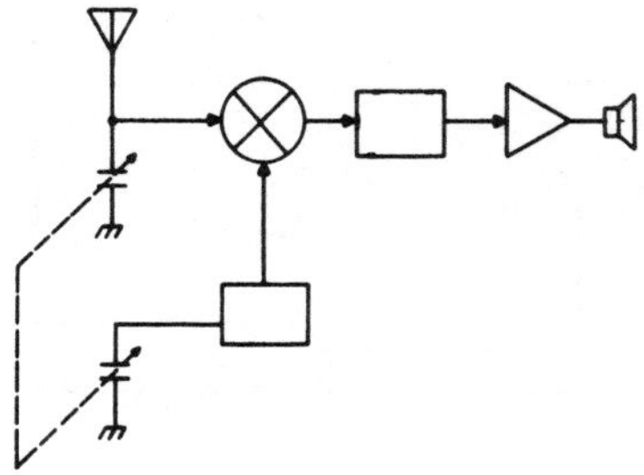

Automatic Gain Control (Agc)

Automatic gain control is necessary in communications receivers because of the wide variations in signal strength encountered at the antenna terminal. While small signals require more amplification in the i-f to produce the same audible output as the stronger signals, the opposite is true in the case of very strong signals, The agc mechanism is intended to follow these variations and properly compensate for them, so that the audible output will remain nearly constant.

The process is further complicated by rapid fading, accommodating for different modes and rates of transmissions, and suppressing the effects of a adjacent channel interference and noise. The agc can be viewed as a control loop as shown in Fig. 18-1.

Its performance characteristics are based on control system theory similar to that presented earlier. The difference is that instead of a phase/frequency-correcting mechanism it uses a gain-correcting mechanism. In a system such as the one in Fig. 18-1 the relative change in output voltage (expressed in millivolts) is proportional to the logarithm of the relative input voltage change (expressed in decibels). This often forgotten fact is usually the missing link in understanding simple automatic gain control systems.

On the practical side, the control voltage produced by the agc system must be applied rapidly to the gain stages at the beginning of each message element received. For receiving a single-sideband signal, a 2 millisecond risetime would be sufficient to overcome any thumping tendency at the beginning of a word. The release requirement is much longer (2 to 3 seconds). The fast-attack slow-release concept has been used extensively by the manufacturers, but the delay timing requirements have been disputed by operators. As a generality, typical circuits having attack times ranging from 2 to 200 milliseconds and release times ranging from 0.5 to 3 seconds have been used extensively.

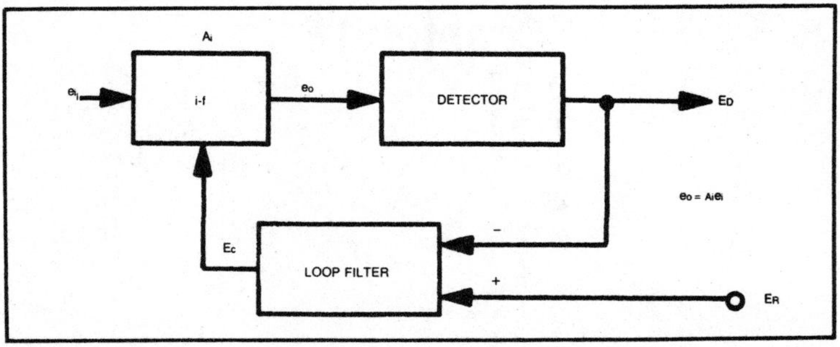

Fig. 18-1. Agc loop.

I-f DERIVED AGC

The i-f derived agc obtains its information at the output of the last i-f stage, usually through a simple diode detector. The averaged dc voltage obtained here is further amplified and applied to all the gain stages in the i-f amplifier, reducing the gain as necessary. Most communications receivers today use i-f-derived agc.

Figure 18-2 shows a practical implementation of an i-f-derived agc as applied to the 9 MHz i-f example communications receiver. The i-f output is detected and applied to a dc amplifier at Q2 which in turn controls the gain of the MC 1590's. The time constants are determined by C_1 and C_2.

AUDIO-DERIVED AGC

Another type of agc is the audio-derived. In this approach the correction voltage necessary to control the gain in the i-f stages is obtained from the audio section of the receiver. The rectified voltage is further amplified and applied to the gain stages of the i-f amplifier much the same way as in the i-f-derived agc.

This method of agc is not very popular in communications receivers because of its relatively poor performance. The narrow bandwidths of the signal at this point in a receiver does not allow for millisecond rise-times, further more in-band intermodulation distortion created in the i-f stages contributes to a phenomenon called "pumping" which is undesirable.

"HANGED" AGC

In a typical agc system, the release time depends on the discharge of a capacitor. Although this is a desired function for SSB or cw reception, the false alarm caused by a noise burst will deactivate the receiver for an unnecessary amount of time, thus the "hanged" agc was created. The "hanged" agc is designed to change the time constants and compensate for noise bursts, while still maintaining the characteristics of the fast-attack slow-release agc for SSB and cw signals. This type of agc can also be viewed as a noise discriminator. Such design usually involves i-f as well as audio-derived agc systems.

THE PIN DIODE ATTENUATOR USED FOR AGC

It is possible to apply agc to the front end of a communications receiver, providing attenuation directly at the antenna. This technique is used successfully to prevent front end overload, considerably reducing intermodulation distortion in this stage. This is accomplished by using a PIN diode attenuator which is used in series with the antenna terminal. The PIN diode will conduct rf only i-f biased properly. Figure 18-3 shows an actual implementation of an audio-derived agc which controls a PIN attenuator, as applied to the 9 MHz i-f model receiver. In this example the audio signal is amplified by U1 and Q1 and detected by diodes D1 and D2. This agc signal is further buffered by Q2 and fed to the base of a current-source transistor Q3. The PIN diode is biased forward from the +12 volt supply. The current will flow through the diode to ground as controlled by the current source transistor Q3 and the current-limiting resistor RL (100mA max), allowing variable attenuation to be achieved at the antenna input. As much as 40 dB of attenuation can be obtained with this approach. A

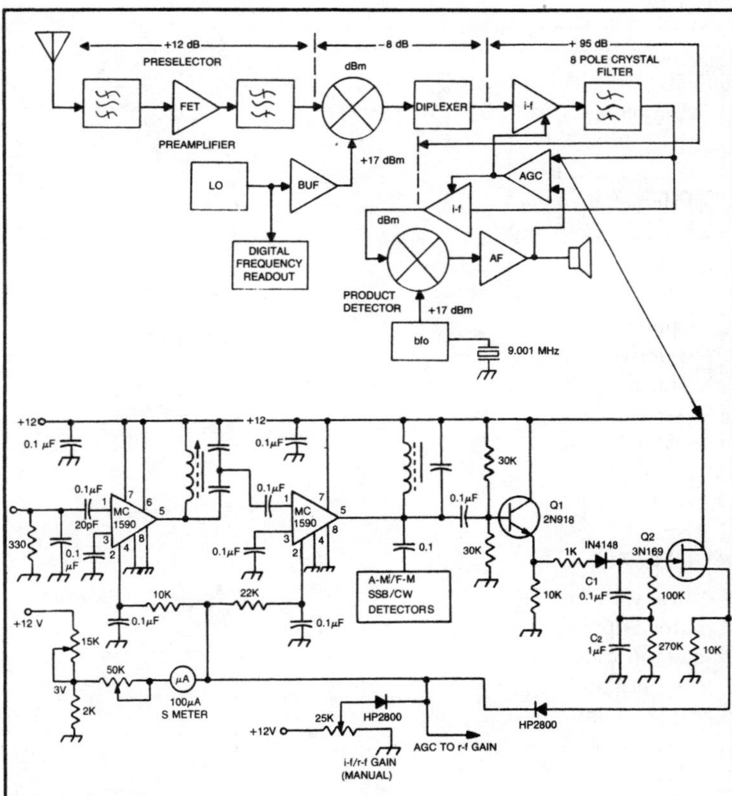

Fig. 18-2. Implementation of an i-f-derived agc.

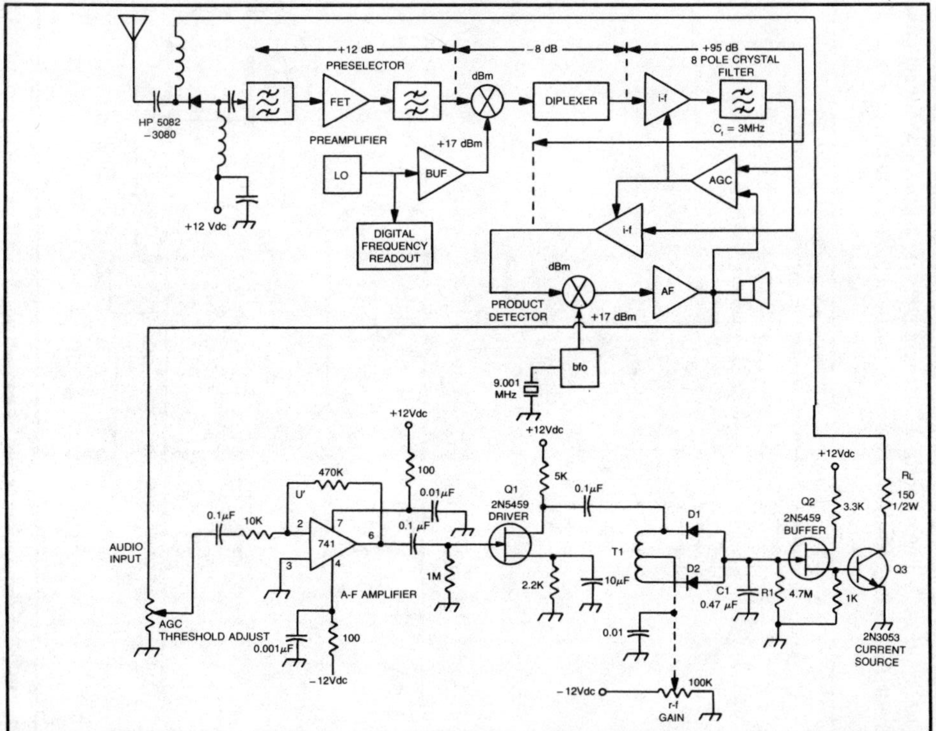

Fig. 18-3. PIN diode agc.

ladder network of several PIN diodes could be substituted for the single diode providing a greater range of attenuation. Transistor Q3 is capable of a continuous collector current of 700 mA and should be heat-sunk if such of installation is contemplated. Figure 18-4 shows a four PIN diode attenuator using the HP-5082-3081. Also, see Fig. 18-5 and Table 18-1.

This approach can provide as much as 80 dB of attenuation, but greater attenuation can not be achieved because of the physical layout which make the "wrap around" leakage phenomenon the limiting factor. PIN diode attenuators can also be used in i-f chains in order to improve upon existing conventional agc's. The technique is, however, somewhat reluctantly used because of the possible intermodulation distortion which it might create.

SPECIAL AGC'S

Another interesting agc method is provided by the rather unusual Plessey, SL 621 IC. This agc generator combines an audio amplifier and two peak detectors with two different time constants, one short and one long. A careful choice of these values will allow the desired independent selection of the attack, hold, and release times. Because of its special design, this system will ignore a noise burst, but will follow a rapidly fading signal, as well as maintain the same agc level for a pause in speech. The performance of the Plessey SL 621 agc generator is shown in Fig. 18-6. It can be seen

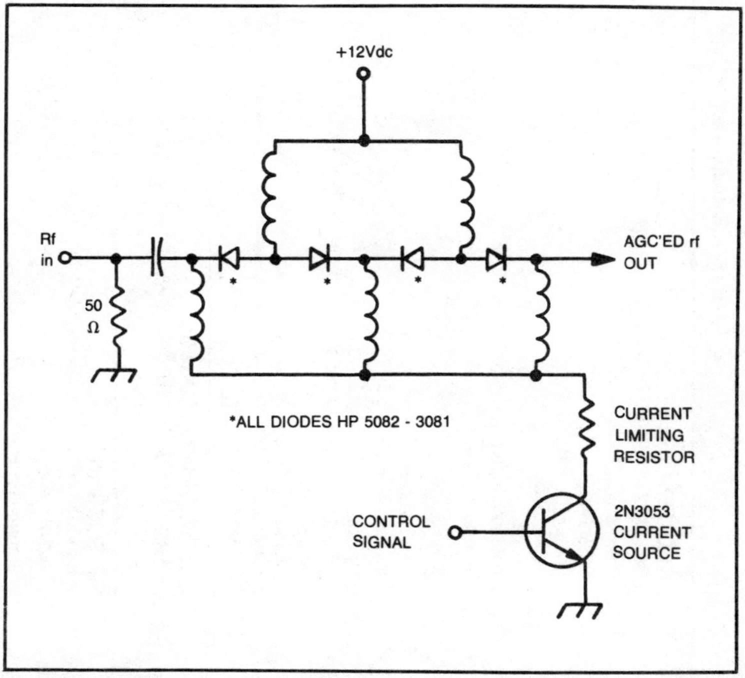

Fig. 18-4. An rf agc using four PIN diodes.

Table 18-1. Specifications of the HF-1030 Communications Receiver.

FREQUENCY

Range: 10kHz — 29.99999 MHz

Resolution: 10 Hz steps

Stability:
(Internal Standard)
1ppm/month
1Hz/°C
1ppm after 15 min warm-up at +25°C

Tuning Modes:
1. Lever control thumb-wheel switches
2. Opto-electrical shaft encoder; 1800Hz and 18kHz/revolution
3. Remote control BCD

Display:
7 digit high intensity red LEDs and thumbwheel switches

Power Interrupt:
Stored frequency data is retained for up to one year. Upon restoration of power, receiver automatically returns to previously tuned frequency

DETECTION CLASSES

Type: AM(A3), CW(A1), SSB(A3J)

RF SECTION

Input: 50 ohms unbalanced, BNC female

VSWR: less than 3:1

Sensitivity: For 10dB S/N
(1-30 MHz typical)

INTERFERENCE IMMUNITY

I-f Rejection: greater than 100dB

Image Rejection: greater than 70dB

Cross Modulation:
Unmodulated wanted signal of 100uV together with a modulated (30% at 1kHz) unwanted signal of 100mV spaced 30kHz apart produces less than 10% cross modulation of wanted signal.

Blocking:
Attenuation of a wanted RF signal of 50uV and caused by an unmodulated unwanted signal of 1V spaced 30kHz away is less than 3dB

Oscillator Re-Radiation: 1.0uV typical

Mean S/N Radio of First Oscillator (typical referenced to 1Hz bandwidth): 90dB, measured 1kHz from carrier

Intermodulation Distortion: (Typical) Third order input intercept point is at least +20dBm for signals separated by 30kHz

Second order IMD is greater than 80dB

INPUT/OUTPUT

Wide Band I-f Output: 40.455 MHz

IF Output: 455kHz; 250mV into 50 ohms (BNC)

Reference Oscillator: Switch select for INT/EXT (Common BNC jack)

AM	(A3)	1.5uV
CW	(A1)	0.3uV
SSB	(A3J)	0.5uV

Preselection:
Automatically selected filter, 10kHz-1.6MHz or 1.6-30MHz; 1.6 and 32 MHz lowpass, plus 40.455 MHz and 40.000 band reject filters

Gain Control:

Type: Automatic and Manual

AGC Range: 120dB, minimum

AGC Threshold: 0.2 uV mimimum; Inputs of 1 uV to 1V produce less than 6dB audio level change

AGC Time Constants: (typical)
SSB: Attack Time: <20ms
Hold Time: adjustable
Release Time: 100ms
AM: Attack Time: 100ms
Release Time: 100ms

I-F SECTION

I-f Frequencies: 40.455MHz and 455kHz

I-f Bandwidths: 40.455MHz ±4kHz at 6dB
455kHz (typical)

CW	0.375kHz at 3dB
	3.0 kHz at 60dB
FSK	1.2 kHz at 3dB
	5.5 kHz at 60dB
SSB	2.2kHz at 3dB
	4.5kHz at 60dB
AM	5.8kHz at 3dB
	14.0 kHz at 60dB

(NOTE: Above bandwidths are standard. Special bandwidths are available at extra cost on special order.)

Ripple: typical 1dB

Output: 1MHz at 1V (1000 ohms)
Input: 1MHz at 1V (1000 ohms)

Audio Outputs:
600 ohm balanced 0dBm nominal, adjustable over a range of −10 to +3dBm

4 ohms at 2 watts with less than 10% distortion

Control: Via 50 pin 3M 3489 Connector
1. BCD (1-2-4-8) 12V CMOS parallel command for Frequency Selection
2. Mode Selection (Ground to Select)

Monitor:
1. AGC output (Analog)
2. BCD (1-2-4-8) parallel output for frequency readout

GENERAL DATA

Temperature Range: 0°C to 50°C Operating
−25°C to 85°C Storage

Relative Humidity: to 95%

Power Requirements:
115/230 VAC ±15%
47-400 Hz at 25 watts
12-30 VDC at 1.2 amp (Amphenol 91/T/3263/9)

Dimensions (mm):
210W X 133H X 368L
(8¾ x 5¼ x 14½ inches)

Weight:
8.2 kg (18 lbs) net
11.4 kg (25 lbs) shipping

Fig. 18-5. Model HF-1030 general coverage communications receiver features a fast switching synthesizer (25 msec switching/settling speed) with a resolution of 10 Hz. The intercept point is + 20 dBm (courtesy of Cubic Communications).

from this figure that the detector with the larger output controls the system. Therefore, if a noise burst was received during speech the long time-constant detector output will dominate, and the noise burst will be ignored. The short time-constant detector will only be activated by the noise burst providing short agc signals, thus contributing to the cancellation. Although the SL 621 is an audio-derived agc system, it qualifies as a "hanged" agc by definition.

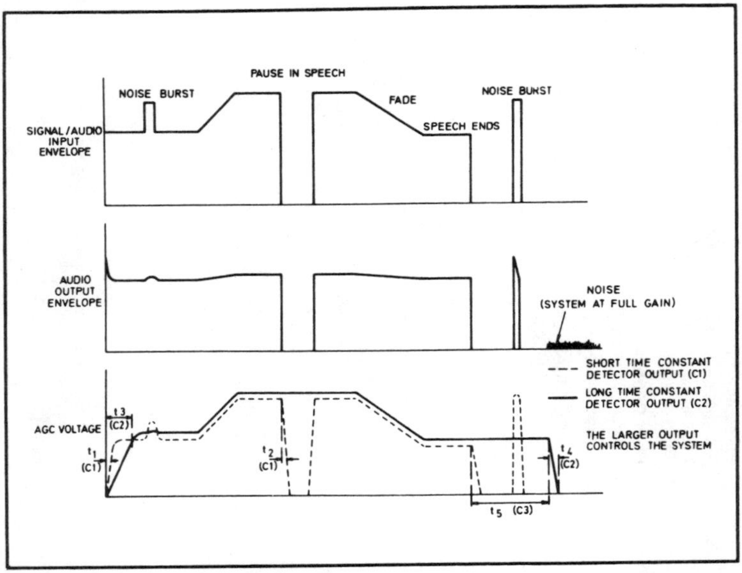

Fig. 18-6. Performance of the Plessey SL 621 IC.

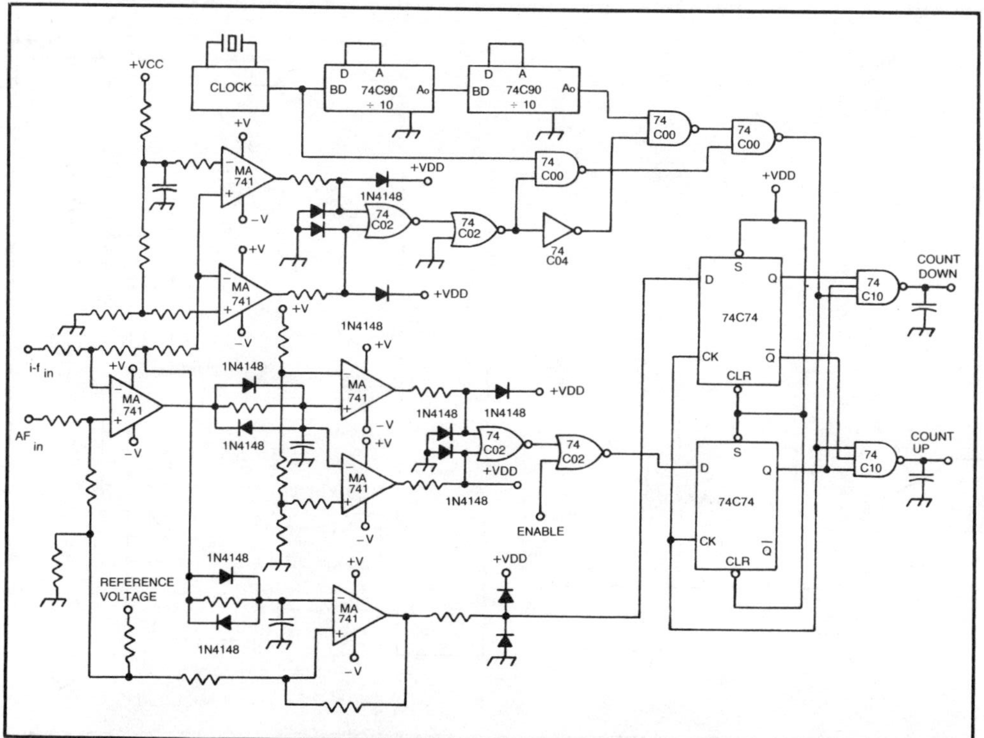

Fig. 18-7. Implementation of a digital agc.

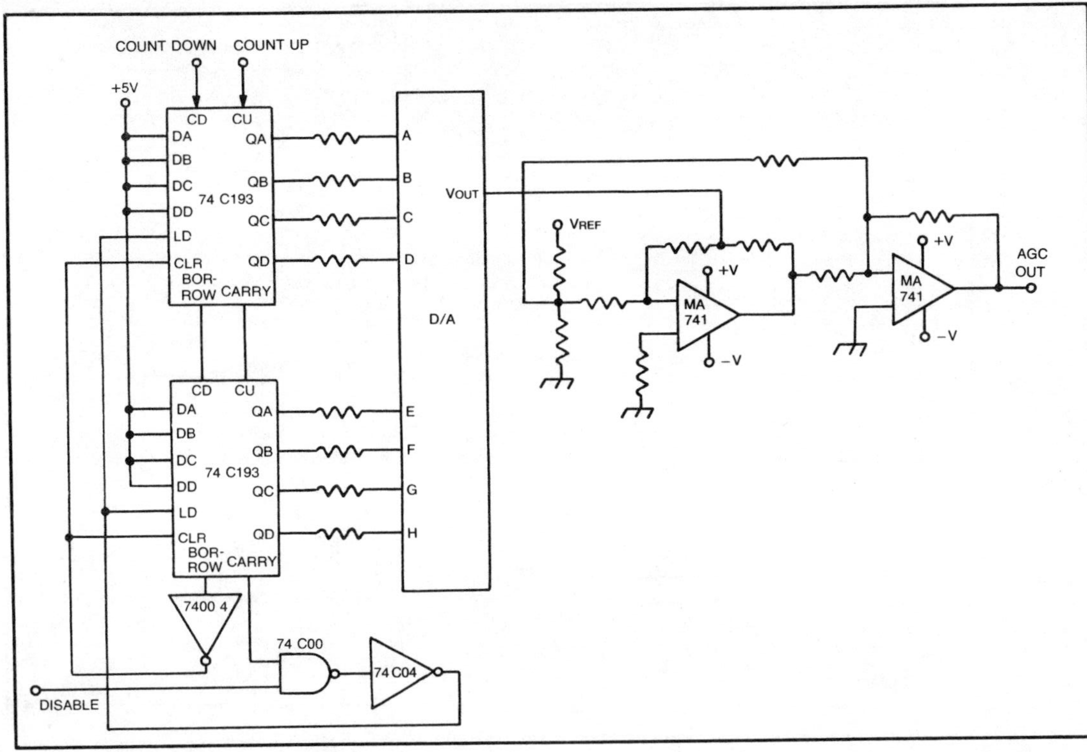

Fig. 18-8. Implementation of a digital agc.

DIGITAL AGC'S

Another agc method which is relatively unknown is the digital agc. An implementation is shown in Figs. 18-7 and 18-8. In this approach, the agc output is a function of an eight-bit digital number which is present at the binary inputs of the digital-to-analog converter. This number is continuously changing, and reflects the counting state of two cascaded up/down counters (74C193) at any given time. If a clock was fed to the counters and the count-up command is activated, the agc voltage at the output of the digital-to-analog converter will increase with every count until reaching the maximum voltage allowed by the number of bits. It is evident that the attack time of this voltage is directly proportional to the frequency of the clock. The higher this frequency is, the faster the maximum digital number (and therefore the maximum agc voltage) is reached.

If the count-down line is activated on the up/down counters, and the clock was slowed down, a reduction in the agc voltage will be accomplished, but this time at a slower rate than it went up. The attack and release times can then be programmed at will by merely changing between two clock speeds when selecting between the count-up and the count-down modes. The two frequencies are usually 40 Hz for the attack time (count up) and 4 Hz for the release time (count down). Signal level information from two points of interest in the receiver (usually the i-f and the audio amplifier) control the gating of the properly chosen clock frequency for the up/down counter, as well as the up or down commands.

This system can be imagined as a continously changing counter which is going up fast and coming down slow, with the agc voltage following this pattern. Because of the complex switching involved in this type of agc, low-current logic should be used in the design. Our example was implemented with CMOS logic for this reason. This relatively expensive agc system has the advantage of being extremely reliable. The precise programmability of the attack and release times also makes it an ideal choice for a communications receiver.

—•••—

Chapter 19

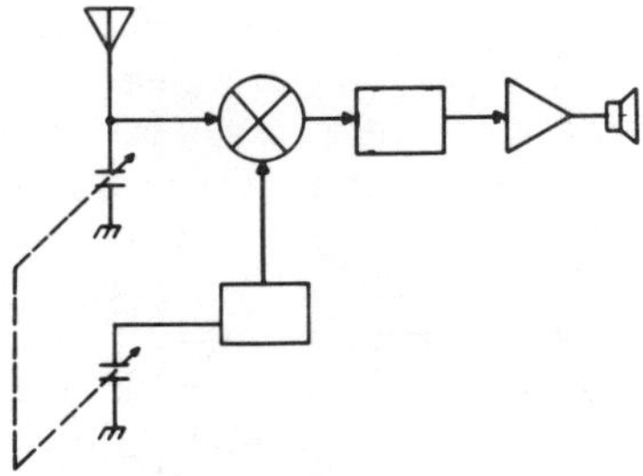

Product Detector and Beat Frequency Oscillator (Bfo)

The last stage of a communications receiver contains one or more detectors, a beat frequency oscillator, and an audio amplifier. While other types of detectors have been discussed earlier in this book, the *product detector* is used in processing single-sideband, continuous wave or frequency shift keying signals. A product detector is actually a mixer which is used in conjunction with the beat frequency-oscillator in order to inject the missing carrier to an SSB signal, or to produce an audible tone when a cw or an FSK signal is received.

Product detectors can take many forms. Figure 19-1 shows a circuit implementation for a diode mixer approach using 1N4148 silicon diodes as applied to the 9 MHz i-f receiver. A low-pass filter is used at the output of the detector in order to reject high-order products generated by the mixer. Shown at B is a circuit implementation for a bfo which can switch between two crystals allowing for upper and lower sideband reception.

Other types of product detectors are also used in communications receivers. Figure 19-2 shows an active product detector which allows for very low intermodulation distortion when using the Motorola MC-1596. These products are typically 60 dB with this approach, a remarkable performance for this active doubly-balanced mixer.

—•••—

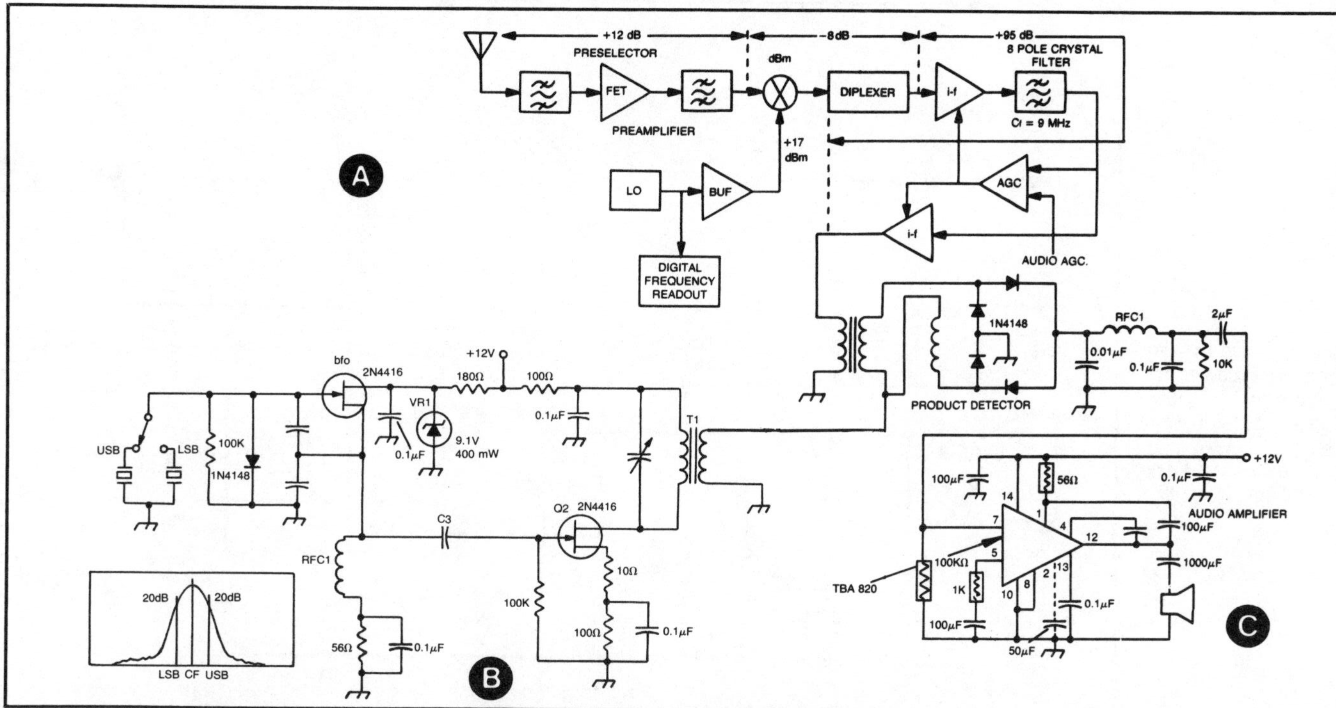

Fig. 19-1. Product detector, beat frequency oscillator, and audio amplifier implementations.

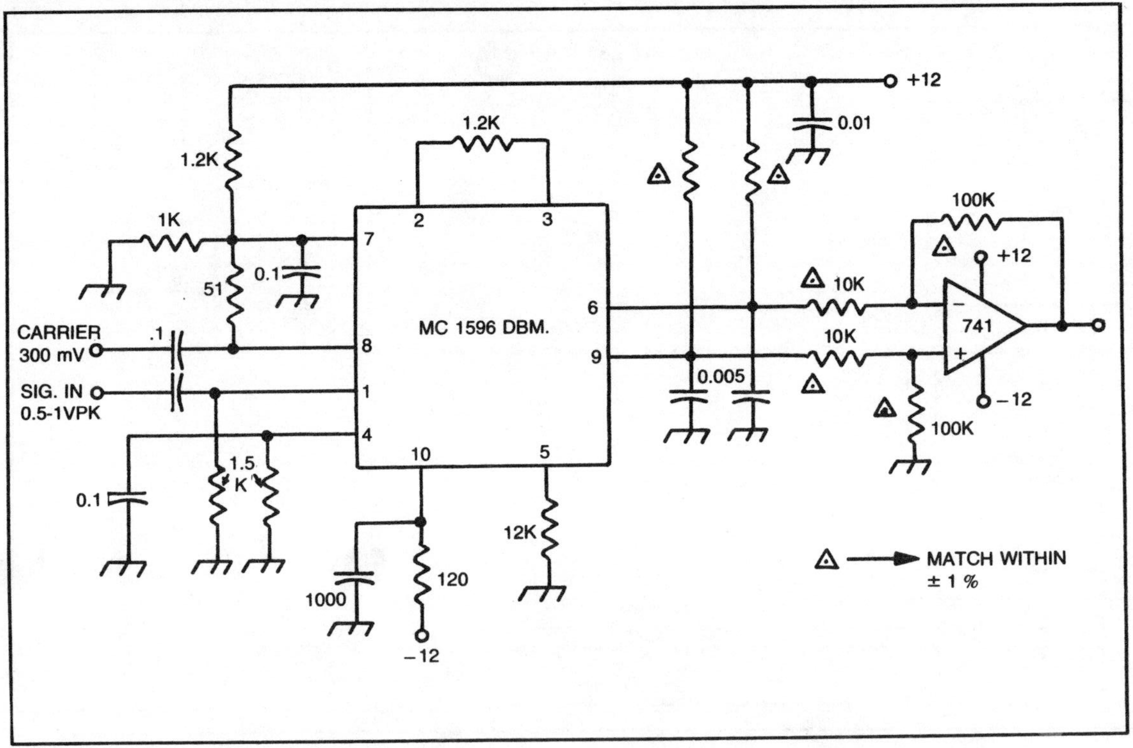

Fig. 19-2. Very low distortion active product detector. This scheme reduces inband intermodulation distortion better than 60 dB.

Chapter 20

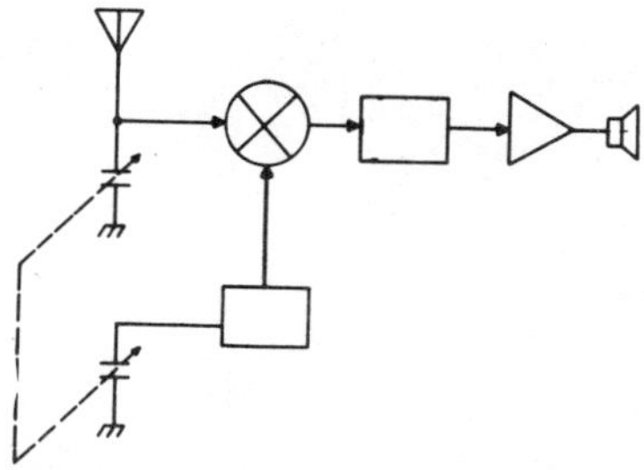

Audio Amplifier Design Considerations

Too many times, designers of otherwise well engineered communications receivers fall into the trap of using inexpensive, readily available audio amplifiers, especially in integrated circuit forms, without considering the harmonic distortion performance of these devices which sometimes exceeds 20%. The use of a good audio amplifier is imperative in the design of a high quality communications receiver. A well designed amplifier capable of continuous power output of 2 to 3 watts, (Institute of High Fidelity-IHF) and exhibiting a total harmonic distortion of no more than 1.5% is highly recommended here. Tailoring of the frequency response to match that of the i-f is also highly recommended. Usually this is achieved by installing a high-pass and a low-pass filter network prior to the amplifier as shown in Fig. 20-1.

Figure 20-2 shows the schematic diagram for a 3 watt audio amplifier which exhibits a typical harmonic distortion performance of less than 1%. This design uses two complementary-symmetry, emitter-follower power transistors. Negative feedback is used in this circuit to reduce distortion and improve stability. Two-tone intermodulation distortion was measured to be better than 70 dB down at 4 watts output using 400 Hz and 2 kHz tones mixed 4:1. Further tailoring of the feedback circuitry might be necessary to achieve this performance.

The intermodulation distortion performance of the audio amplifier is just as important as the intermodulation distortion of the rest of the receiver if performance is to be maintained throughout. This statement is also true for the transducer, the speaker, or headphones used. When building this amplifier, the final transistors should be heat sunk and electrically insulated from ground.

Shown in Fig. 19-1 is an integrated circuit audio amplifier using the TBA 820 which is rated at 2 watts. This design exhibits acceptable perfor-

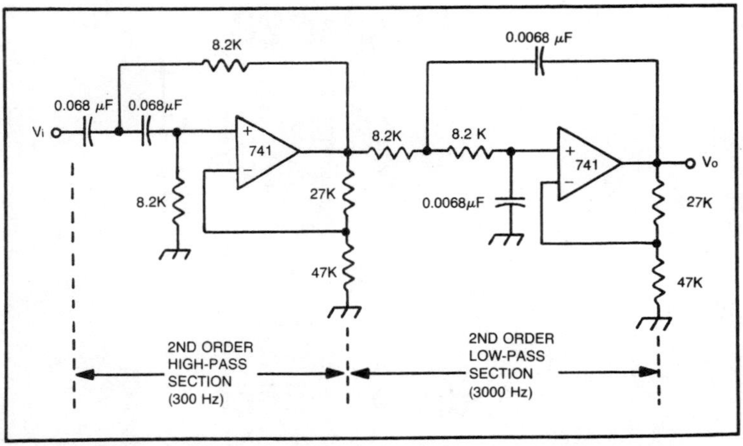

Fig. 20-1. Tailoring frequency response of an audio amplifier in a communications receiver.

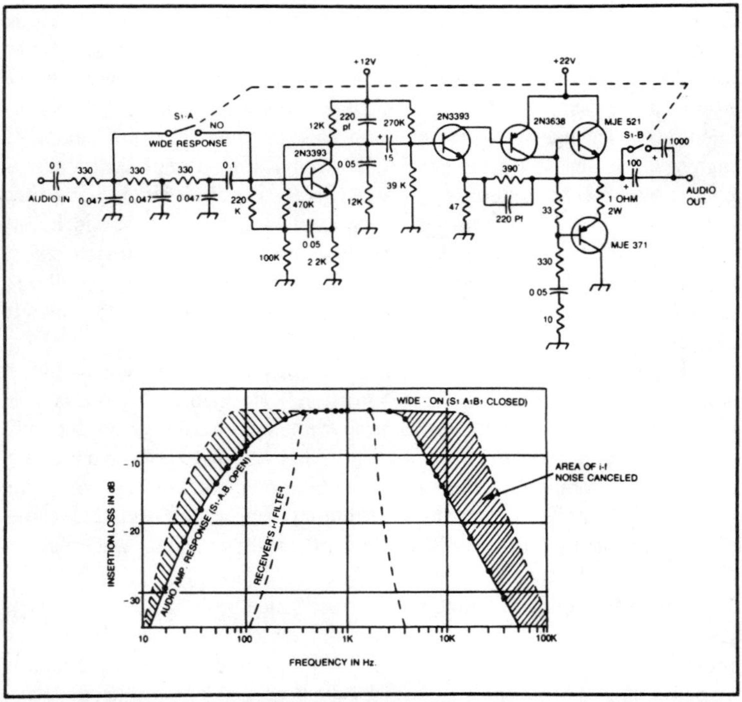

Fig. 20-2. Practical low distortion audio amplifier for a communications receiver showing frequency response. This design is tailored so that frequencies above 3 kHz are greatly attenuated, (S1-A, B, open) providing effective cancellation of noise generated in the stages following the i-f filter. A wider response can be achieved by closing switch S1-A,B as shown.

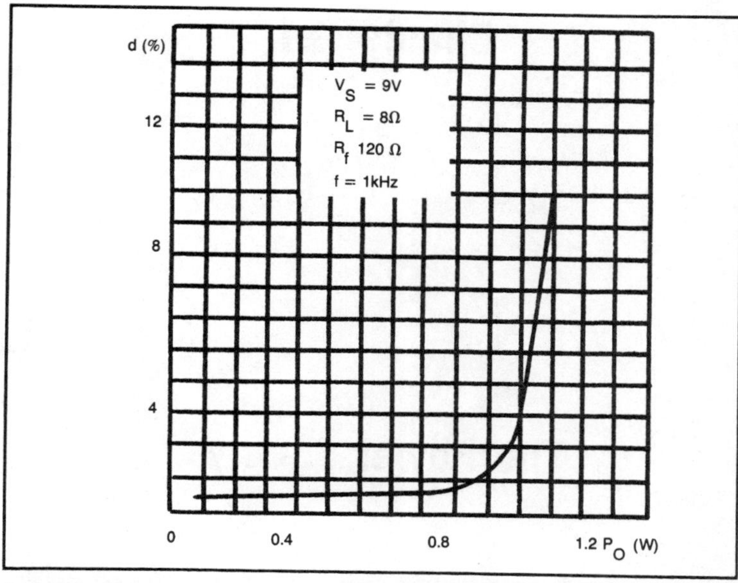

Fig. 20-3. Distortion performance of the TBA 820 audio amplifier versus power output.

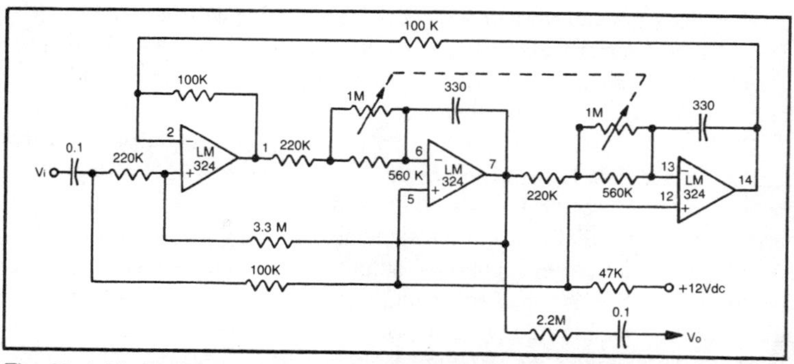

Fig. 20-4. Variable-frequency audio active filter. Range is 500 Hz to 2 kHz.

mance below 1 watt. Above this level, harmonic distortion increases rapidly accounting for poor performance especially when trying to copy weak signals. Figure 20-3 shows the distortion versus output power performance of the TBA 820.

Additional selectivity can be obtained in the audio portion of a communications receiver. This is accomplished through the use of active filters such as the one shown in Fig. 20-4. This is a state-variable filter using three operational amplifiers in a unity-gain configuration. This design provides sharp tuning for cw signals over the frequency range of 500 Hz to 2 kHz.

—•••—

Chapter 21

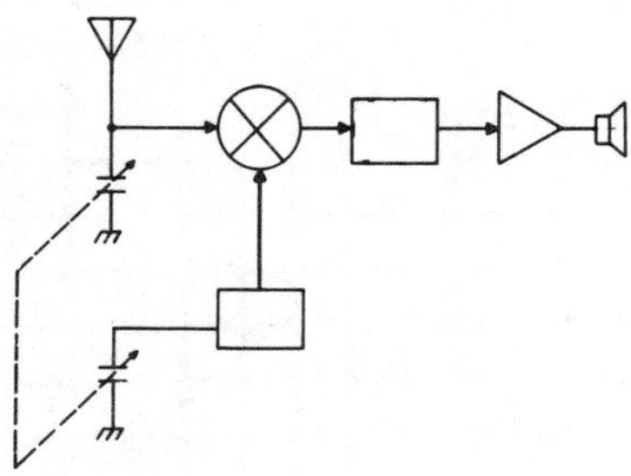

The Power Supply

The power supply should not be neglected when designing a high-performance communications receiver. High-efficiency full-wave rectification should be used if possible, coupled with double regulation and adequate filtering. It is not uncommon to find two series regulators feeding each other in a high-performance power supply as shown in Fig. 21-1. Switching power supplies should be avoided because of the noise generated that can easily pollute the receiver's sensitive circuitry.

I am against a single power supply in a communications receiver, because of the FM-like phenomenon resulting from it. Many low-cost communications receivers power their audio amplifiers as well as their variable tuning oscillators from the same power supply. Because of inefficient design in the power supply, the output voltage to these elements will vary as peak current is required for audio signal reproduction. The result is pulling of the variable tuning oscillator which is actually frequency modulated by the sound of the receiver.

This phenomenon is very annoying and can only be eliminated by rebuilding the power supply. When designing a receiver, use separate power supplies for the Vfo and i-f amplifiers. The audio amplifier should have its own power supply. Digital counters should also be powered from a separate supply, reducing the chance of digital noise feedback into the rf circuits.

Ground loops are also a consideration. When designing with multi-power supplies, there is no known way of predicting ground loops in a communications receiver, and they can be the final obstacle in achieving high performance. They manifest themselves by introducing severe amounts of hum in the output of the receiver. Good grounding techniques may or may not solve the problem. Although theoretically understood,

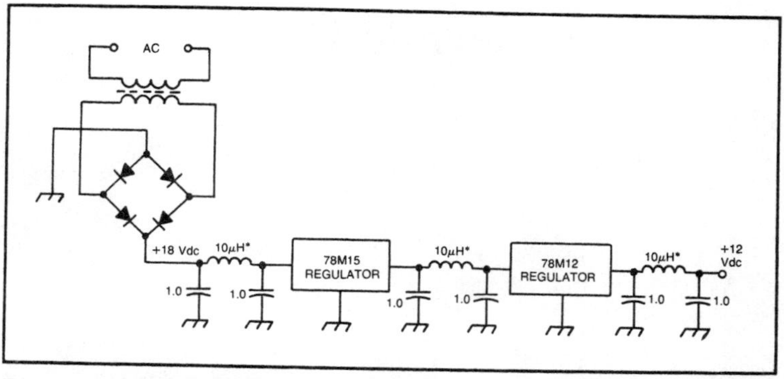

Fig. 21-1. Doubly-regulated power supply is used to insure superior performance in a communications receiver.

ground loops will almost invariably impact the performance of a newly designed receiver. Removing them has always proved to be a time consuming trial and error process.

—•••—

Chapter 22

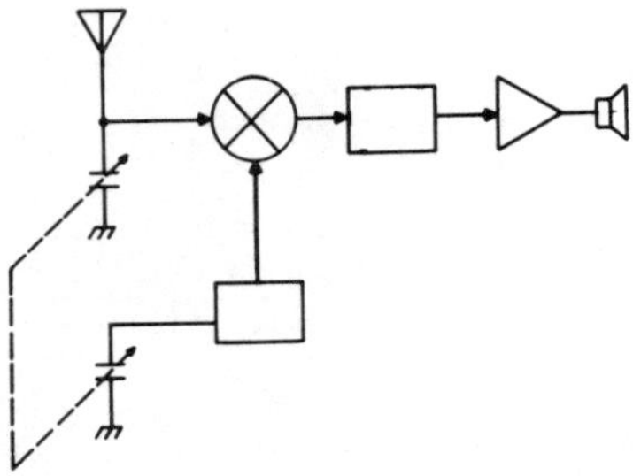

Putting It All Together

Designing and building a high performance communications receiver is probably one of the hardest tasks, an individual or a company can undertake. Although it is relatively easy to put together building blocks and combine them to achieve electrical performance, the implementation of these blocks in the total ensemble of the package can make the difference between failure and success.

The actual design of a communications receiver involves many sciences from mechanical to digital. Figure 22-1 shows the actual implementation of a high performance communications receiver which I designed and built. Electrically the system is a double conversion approach with a first i-f at 75 MHz and the second i-f at 455 kHz. A very high level mixer is used in the front end allowing for an intercept point of +22 dBm. A classic Vfo is used in conjunction with crystal oscillators providing a resolution of 2 kHz per revolution of the main knob. This, combined with narrow switchable filters makes for an extremely sharp communications receiver. To provide maximum isolation, the Collins mechanical filters used in the second i-f are switched with relays. Frequency is read through a digital display with a resolution of 100 Hz. The layout of the receiver shows a compartmented approach.

All circuits are built on 3½ × 5 inch PC boards and gold-plated connectors are used throughout. The advantage of this design is mainly that circuit modifications and improvements are possible without altering the actual receiver.

Over the years, many modifications have been implemented in a continuous search for technical excellence. Mechanically, the receiver is made of anodized black aluminum. All parts have been extensively machined and there are 186 screws in the system. The gear drive exhibits no backlash because of its different design. Light magnesium gears are used throughout. The variable frequency oscillator is driven by this gear train

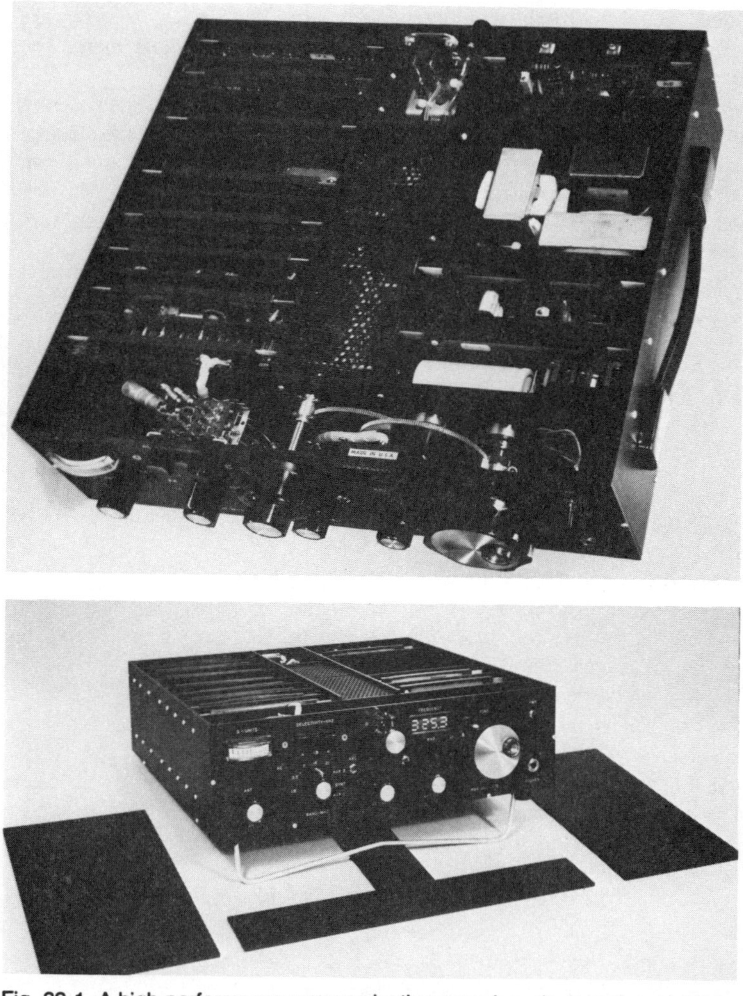

Fig. 22-1. A high performance communications receiver designed and built by the author.

and is located in a closed compartment in the middle row close to the variable tuning oscillator (board) which can also be unpluged. The receiver has three tops, a T-like top which covers the mechanics, and two hinged side covers which cover the electronics. Upon closing the covers, each circuit board is fully isolated in its own housing because of the walls between the compartments. The electrical interconnections between the boards are provided through two motherboards located on the bottom of the receiver. All interconnect circuits are either shielded or etched into the motherboards. The bottom cover is only one-quarter of an inch from the circuits, providing extra shielding. There are two extender boards provided, should any troubleshooting be necessary. When not in use, they are packed in a plastic bag and housed in the transformer compartment. The receiver weighs 15 lbs. and is fully portable.

Designing and building a high performance communications receiver is a good test of technical maturity for any individual who dares to attempt it. It is my opinion that such a multi-science test should be of significant value in asserting the engineering level of new college graduates, who will be faced with signal processing problems much similar to those encountered in a communications receiver in their future careers.

—•••—

Chapter 23

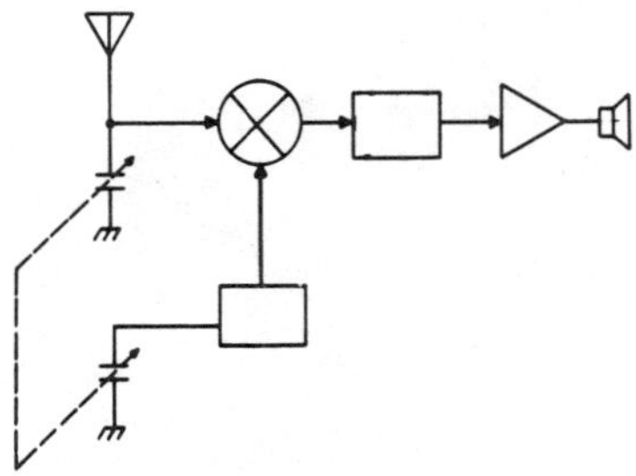

Packaging and Mechanical Considerations

Today, the electrical performance of a communications receiver can be mathematically predicted with the help of computers. As a result, the amount of circuit design required by a manufacturer is minimal. Of much more significance is the amount of performance that can be achieved for the price. The question remains how to build a high performance low cost receiver. The answer lies in how the receiver is setup mechanically, or how it is packaged. Designing a communications receiver with stringent cost and labor boundaries automatically leads to the idea of using plug-in modules, as previously discussed.

These modules would functionally resemble the block diagram of the receiver, and can be individually manufactured and tested before integration into the receiver's frame. Fault isolation is automatically achieved with the use of diagnostic computers, as shown in Figs. 23-1 and 23-2, thus reducing the manufacturing cost and allowing for more performance to be incorporated in the design for the same price. Figure 23-3 shows the extensive use of module boards in the Norlin SR-2090 communications receiver.

Because of the performance goals required in a communications receiver, the problem of interconnecting the modules economically and efficiently becomes a major cost element of the design. Demanding shock, vibration, and temperature requirements, as well as electromagnetic-interference (EMI) standards call for new and economical ways of packaging. The problem is further aggravated by the highly digital nature of today's communications receiver. As many as a hundred control lines are sometimes needed to allow functional access to the various modules from the main microprocessor as well as from the commands and interfaces located on the front and back panels. The digital noise generated by these lines can greatly interfere with the performance of the receiver. Introducing lowpass filters in each one of the lines would prove too excessive for the

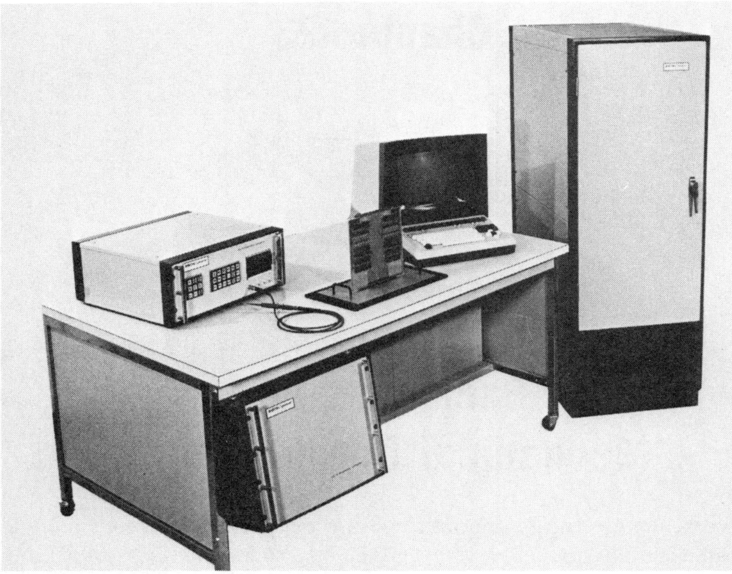

Fig. 23-1. Diagnostic computer used for automatic testing of module-boards.

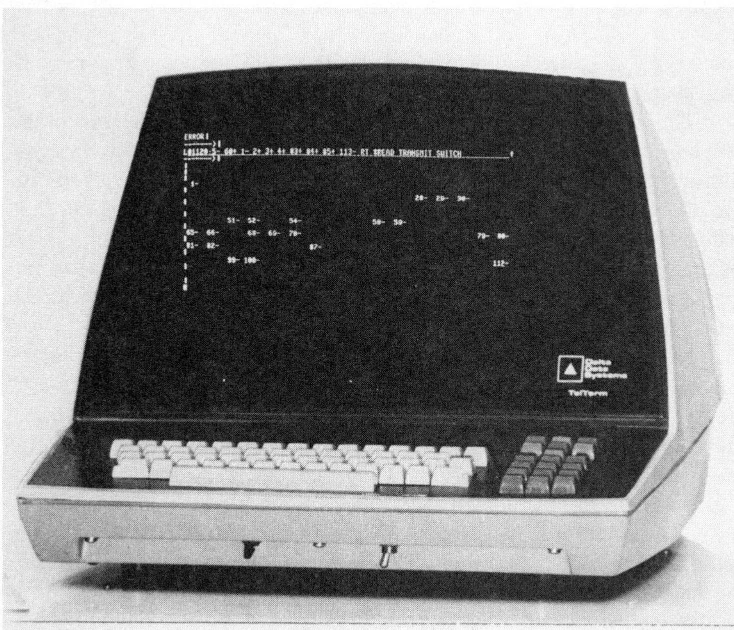

Fig. 23-2. Actual display of faults on the crt of a diagnostic computer.

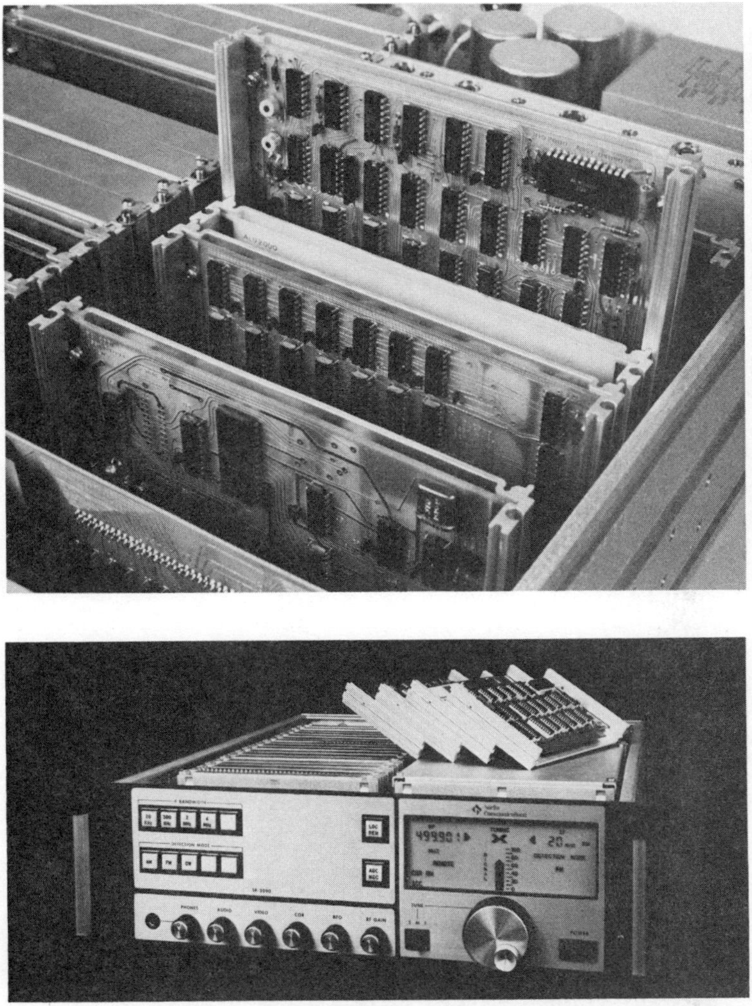

Fig. 23-3. Mechanical layout of the SR-2090 communications receiver shows a modular approach (courtesy of Norlin Communications).

cost of the receiver, while the use of ceramic capacitors to by-pass each circuit would only provide suppression at the self-resonant frequency of the capacitors. The economical solution to the problem is the use of a special flexible flat cable which is wrapped in metal foil. This provides a cost effective solution as shown in Fig. 23-4. It has been found that the filtering provided by this type of interconnection technique costs five times less than conventional line filtering. The method has been successfully applied in production by Watkins-Johnson Co. in an attempt to manufacture a low-cost, high-performance communications receiver, the WJ-8718.

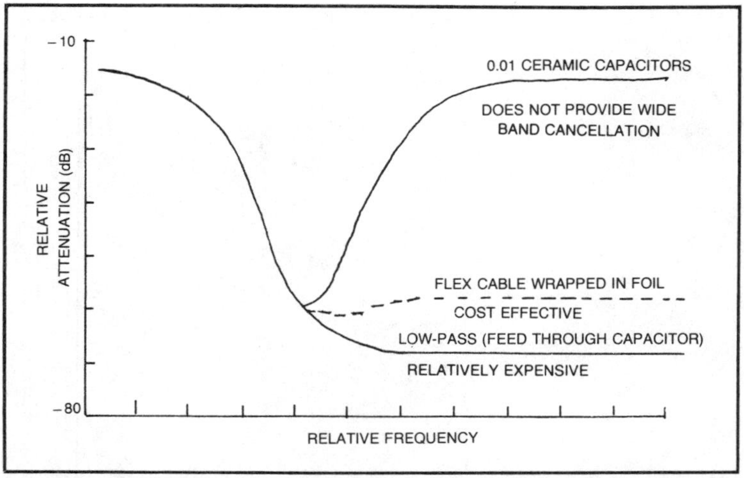

Fig. 23-4. Relative attenuation and cost of different methods of noise cancellation for interconnect lines in a communications receiver.

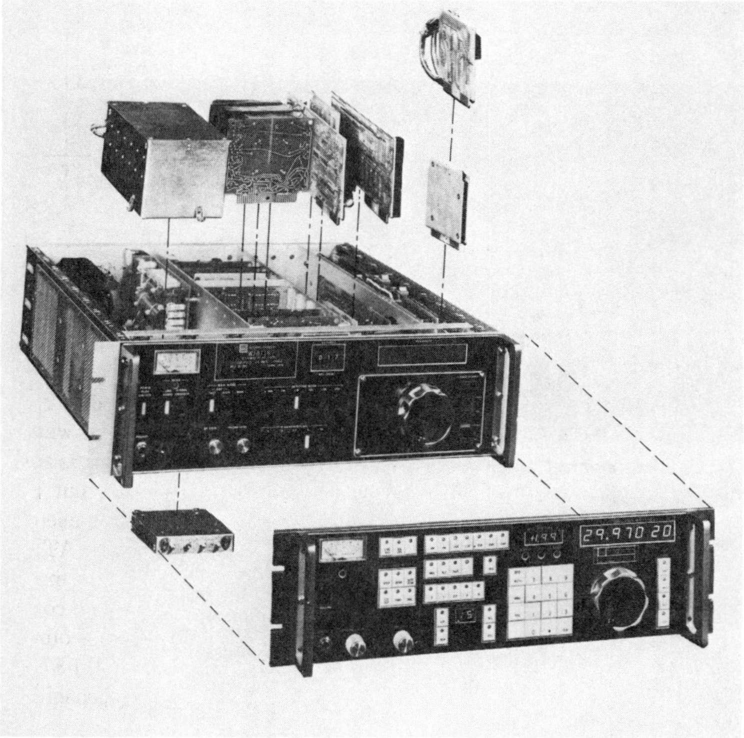

Fig. 23-5. Packaging of the Watkins-Johnson WJ-8718 HF receiver (courtesy of Watkins-Johnson Company).

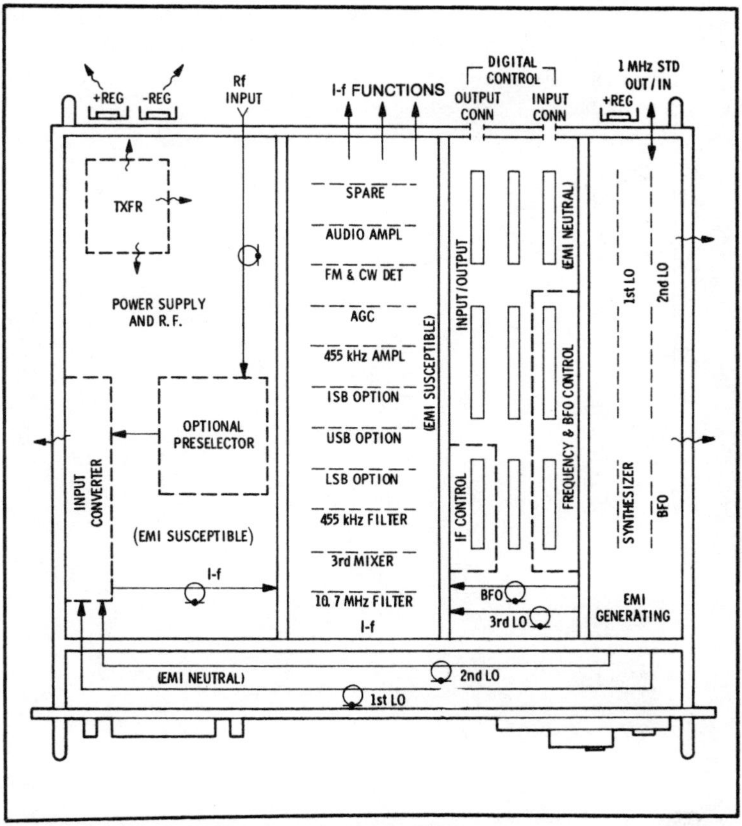

Fig. 23-6. Mechanical layout of the WJ-8718 HF receiver.

Watkins-Johnson Co. in an attempt to manufacture a low-cost, high-performance communications receiver, the WJ-8718.

Additional manufacturing cost reductions have been obtained by the elimination of time consuming milling and tapping of the chassis, while substituting punching and bending of sheet metal parts. The use of nuts and washers has been reduced to a minimum while emphasis was put on efficient ways of interconnecting the functional blocks. Extensive use of automated board assembly as well as automatic test equipment (ATE) makes the manufacturing of the WJ-8718 HF receiver one of the most automated production processes of its kind, allowing for a true low-cost, high-performance communications receiver at a fraction of the cost of other comparable receivers. Figure 23-5 shows an exploded view of the WJ-8718 receiver at the subassembly level. Shown in Fig. 23-6, is the mechanical layout of the receiver.

In conclusion, receivers can be built today with exceptional performance, but the manufacturing cost can make them prohibitively expensive.

The greatest task remains how to get the most performance for the lowest cost. While electrical design remains important in this regard, the solution lies in using more ingenious packaging techniques, as well as the automation of production lines and testing techniques.

—•••—

Chapter 24

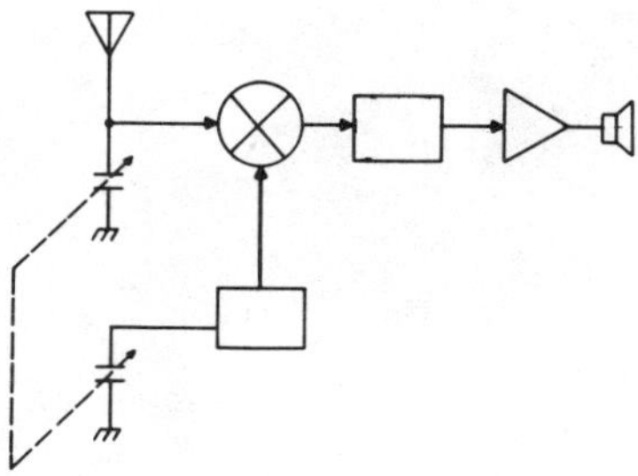

Radio Receiver Technology Today

This book spans about 80 years of radio receiver technology. While this is a relatively short period, the advent of communications receivers has had its technological impact. One may ask: Where are we today?

While the superheterodyne is still the most popular approach, with an i-f as high in frequency as 2 GHz, the trend is toward even higher i-f's. While selectivity is still achieved at lower frequencies such as 455 kHz. New technological discoveries in the consumer receiver market will definitely impact the commercial and military communications receiver.

The development of low cost microprocessor-controlled synthesizers, such as the one used in the GE 4870 clock radio shown in Fig. 24-1, will allow for more sophisticated communications receivers at lower costs in the near future.

Large-scale integrated circuits (LSI) are being developed to replace discrete functions not only in the i-f area but also in the front end and mixer areas. Extremely low-noise amplifiers using GaAs FET technology have been recently introduced, allowing receivers for the first time to listen deeper into space for possible signs of extraterrestrial life. In today's technology, such a specialized communications receiver can provide useful reception of intelligent signals from a planet located some 15,000 light-years away.

Although this may seem as a great distance, it is only half way to the center of the Milky Way galaxy. It has been established that such a receiver will cover a preferred frequency range between 1400 MHz and 1727 MHz, the "water hole"* frequency.

*Oliver, B.M., and Billingham, J.: Project Cyclops, a design study of a system for detecting extraterrestrial intelligent life. NASA CR 114445, 1972.

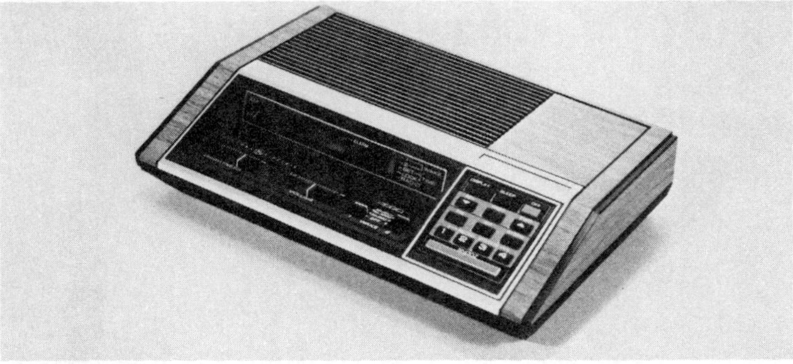

Fig. 24-1. Technologies developed for the consumer market should provide new, low cost solutions for the communications receiver of tomorrow. The GE 4870 clock radio is one of the first microprocessor-controlled, synthesized consumer radio, allowing for precise tuning of the AM and FM bands. Four programmable memories are also provided for instant recall at predetermined times, along with a variety of other microprocessor-controlled functions.

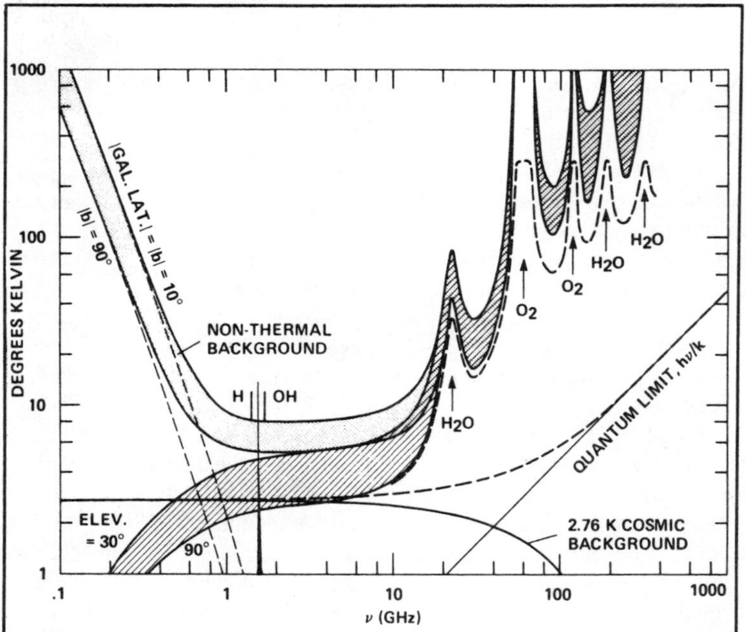

Fig. 24-2. The "Water Hole" window provides the best overall noise performance for an interstellar communications receiver. This frequency band was also chosen as a meeting point for extraterrestrial civilizations which base their existence on water. The signal shown represents the hypothetical strength of a transmitter similar to that of the Arecibo Observatory, located some 60 light years away and aimed at us.

Fig. 24-3. Aerial view of the world's largest Radio Observatory, Arecibo in Puerto Rico where Extraterrestrial Intelligence (ETI) searches are being conducted. The Arecibo Observatory is part of the National Astronomy and Ionosphere Center which is operated by Cornell University under contract with the National Science Foundation.

The rationale behind this choice is based on the ideas that:

- Life in the Universe is a function of water.
- An extraterrestrial civilization attempting to communicate with us would transmit in the water frequency band which is located between the spectral line of hydrogen (1420 MHz) and the spectral lines of hydroxyl (1612 to 1727 MHz). Although other frequencies have been previously suggested, it has been recently discovered that this band of frequencies is the quietest in the spectrum as shown in Fig. 24-2.

Existing radio observatories such as the one shown in Fig. 24-3 are already searching the universe at the mentioned frequencies. An organized effort has been initiated in both the United States and the Soviet Union to systematically listen for extraterrestrial signals*.

It is understood that the Soviet Receiving System will be dedicated for listening to wide-band omni-directional pulses. Several receivers located geographically apart will be looking for time of arrival coincidence of the pulses in order to identify their extraterrestrial nature. In the United States, the approach involves narrow-band signal processing with the help

*Search for Extraterrestrial Intelligence (SETI)

of highly directive antenna beams. While this method provides a superior signal-to-noise ratio, the probability of intercept is greatly reduced due to the relatively narrow antenna beamwidth and the narrow-band processing. In order to increase the probability of intercept, new concepts in signal processing have evolved. Recent developments in cryogenically cooled up-converters with an i-f at 20 to 30 GHz and using Masers** as first i-f amplifiers, will provide exceptionally low noise and instantaneous bandwidths covering the "water hole" frequency range. Instantaneous detection was also achieved recently when, for the first time, interaction of light and sound has been implemented in the Bragg Cell Receiver.

—•••—

**Microwave Amplification by Stimulated Emission of Radiation. Device similar to a laser which amplifies coherently generated microwave emissions.

Chapter 25

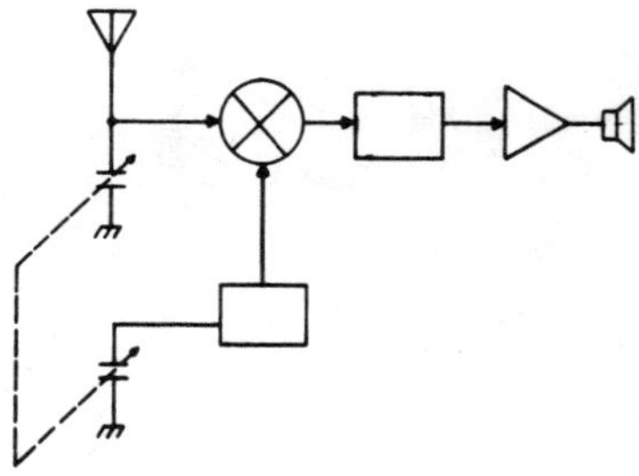

The Bragg-Cell Receiver

The phenomenon of interaction between sound and light was suggested by Louis-Marcel Brillouin* some sixty years ago. This theory was based on the hypothesis that monochromatic (single frequency such as red, green, etc.) Light would be diffracted in the presence of sound. This principle was proven in 1932 when the diffraction effect was observed in an experiment involving a source of filtered light that illuminated a column of water into which sound has been generated. Proper adjustment of the angle of incidence of the light source, allowed for the first-order diffraction line to become more intense while providing cancellation of the other lines. This angle was later called the Bragg angle and the phenomenon constitutes the basis for the acousto-optical (A/O) receiver, or as it is sometimes called, the Bragg-cell receiver.

In a Bragg-cell receiver, such as the one shown in Fig. 25-1, several rf signals of interest which are present at the antenna are converted to a wide-band i-f (40 MHz) centered, in our case, at 70 MHz. This is a typical superheterodyne process, and the signals contained within the i-f are further amplified by a high-gain high-dynamic range i-f amplifier, and are finally applied to the Bragg-cell through a piezoelectric transducer which is bonded to it. See Fig. 25-2.

The Bragg-cell is a block of very pure crystalline material, usually quartz or lithium niobate, approximately 1 cm × 1 cm × 10 cm in size. When excited by electrical signals within the pass-band of its transducer, the Bragg-cell is internally exposed to sound waves corresponding to the signals received.

*Louis-Marcel Brillouin (1854-1948). A French physisist, famous for his ferromagnetic theory, perpetual motion opposition, and Brillouin scattering phenomenon.

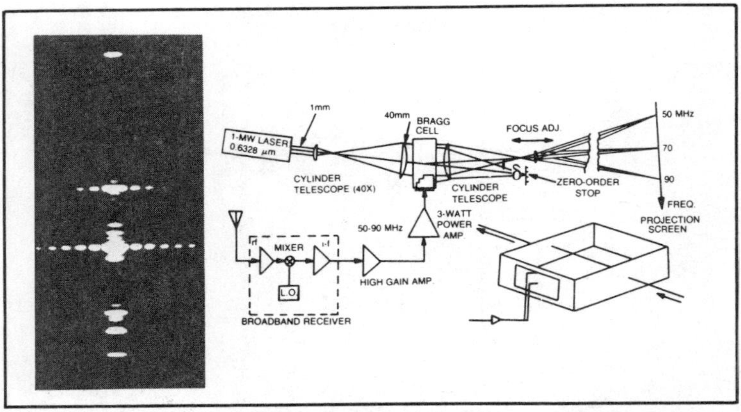

Fig. 25-1. The Bragg-Cell receiver principle, and typical display of received signals.

Fig. 25-2. Bragg-Cell signal processing is being developed by GTE-Sylvania for a variety of applications. The beam from a Helium-Neon laser is diffracted by the acoustic wave in the Lithium Niobate Bragg-Cell (courtesy of GTE-Sylvania).

If a beam of monochromatic light such as that produced by a laser is injected at the certain angle into the Bragg-cell, an instantaneous deflection will occur for each of the sonic signals traveling through it. This happens because the acoustic energy in the material causes changes in the refractive index between the peaks and valleys of the wave. The resultant diffraction pattern is an instantaneous display of the signals which can be viewed panoramically on a screen within the range of the signals applied to the transducer. The deflection angle and the intensity of the light beam are proportional respectively to the frequency and power of the signals being received.

The Bragg-cell receiver can be viewed as a parallel-processing device which acts as an optical Fourier transformer that continuously converts the time varying input into discrete frequency sets represented by separate instantaneous beam diffraction angles corresponding to each frequency.

The instantaneous nature of the Bragg-cell receiver allows for a high probability of intercept (POI) since many signals can be "viewed" at the same time, as in the case of a radio telescope. Detection of the displayed signals is implemented by focusing the multiple light beam on a linear array of PIN photodiodes which are spaced very closely together. The position of each detector corresponds to a specific frequency within the band-pass of the receiver allowing for parallel frequency detection of all signals within that band. As many as a thousand detectors could be incorporated into such a signal collection installation, allowing for a typical frequency cell resolution of 1 MHz or less.

While not completely understood, from an application point of view, this type of receiver is envisioned as an instantaneous spectrum analyzer, identifying signals of interest and directing conventional digitally-programmable receivers to the particular frequency for narrow-band analysis.

There are few Bragg-cell receivers in existance today. Figure 25-3 shows a Bragg-cell channelized receiver developed by GTE-Sylvania, and shown in Fig. 25-4 are some commercially available Bragg-cells manufactured by IntraAction Corp. The typical performance of a Bragg-cell receiver

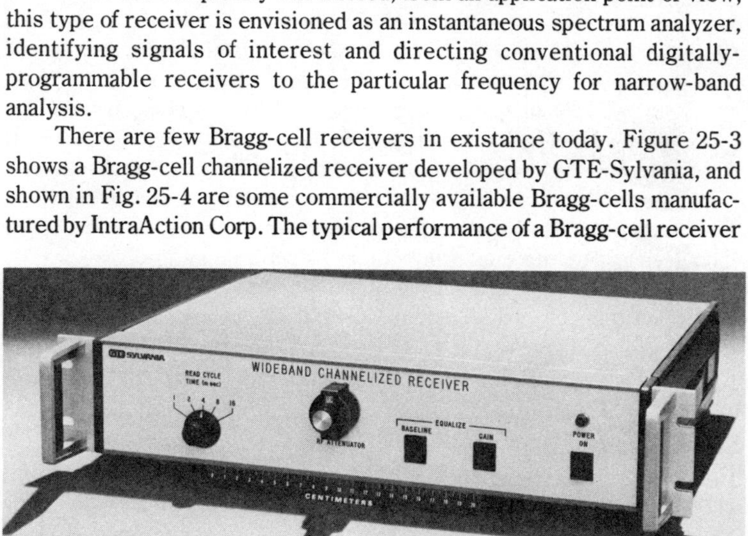

Fig. 25-3. A Bragg-Cell receiver developed by GTE-Sylvania (courtesy of GTE-Sylvania).

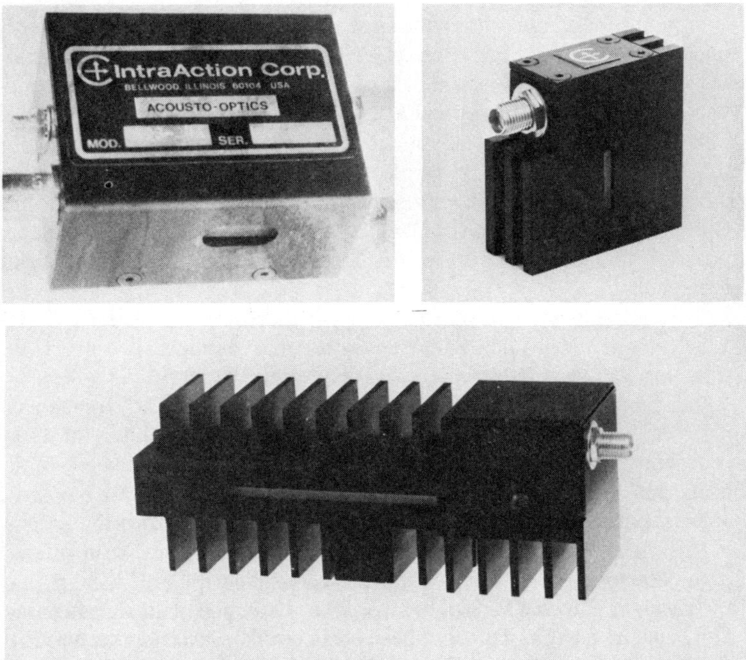

Fig. 25-4. Several commercially available Bragg-Cells (acousto-optic deflectors), which can provide a high number of resolvable positions (as many as 1000 spots) (courtesy of IntraAction Corp.).

is shown in Table 25-1. This table shows that one of the disadvantages of the Bragg-cell receiver is its limited dynamic range.

Other disadvantages of this type of receiver are its mechanical complexity associated with the packaging of the laser source, optical train and the detector array. New approaches in Bragg-cell receiver technology are looking at replacing the relatively bulky neon-helium laser with solid state devices which can be incorporated together with the Bragg-cell, lenses, and photosensors on the same substrate as shown in Fig. 25-5.

In conclusion, acousto-optical signal-processing technology has been emerging from the experimental laboratory development phase into an

Table 25-1. Typical Performance Characteristics for a Bragg-Cell Receiver.

Rf Bandwidth	1 GHz
Dynamic Range	30-45 dB
Sensitivity (S/N = 1)	−90 dBm
Frequency Resolution	1 MHz
Frequency Accuracy	± 1.5 MHz
Time of Arrival Resolution	>1 μsec

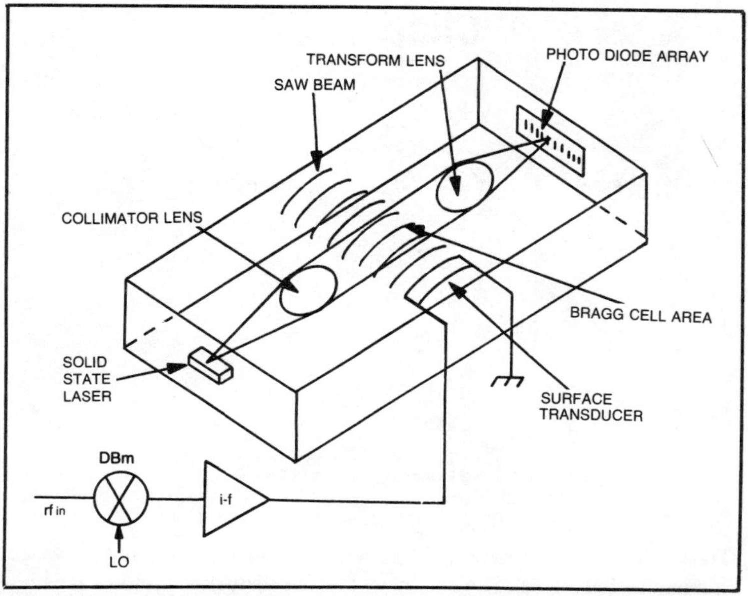

Fig. 25-5. Integrated acousto-optic receiver. A solid-state laser, optical elements, the Bragg-Cell, and the array of photo detectors are all incorporated on the same substrate, significantly reducing the size of the receiver.

advanced process of signal collection with wide band applications. While still in its infancy, the Bragg-cell receiver should provide an improved signal processing tool in a dense rf spectrum.

—•••—

Chapter 26

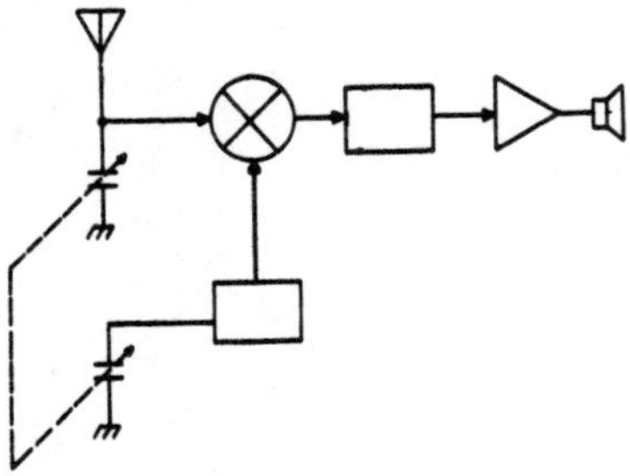

Future Trends

Today, the classical receiving system is changing rapidly, due to advances in many technologies. The communications receiver of tomorrow will be increasingly called upon to assume demanding tasks of signal processing in a complex environment. Extremely high speed frequency scanning and processing of the received data must take place in order to insure the highest probability of intercept. Extensive use of microprocessors will allow for immediate digestion and interpretation of the received data. Digital filters* are also being implemented in conventional i-f amplifiers, and as their sampling frequency is increased, they are considered as total signal processors in a no-conversion, TRF-type receiver.

New developments in analog and digital filters should allow for the realization of single up-conversion receivers which can achieve the required final selectivity in the first i-f. Extremely high dynamic ranges will be required to cope with the future's densely-populated rf environment. It is possible that the receiver of the future will be a hybridized approach involving several generic receiver types such as the superheterodyne, tuned radio frequency (TRF), and/or Bragg-cell, providing an increased probability of intercept over the conventional types.

In retrospect, it is hard to believe that only eighty years ago, the first intelligent signal was heard with a coherer-type receiver. This book had attempted to cover technical achievements in radio receiver design over this relatively short period of time. It is hoped that the book will serve as stimulus for creativity in those interested in technical excellence.

• • • — • —

*Digital Filters use discrete time samples of the input signal to output similar samples as modified by arithmetic, computer-like operations. The result is that an analog filter is implemented digitally, allowing for changes in parameters to be easily obtained by modifying the software.

Index

Edited by Roland Phelps